Innovations in Transport

Innovations in Transport
Success, Failure and Societal Impacts

Edited by

Bert van Wee

Professor of Transport Policy, Delft University of Technology, the Netherlands

Jan Anne Annema

Associate Professor of Transport Policy, Delft University of Technology, the Netherlands

Jonathan Köhler

Senior Scientist, Fraunhofer ISI, Germany

Cheltenham, UK • Northampton, MA, USA

Published by
Edward Elgar Publishing Limited
The Lypiatts
15 Lansdown Road
Cheltenham
Glos GL50 2JA
UK

Edward Elgar Publishing, Inc.
William Pratt House
9 Dewey Court
Northampton
Massachusetts 01060
USA

A catalogue record for this book
is available from the British Library

Library of Congress Control Number: 2022944514

This book is available electronically in the **Elgar**online
Economics subject collection
http://dx.doi.org/10.4337/9781800373372

ISBN 978 1 80037 336 5 (cased)
ISBN 978 1 80037 338 9 (paperback)
ISBN 978 1 80037 337 2 (eBook)

Printed and bound in Great Britain by TJ Books Limited, Padstow, Cornwall

Contents

Contributors

Jan Anne Annema, Associate Professor in Transport Policy, Delft University of Technology, the Netherlands.

Yashar Araghi, Research Scientist, Netherlands Organisation for Applied Scientific Research (TNO), The Hague, the Netherlands.

Bart van Arem, Full Professor of Transport Modelling, Delft University of Technology, the Netherlands.

Marlous Arentshorst, Copernicus Institute of Sustainable Development, Utrecht University, the Netherlands.

Wouter Boon, Associate Professor in Innovation Studies, Copernicus Institute of Sustainable Development, Utrecht University, the Netherlands.

Long Cheng, Postdoctoral Researcher in Transport Planning, Ghent University, Belgium.

Gonçalo Homem de Almeida Correia, Associate Professor of Multimodal Urban Transport and Smart Mobility, Delft University of Technology, the Netherlands.

Jannes Craens, Partner, ScaleUp Capital, Eindhoven, the Netherlands.

Ron van Duin, Professor in Port & City Logistics, Rotterdam University of Applied Sciences, the Netherlands and Assistant Professor in Logistics, Delft University of Technology, the Netherlands.

Erik Figenbaum, Chief Research Engineer, Department of Technology, Institute of Transport Economics, Norway.

Koen Frenken, Full Professor in Innovation Studies, Copernicus Institute of Sustainable Development, Utrecht University, the Netherlands.

Cyriac George, Research Scientist, Department for Mobility, Institute of Transport Economics, Norway.

Jonathan Köhler, Senior Scientist, Fraunhofer ISI, Germany.

Ove Langeland, Senior Research Political Scientist and Sociologist, Department for Mobility, Institute of Transport Economics, Norway.

Toon Meelen, Assistant Professor in Innovation Studies, Copernicus Institute of Sustainable Development, Utrecht University, the Netherlands.

Karla Münzel, Postdoctoral Researcher, University of Twente, the Netherlands.

Bonno Pel, Researcher/Lecturer in Governance of Sustainability Transitions, Université Libre de Bruxelles, Belgium.

Walther Ploos van Amstel, Professor in City Logistics, Amsterdam University of Applied Sciences, the Netherlands.

Hans Quak, Professor, Smart Cities and Logistics, Breda University of Applied Science, the Netherlands and Senior Scientist, Netherlands Organisation for Applied Scientific Research (TNO), the Netherlands.

Henrike Rau, Professor of Social Geography and Sustainability Research, LMU Munich, Germany.

William Riggs, Professor of Management and Engineering, University of San Francisco, USA.

Joachim Scheiner, Professor in Transport Planning, TU Dortmund University, Germany.

Kunbo Shi, Postdoctoral Researcher, in Social and Economic Geography, Ghent University, Belgium.

Maaike Snelder, Associate Professor, Transport Network, Delft University of Technology, the Netherlands and Principal Scientist, Netherlands Organisation for Applied Scientific Research (TNO), the Netherlands.

Qi Sun, PhD candidate, Eindhoven University of Technology, the Netherlands.

Lóránt A. Tavasszy, Professor in Freight and Logistics, Delft University of Technology, the Netherlands.

Wijnand Veeneman, Associate Professor, Governance of Infrastructures, Delft University of Technology, the Netherlands.

Bert van Wee, Professor in Transport Policy, Delft University of Technology, the Netherlands.

Isabel R. Wilmink, Senior Research Scientist, Netherlands Organisation for Applied Scientific Research (TNO), The Hague, the Netherlands.

Frank Witlox, Senior Full Professor in Economic Geography, Ghent University, Belgium.

Giovanni Zenezini, Assistant Professor in Supply Chain Management and Project Management, Turin Polytechnic University, Italy.

Preface

Innovations have played a very important role in the transport sector and will continue to do so in the future. There are many papers and books about innovations, focusing on theories, concepts, frameworks, impacts and so on, and the transport sector has received a lot of attention in that literature. But to the best of our knowledge, a book that introduces a selection of key theories of frameworks and applies these and other theories/frameworks to (candidate) innovations does not exist. This book aims to fill this gap, focusing on complex public–private innovations only. We hope it will stimulate BA, MSc and PhD students, other researchers, policymakers and many others to better understand the success or failure of innovations and factors contributing to this success or failure, and the possible societal impacts of these innovations.

We thank Edward Elgar Publishing for agreeing to publish this book open access as well as in hard copy. And we thank TRAIL Research School, the Netherlands for their financial support.

Bert van Wee
Jan Anne Annema
Jonathan Köhler

1. Introduction to *Innovations in Transport*

Bert van Wee, Jan Anne Annema and Jonathan Köhler

TRANSPORT INNOVATIONS AND IMPACTS ON SOCIETY

Innovations induce important societal changes, in general, and also in the transport system. In this book we define transport innovations as 'new elements of the transport system that are implemented in the real world'. In other words, a new element that has not been implemented yet is not presently an innovation, but a potential or candidate innovation. We define the transport system as 'the comprehensive system of infrastructures, vehicles, fuels/energy supply, services, prices, travel times and effort to move people or goods'. We label infrastructures, vehicles, fuels/energy supply, and services as the main components of the transport system. 'Effort' includes all components relevant for the resistance of travelling, additional to travel time and travel costs. It includes factors like perceived safety, mental and physical effort, and the (dis) joy of travelling by various modes of transport. Consequently, transport innovations can include all those main components of the transport system, as well as their mutual interactions. And innovations can be a completely new form of infrastructure, vehicle or energy supply, or completely new services, but also a part of one of these components. For example, we consider the high-speed train as an innovation that is part of the rail system. And we consider the electric vehicle as an innovation within the car system. Websites that allow for the online comparison of airline tickets are also innovations, coming within the overall category of information provision and ticketing.

In some places in this book a distinction is made between radical and incremental innovations. It is not straightforward to assign innovations to either of these two categories. Take the introduction of the three-way catalytic converter to cars in the 1980s as an example. This technology did not have an impact on the functional characteristics of cars, so from the perspective of the car as a mode of transport it can be seen as an incremental innovation. But from the perspective of the exhaust system of cars it certainly is a radical innovation because, unlike exhaust systems without a three-way catalytic converter, it does (strongly) reduce exhaust emissions, and it changed the motor management system, replacing carburettors with injection systems. In addition, it requires unleaded petrol, inducing fuel changes. So, the distinction between radical and incremental innovations depends on the perspective one takes with respect to innovations.

Important examples of transport infrastructure innovations implemented in the past are motorways, rail, high-speed rail, maglev rail, modern airports, and cycle lanes. In the category of vehicles, examples are all vehicle types (cars, trains, bikes, ...) as well as major changes within these vehicle types, such as e-bikes, or electric cars. The introduction of liquefied natural gas (LNG) and unleaded petrol, as well as electric cars, are examples of fuel/energy innovations.

The fact that we mention the electric car as both a vehicle- and an energy-related innovation makes clear that some innovations combine multiple components of the transport system. In the case of electrical vehicles, the component of infrastructure is also included: electric cars need charging infrastructure.

In the area of services, innovations include shared bikes and cars, smart cards for public transport, and mobility as a service (MaaS). Innovations in the area of ticketing and information are important service innovations: nowadays it is much easier to find timetables of public transport and airline services, and book tickets, than a few decades ago.

In the area of goods transport, the container is a radical innovation, allowing for the combination of many product types and shippers on one vessel, and reducing barriers for multimodal goods transport (van Ham and Rijsenbrij, 2012; Lau et al., 2013).

The concept of transport innovations is strongly related to time and place: it is important where and when new elements are introduced. For example, building cycle lanes in a city without any bike infrastructure is an innovation for that city, whereas it is not in cities with a long cycling tradition.

Transport innovations implemented during the past two hundred years had many wider impacts on societies, as well as dramatically changing the transport system. Improved transport systems made the implementation of complex supply chains possible, leading to relocations of production in its various stages. And improved transport systems fuelled urban sprawl, just to mention a couple of important societal changes. A part of these changes is that transport has become much less expensive over the past centuries (Filarski and Mom, 2008), which has contributed to increasing welfare levels. The changes in the transport system made it possible for people to reach many more destinations, extending their activity spaces dramatically. The theory of constant travel time budgets illustrates that over a large group of people, such as all inhabitants of a country or state, the average time spent travelling is quite stable at around 60 to 75 minutes per person per day (e.g. Mokhtarian and Chen, 2004). Therefore, innovations that made transport faster made people travel further within roughly the same time, resulting in important accessibility gains (but also higher environmental impacts).

SUCCESS OR FAILURE OF POTENTIAL TRANSPORT INNOVATIONS, AND SOCIETAL IMPACTS OF TRANSPORT INNOVATIONS

Looking at the past two centuries, we can conclude that the transport system is changing constantly. So the transport system will very likely also change in the future. But it is difficult to accurately forecast which potential innovations might be successfully implemented in the future and what the impacts of these implementations on societies might be. It is also very difficult to estimate the importance of public policies, and the research and development of companies, for the success or failure of potential innovations, as well as their societal impacts. Should policy makers and planners develop policies and plans for new potential innovations or not? If so, which? Should they, for example, plan new infrastructure for e-bikes, or for the Hyperloop? If so, where and when? Should they adapt cities to accommodate automated vehicles? Should they reduce parking spaces in residential areas to be built, because people might increasingly share cars? For companies it is very difficult to accurately forecast the return on R&D investments in general, including in the area of transport innovations. This applies even more to radical innovations (as opposed to incremental ones): the list of failed potential transport innovations is much longer than the list of successfully implemented transport innovations (Filarski and Mom, 2008). So, activities in the area of innovations, both for companies and for policy makers, are often in the category of 'high risk, high gain'. Many companies need innovations to survive in the longer term, so they need to take onboard the high risk of related R&D investments. And policy makers know that the transport system will change in the next hundred years, but not clearly how. In other words, uncertainties are very large, leading to investments that are high risk in terms of time and money, but if successful, transport innovations might significantly change the transport system and society.

AIM AND SCOPE OF THIS BOOK

We are not under the illusion that we can greatly reduce those uncertainties. But in producing this book we do have the ambition to be helpful in dealing with those uncertainties.

More specifically, this book aims to help understand the success or failure of potential transport innovations, as well as their societal impacts.

Here we define 'success' as the real-world implementation of potential innovations. We realize this is a simplistic definition of success – innovations might be implemented, but we might regret this implementation with hindsight because the disadvantages outweigh the advantages of the real-world innovations. We argue that another definition of success in this case could be that potential innovations not only need to be implemented in the real world but they also need to have net benefits. We do not dig into the evaluation of all pros and cons; this is beyond the scope of this book.

We put the topic of transport innovations into the context of (changing) societies. What works in one context does not necessarily work equally in another. For example, rail infrastructure might serve high-income groups in one country, increasing inequalities, and low-income groups in another country, reducing inequalities. And societies are constantly changing, examples of important changes being digitalization, population decline or growth, increasing focus on sustainability, the aging population and globalization. All these changes can have important impacts on the transport system in general, as well as on the success or failure and societal implications of potential transport innovations.

The book is limited to complex innovations in which both public and private actors are involved. Without being able to draw a sharp line – it can be difficult to say beforehand what the role of different actor types in the real-world implementation of potential innovations might be – we can give some examples. Further implementation of MaaS or hydrogen cars, and hydrogen airplanes and the Hyperloop, typically will never happen without the involvement of public and private actors, but a new ticketing platform for airline tickets probably can be implemented without public involvement.

To reach our aims we combine the insights from several disciplines. Innovation and transportation sciences provide the most important theoretical underpinnings. It is important to realize that innovation sciences combine and integrate several other disciplines, such as evolutionary economics, business economics, managerial sciences, psychology and history.

The target audience for this book includes academics in the area of transport innovations at the strategic level, students, companies in the area of transport innovations, policy makers and planners.

Part I of the book introduces several frameworks that help to understand the success or failure of potential innovations in general, as well as societal implications. Part II then discusses several potential transport innovations, partly making use of the insights of Part I.

POLICIES RELEVANT FOR INNOVATIONS

Before we continue with Part I, we will explain the role of public policies, because of the focus of this book on complex public–private innovations. Table 1.1 shows the dominant relationships between policy instruments and technological and service innovations. In addition, it shows how these instruments can influence other determinants for the impact of the transport system on the environment, safety and accessibility. These determinants indirectly are relevant for the effect of policy instruments on innovations, as we will explain below.

Table 1.1 shows that in general (not limited to policies relevant for innovations) policy makers have multiple categories of instruments available, a first category being regulations. Emissions and safety standards for vehicles and maximum speeds on roads are important examples. A second type of policy instrument concerns prices, such as levies on fuel or subsidies on public transport. Next, via land-use planning they can influence travel behaviour, example policies being those that influence densities, the degree of mixing of land uses, and distances between public transport access points and residential, work and other areas (e.g. Ewing and Cervero, 2010). The fourth category encompasses infrastructure policies, mainly those pertaining to building and maintaining roads, railways, ports, airports, canals and other infrastructures. Next, many policy makers develop specific public transport policies, in addition to the four policy categories above, such as defining routes and service levels for public transport services and rules for tendering public transport services. Via marketing, information and communication they can indirectly (try to) influence the behaviour of travellers and other actors, such as public transport companies. As Table 1.1 makes clear, these policy types can be used to stimulate or discourage innovations, and to change transport volumes (expressed by the number of passenger kilometres or the number of trips), modal split (volumes split by travel mode – in the case of passenger transport: aircraft, car, public transport, bike, motorized two-wheelers, and other; in the case of goods transport: shipping, barge, lorry, train, air), the efficiency of using vehicles (expressed in load factors in the case of goods transport and vehicle occupancy rates in the case of passenger transport), and the way of using vehicles (speeds, acceleration, braking). It is beyond the aim of this chapter to explain all the instruments in detail. We limit ourselves to the options to use policy instruments to stimulate or discourage innovations.

Not all these policies are important from the perspective of innovation. Table 1.1 shows that regulations, prices, infrastructure and public transport policies can be used to stimulate or discourage innovations, both technological and service innovations. To illustrate we give some examples. Via regulations

Table 1.1　　*Dominant relationships between policy instruments, innovations and determinants for the impact of transport on the environment, safety and accessibility*

Policy instruments	Technological innovations	Service innovations	Transport volume	Modal split	Efficiency of using vehicles	Driving behaviour
Regulations	*	*	*	*		*
Prices/subsidies	*	*	*	*	*	*
Land-use planning			*	*		*
Infrastructure	*	*	*	*		*
Public transport policies	*	*				
Marketing, information and communication			*	*	*	*

Source: Updated from van Wee (2009).

policy makers can allow or forbid innovative technologies, such as hydrogen cars, private small aircraft or electric scooters. Or they can require innovative services like ticketing, information and payment services for public transport. Next, they can stimulate or discourage technical and service innovations via price instruments. They could give tax benefits or subsidies to innovations they want to support, and put levies on innovations they do not want. An example of the latter category is (higher) taxes on vehicle types they prefer to not be sold frequently, for environmental or safety reasons. They can subsidize information provision services for public transport that cannot survive on a commercial basis only. Infrastructure policies could be the decision to build innovative infrastructure types, such as high-speed rail lines, charging infrastructure for electric vehicles, or maybe maglev rail or Hyperloop infrastructure in the future. Once these infrastructures are in place, services using them will be offered. Through specific public transport policies, such as tendering procedures, policy makers can force or stimulate public transport service providers to use innovative technologies like electric buses or implement innovations in information and ticketing services.

THE CHAPTERS IN THIS BOOK

We finally give a brief overview of the chapters in this book.

Part I   Frameworks for Analysing Transport Innovations

Part I looks at frameworks for analysing innovation in transport. It does not attempt to be comprehensive, but considers the state of the art in innovation theories applied to sustainable transport, with an emphasis on approaches to understanding behaviour and an example of current modelling ideas.

In Chapter 2 Pel discusses the use of transitions theory in transport innovation. The author considers moving from ecological modernization to 'system innovation', from optimization to experimentalism, from sustainable technologies to socio-cultural transformation and from innovation to exnovation (the decline of established systems and technologies). He shows how transitions theory puts innovation success, failure and societal impacts in a broader perspective, and how it takes the pursuit of 'sustainable' transport innovation well beyond the development of clean technologies.

Chapter 3 written by Zenezini and Tavasszy provides an example of how current developments in transport modelling in the use of an agent-based modelling (ABM) approach can be used to analyse micro-level decisions and behaviour. The authors address the reorganization of logistics processes, usually neglected in innovation studies of transport, by representing transport systems as business ecosystems. They introduce a new ABM framework for freight transport innovations that helps to fill this gap.

In Chapter 4 Rau and Scheiner review the use of travel biographies. Investigations of mobility biographies in various disciplines and in interdisciplinary collaborations have cast new light on changes in how people travel that relate to specific life events and phases. Recent innovations in mobility biographies research (MBR) add that individuals cannot make autonomous choices but that their decisions are embedded in complex contextual conditions. Practice-theoretical approaches imply that wider material and social conditions (including transport innovations) represent constitutive elements of practices in transport behaviours. Taking individuals' and policy stakeholders' subjective representations of transport issues seriously (as opposed to simply relying on 'matters of fact') is another approach that may help to identify opportunities for change. MBR contributes to creating such knowledge by studying stability and change in mobility over the life course.

Research has shown that travel behaviour can be influenced by economic and social nudges. In Chapter 5 Riggs explores how individuals' decisions are framed by the built environment as well as by behavioural factors. It explores how transportation nudges constitute a growth area for transportation planners and engineers, particularly as travel choices and trip complexity increase with smart, connected and automated mobility services.

In Chapter 6 Annema reviews theories of innovation applied to transport. The theories considered are (1) the opportunity vacuum as a conceptual model for the explanation of innovation, (2) the technological innovation systems (TIS) approach, (3) the political economy of transport innovations, (4) the multi-level perspective (MLP) on transitions, (5) Rogers's theory on diffusion of innovations, (6) the hype cycle and (7) theories and frameworks that aim to explain the role of industries in innovations. Modern innovation theories – broadly speaking – reject the idea of the possibility of an easy innovation fix. Innovations in transport involve many actors with large and opposite interests. Theories can provide the basis for policy development and innovation management. They can help to understand the social and technical processes of innovation in transport. They can help to understand which factors could positively contribute to the success of a specific innovation. They can also help to understand time and place in innovation adoption.

Part II   Transport Innovations

Part II first presents three chapters on technological innovations that are already partly implemented: vehicle electrification, e-bikes, and light electric vehicles in city logistics. In Chapter 7, Langeland, George and Figenbaum discuss the case of vehicle electrification in Norway. Norway is a very interesting case when it comes to vehicle electrification because of its worldwide leading position with respect to the market penetration of battery electric vehicles. The authors evaluate how suitable the technological innovation systems (TIS) framework is for understanding emerging technologies such as battery electric vehicles (BEVs) and fuel cell electric vehicles (FCEVs). The content of the chapter in that respect cuts across the themes of Parts I and II as it introduces and evaluates the TIS framework while also using BEVs and FCEVs as a case study.

Sun, in Chapter 8, then elaborates on the success and failure plus impacts of the e-bike. Departing from the transitions theory perspective of Pel (Chapter 2) and other theories partly presented in Chapter 6 by Annema, Sun explains why this transport innovation has gained a high level of popularity. An important conclusion is that environmental and social impacts to a large extent depend on the impacts of e-bikes on travel behaviour, including substitution between modes. These impacts are discussed making use of the mobility biographies approach as presented by Rau and Scheiner in Chapter 4 of this book.

In Chapter 9 van Duin, Ploos van Amstel and Quak next focus on light electric vehicles for city logistics, making use of the TIS framework of Langeland, George and Figenbaum presented in Chapter 7. Because the framework didn't provide them explicit insight as to whether the factors presented in the TIS framework are success or failure factors, strengths or weaknesses, opportunities or threats, they suggest an additional framework to find strategies for how to proceed with the outcomes of the TIS framework by categorizing them as strengths, weaknesses, opportunities and threats.

In Chapter 10 Snelder, Homem de Almeida Correia and van Arem discuss automated driving, an innovation that at the time of writing this book (2022) has been partly implemented but is only in its early stages. Fully automated driving is a candidate innovation. The authors discuss the possible or likely first-order (travel related), second-order (vehicles, land use and location choices, and infrastructure), and third-order (societal relevant effects, mainly environmental and safety effects) implications and how these relate to success or failure of future developments of automated driving. They conclude that the future of (higher levels of) automated driving is quite uncertain, and that this future depends on related fundamental challenges with respect to human factors, technology, infrastructure and legislation.

After these more hardware-related innovations, chapters 11 to 13 focus on service innovations: carsharing, MaaS and e-shopping. In Chapter 11 Münzel, Arentshorst, Boon and Frenken explain the success and failure plus societally relevant impacts of carsharing. Based on interviews with stakeholders they conclude that stakeholders agree on the importance of developing visions for the future transportation system of which carsharing is an integral part, setting up an information campaign, and implementing measures supporting carsharing in municipalities. Controversial issues include the sharing of data among industry providers, collaboration for setting up an aggregated overarching booking platform, and changes in the national taxation of car ownership. The results show that measures supporting the carsharing niche to grow incrementally are evaluated as feasible and desirable while more disruptive, potentially higher-impact measures changing the car ownership regime are less popular across the consulted stakeholder groups.

In Chapter 12 Veeneman focuses on mobility as a service (MaaS), a service innovation that has received a lot of attention from researchers and policy makers during the past five years or so. He departs from the transitions theory perspective of Pel (Chapter 2). Making use of real-world cases he shows how MaaS developed (1) from a specific service to help travellers deal with a fragmented landscape of mobility towards an incremental system innovation, (2) from a singular private innovation to be implemented to an intricate effort from both governmental and private entities, (3) from a new additional service to a rethink of governance, and (4) from a system improvement to a system innovation.

In Chapter 13 Shi, Cheng and Witlox discuss e-shopping, including the relationships between e-shopping and travel behaviour. Departing from the multi-level perspective of Geels (2011) (see also Chapter 6 by Annema in this book), they provide a conceptual framework to understand the emergence and growth of e-shopping and then consolidate the existing literature to analyse in particular its transport and social implications and associations with the built environment.

Next, in Chapter 14, Araghi and Wilmink elaborate on the success or failure and impacts of a candidate innovation that potentially could be implemented in the more distant future, after 2030: the Hyperloop. The content of this chapter, like that of the chapter by Langeland, George and Figenbaum, cuts across the themes of Parts I and II of this book, because it starts with the notion that the Hyperloop is a disruptive innovation, and a framework for such innovations is lacking. The authors therefore first develop a conceptual disruptive innovation framework, setting off from the disruptive innovations theory of Christensen et al. (2015; 2018). They conclude that it is unlikely that the Hyperloop will be disruptive in the short term (i.e. the coming decade), but that the need to reduce the carbon footprint of the transport sector and possible breakthroughs in reducing the infrastructure expenses (and subsequently travel costs) of the Hyperloop may be helpful for its realization and the potential disruption of the transport sector.

Finally, in Chapter 15, Craens, Frenken and Meelen take a completely different angle: they depart from a mission-oriented innovation policy. A key difference between this chapter and the other chapters in Part II of this book is that no specific innovation is the centre of attention, but rather a mission of policy makers, in this case the Swedish 'Vision Zero' approach to traffic safety, aiming to greatly improve road safety. They explain what the Vision Zero policy entails, how stakeholders dealt with 'transformational failures', and what made the policy a success. They end with lessons for the development of new mission-oriented innovation policies to address societal challenges.

After reading all the chapters it becomes clear that several in Part II make use of theories and frameworks presented in Part I. But the links between both parts are not very strong, one reason being that Part I unavoidably presents only a selection of relevant frameworks. There are many more, and the authors of the chapters of Part II were free to choose the theoretical underpinnings for their chapter without being limited to the frameworks presented in Part I. Another reason is that some chapters do not set off from an existing innovations framework at all, but rather take a more pragmatic stance to discuss the innovations at stake.

Unavoidably this book only discusses some of the more important examples of transport innovations, and there are many more that could play a substantial role in the future, examples being drone deliveries, hydrogen aircraft, e-fuels, non-road vehicle automation, mobility hubs and shared taxis. So this book is not aimed to be a handbook providing all the important theories available to help understand complex public–private innovations, nor one that discusses all dominant (candidate) transport innovations. Rather, it is meant to be a source of inspiration to help understand the success or failure of complex public–private (candidate) transport innovations and their societal impacts.

Because students studying complex public–private innovations in the transport system are a very important category of users of our book, we end with some reflections on how students can best use this book. Firstly, in searching for theories and theoretical frameworks students are encouraged to not only read the theoretical chapters of Part I plus a few additional theoretical contributions presented in Part II. There are many more theories and theoretical frameworks that can be useful to understand the success or failure and the impacts of innovations. Two of us have been teaching these topics to master's students at Delft University of Technology for many years, and students were able to find many more theories and frameworks. Secondly, we encourage students to study the innovations presented in Part II from the perspective of theories other than those presented in this book, which may result in gaining additional insights. Finally, and most importantly, we hope this book encourages students to study innovations other than those presented in Part II, and we hope they are inspired by the chapters presented in this book.

REFERENCES

Christensen, C.M., McDonald, R., Altman, E.J. and Palmer, J.E. (2018), Disruptive innovation: An intellectual history and directions for future research. *Journal of Management Studies*, 55(7), 1043–1078.

Christensen, C.M., Raynor, M. and McDonald, R. (2015), What is disruptive innovation? *Harvard Business Review*, 93, 44–53.

Ewing, R. and Cervero, R. (2010), Travel and the built environment. *Journal of the American Planning Association*, 76(3), 265–294.

Filarski, R. and Mom, G. (2008), *Van transport naar mobiliteit: De transportrevolutie 1800–1900*. Stichting historie der techniek. Zutphen, the Netherlands: Walburg Pers.

Geels, F.W. (2011), The multi-level perspective on sustainability transitions: Responses to seven criticisms. *Environmental Innovation and Societal Transitions*, 1(1), 24–40.

Lau, Y.-Y., Ng, A.K.Y., Fu, X. and Li, K.X. (2013), Evolution and research trends of container shipping. *Maritime Policy and Management*, 40(7), 654–674.

Mokhtarian, P.L. and Chen, C. (2004), TTB or not TTB, that is the question: A review and analysis of the empirical literature on travel time (and money) budgets. *Transportation Research Part A*, 38(9–10), 643–675.

Van Ham, H. and Rijsenbrij, J. (2012), *Development of Containerization: Success through Vision, Drive and Technology*. Amsterdam: IOS Press.

Van Wee, B. (2009), Transport policy: What it can and can't do? Paper presented at the European Transport Conference, Noordwijkerhout, 4–6 October 2009.

PART I

Frameworks for analysing transport innovations

2. A transitions theory perspective on transport innovation

Bonno Pel

1. INTRODUCTION: SUSTAINABILITY TRANSITIONS AND THE SEARCH FOR SYSTEM INNOVATION

Transport innovation is a shared interest of transport engineers and social scientists, policy-makers and enterprises, historians and futurologists. Accordingly, there is a multitude of transport innovation definitions, each with their particular exemplars. Archetypical visualizations are the flying cars and the elevated monorail tracks – reaching for the sky. Research on transportation and mobility has well surpassed this obsession with technological novelty, however. Strategies towards sustainable mobility typically take integrative approaches in which travel behaviour is a key dimension (Banister, 2008). Beyond technological fix approaches and beyond predict-and-provide logics, comprehensive, 'joined-up' transport strategies have become leading strategic orientations (Meyer and Miller, 2001; Berger et al., 2014). Meanwhile, the very understanding of 'sustainable' transport has undergone a certain evolution: rather than reducing it to matters of ecological efficiency and clean energy, transport is increasingly often taken as a more fundamental issue (van Wee, 2011) of freedom, social-spatial development and power. Framed more broadly as mobility, it can even be considered the key organizing principle of contemporary social order (Urry, 2012).

The above developments signal an overall shift towards systemic outlooks on transport dynamics and policy. The scope and significance of transport *innovation* is changing accordingly: in this book, transport innovations are defined as 'new elements of the transport system that are implemented in the real world'. This definition indicates first of all that a broad range of transport innovations could be worthwhile pursuing, as more or less pivotal system elements. It also indicates, secondly, the relevance of comprehensive transport innovation strategies that target several interrelated system elements.

Taking a systemic perspective on transport innovation, this chapter discusses the specific understanding of system innovation as brought forward in transitions theory (Köhler et al., 2019). The key idea in transitions theory is that sustainable mobility – like current persistent sustainability challenges more generally – calls for system-level, multi-dimensional and organization-transcending innovations. Such system innovations result from 'cascades' of mutually reinforcing innovations in technologies, consumer choices, business models, institutional arrangements, et cetera (Rotmans, 2005). Depending on their interactions with broader societal changes over longer periods of time, such system innovation processes could ultimately contribute to sustainability transitions, commonly defined as 'long-term, multi-dimensional, and fundamental transformation processes through which established socio-technical systems shift to more sustainable modes of production and consumption' (Markard et al., 2012).

The associated 'transitions theory' combines insights from Science & Technology Studies, evolutionary economics, institutional theory, sociology, and governance studies, amongst others. It comprises well-established frameworks like the multi-level perspective (MLP), strategic niche management (SNM), transition management (TM) and technological innovation systems (TIS; see Chapter 6), but it cannot be reduced to any of those frameworks. This chapter bypasses discussions on the scope and limitations of particular models (e.g. Temenos et al., 2017 on the MLP), which as such obscure their common focus on system innovation. It will capitalize on the interdisciplinary diversity of the transitions research community. Each of the aforementioned frameworks captures certain aspects of these transformation processes: whilst the analytical model of the MLP sets the scene of long-term system evolution, SNM, TIS and TM are providing more hands-on innovation management insights. Where TIS and SNM provide specific guidance on the cultivation of sustainable technologies and innovations, TM pays relatively more attention to the associated issues of transformative governance. Meanwhile, MLP and TM stand out as the frameworks in which the distinctly *systemic* outlook of transitions theory – and its divergence from conventional innovation management – is the most pronounced. Key defining characteristics of this (broadly defined) 'transitions theory' are:

1. The *long-term, evolutionary* perspective. Informed by evolutionary economics and the history of technology, transport innovations are considered as incremental extensions of, or radical breaks with, technological paradigms and dominant socio-technical structures ('regimes'). Regime shifts (e.g. the build-up towards or break away from car dependency) are considered to take periods of 30–40 years.

2. The *multi-dimensional* outlook. Broadening the evolutionary economics focus on technologies with insights from Science & Technology Studies, institutional theory, sociology, and governance theory, transport innovation is considered to involve various social, institutional, economic and cultural innovations as well.
3. The focus on *system innovation*. Informed by insights from complexity theory, innovations are considered for the reinforcing and dampening feedbacks they give to ongoing processes of change ('modulation'). Placing local and isolated innovation processes in a 'bigger picture' (Smith et al., 2010) of systems-level innovation, this also problematizes the associated governance philosophies: ecological modernization strategies and procedures of evidence-based policy are mistrusted for their reductionist tendencies.

The latter point is important. As further discussion will bring out, transitions theory comes with outspoken critiques on various 'conventional' innovation approaches that pursue innovation along a logic of incremental and well-controlled optimization. This rational optimization model continues to prevail in transport innovation, as it fits well with the positivistic scientific traditions of transport engineering, transport planning and transport economics. By contrast, this chapter highlights how transitions theory argues for an altogether different innovation logic. Other than seeking to improve or complement established repertoires of rational innovation management, it introduces a different innovation logic and a different governance philosophy that puts innovation in the service of structural societal transformation. In other words, transitions theory mainly adds to transport innovation by reframing its *purposes*. Its more particular implications will be brought out along the following research question: How does the transitions theory perspective enrich existing insights on the success, failure and societal impacts of innovations in transportation?

The question is answered along four basic aspects of transport innovation. The transitions-theoretical perspective challenges conventional insights regarding the rationale, governance philosophy, relevant dimensions and scope of transport innovation. It proposes to move from ecological modernization to 'system innovation' (section 2), from evidence-based improvement to experimentalist governance (section 3), from sustainable technologies to social-institutional transformation (section 4), and from innovation to exnovation (section 5). The concluding section summarizes how transitions theory puts innovation success, failure and societal impacts in a broader perspective, and how it takes the pursuit of 'sustainable' transport innovation well beyond the development of clean technologies. After highlighting the added insights of the transitions-theoretical 'bigger picture', the chapter also provides some critical reflections on its limitations (section 6).

2. RATIONALE: FROM ECOLOGICAL MODERNIZATION TO 'SYSTEM INNOVATION'

As indicated in the introduction, transport innovation is a particularly influential example of the modernist beliefs in progressive innovation that 'pushes the boundaries' of what can be realized in society. It is therefore unsurprising that transport innovation has also become a key area of ecological modernization, that is, the conviction that environmental challenges can be resolved without revolutionizing the current institutional arrangements of society (Hajer, 1995). Transitions theory indeed considers transport innovation as a kind of ecological modernization – stressing however that such ecological modernization should be fundamental and transformative.

The recurring concern in transitions theory is precisely the difficulty to achieve more than shallow, incremental innovations. Through its typical cross-sector view, transitions theory considers transport as one field of environmental policy alongside others. Discussing the similarities with developments in energy, water management, agriculture, waste management and healthcare, Rotmans (2005) critically observed how sustained and extensive transport innovation efforts had eventually just hardly delivered on their promises of ecological modernization. Providing only temporary relief and local stopgaps, many of the innovative solutions even seemed to reinforce the 'persistent problems' of environmental degradation and congestion. In order to address the societal roots of the problems, ecological modernization would have to reach well beyond incremental tinkering and undertake innovation on the appropriate systemic level: 'system innovation', comprising innovations in technologies, processes and behaviours and crucially involving organization-transcending innovations on the level of administrative routines, policy paradigms, planning doctrines, appraisal procedures and consumer cultures.

This strategic orientation towards *system* innovation implies several more specific advances beyond conventional 'ecological modernization' modes of innovation. First of all, it challenges the understanding of transport innovations as ways to reduce transport externalities.

From Reduction of Externalities to 'Regime Shift'

It is quite common to consider mobility/transport as socially and economically desirable practices of which the negative externalities should be dampened. This is in any case the angle taken by transport economics, just as the notion of ecological modernization is premised on the idea that externalities can be internalized, dampened and compensated. Illustrative examples of such externality-dampening innovations are the advances made in combustion efficiency, or the long history of traffic safety innovation – from bumpers and safety belts to the current automated driver assistance systems. Yet whereas these innovations are typically meant to dampen side-effects, system innovation seeks to transform the underlying societal structures through which the externalities keep coming back: transitions theory claims that systemic mobility problems call for similarly systemic solutions and innovations. This understanding coincides with the more longstanding literature in transportation research on car dependency (Newman and Kenworthy, 1989). Transitions theory focuses similarly on the systemic feedbacks between land-use structures and transport behaviours, searching for innovations that somehow intervene in the ensuing path dependency (Switzer et al., 2013). Building on insights on the history of technology, it develops a diagnosis of car dependency that reaches beyond spatial-economic factors – it is also understood to comprise the social, infrastructural, technological obduracy of derivative practices like underground parking (Stanković et al., 2021), the extensive supply chains of the automotive industry (Wells and Nieuwenhuis, 2012), the guidelines and expertise of dedicated governmental organizations (Geels, 2007), and the cultural-ideological emergence of accelerated lifestyles (Urry, 2004). The key consideration is that transport innovations should be somehow targeting the extensive socio-technical 'regimes' (Geels, 2005) through which transportation has become so structurally unsustainable – the 'rules of the game' that are typically taken for granted in approaches aiming to dampen externalities.

From System Improvement to System Transformation

The search for 'regime shift'-inducing innovation comes with a particular suspicion towards incremental innovations. Efficiency improvements are hardly considered as innovation success. Rotmans (2005) is particularly outspoken about the pseudo-solutions offered through 'end of pipe' solutions and temporary fixes. Ultimately, such solutions will only allow the car dependency 'regime' to further expand. This critique of system improvement approaches follows the transportation science arguments against one-dimensional transport policy strategies (Berger et al., 2014), and the underlying analyses of 'induced travel' and 'rebound effects'.

Beyond this analytical attentiveness to unintended side-effects, transitions theory takes a particular critical-political view on the matter: incremental improvements are rejected as reinforcements of an unsustainable transport system and as strategic moves of incumbent actors to maintain their dominant positions. Wells and Nieuwenhuis (2012) indicate for example how much innovation is still guided by an automotive industry in search of continuity.

Focusing on these power structures, transitions theory is very concerned with the cultural-political processes through which emergent radical innovations ('niches') become watered down and 'captured' by incumbent actors (Geels et al., 2012), or end up by the wayside as they fail to fit in with the prevailing socio-technical system. In their seminal article on SNM, Kemp et al. (1998) have thus famously asked why many of the promising sustainable technologies eventually do not make it from the car exhibition to the showroom, running up against combinations of rigid supply chains, risk-averse enterprise strategies, outdated regulations, established expertise and cultural conservatism. Analysing the turbulent dynamics of transitioning processes, transitions research typically explores how transport innovations – through various actors' attempts to push the process in desired directions – tend to hover between incremental system improvement and more radical system transformation. Typical empirical inquiries are: How could innovation in traffic management bring more than mere 'draining with the tap flowing' (Pel and Boons, 2010)? Could the electrification of mobility bring along changes in travel demand as well (Geels et al., 2012)? How and why are the advances towards sustainable surface mobility about to be overtaken by the rise of personal aeromobility (Cohen, 2010)? These inquiries mark how transitions theory takes a particularly critical and politically oriented view on innovation success.

From Impact Assessment to Institutional Change

Insisting on system-transformative innovation, transitions theory not only introduces a critical angle on efficiency improvements but also shifts attention away from the very assessment and appraisal of concrete environmental, social and economic impacts. Key considerations are that such quantitative assessments often provide only snapshots of change processes, that they tend to be narrowly scoped (cost–benefit analysis, notably), that powerful actors tend to 'game' the measurements (VW and Dieselgate), and that performance indicators tend to be shaped by the expertise and political orientations of the prevailing 'regime'. Transitions approaches focus instead on *learning* effects, and on *systemic* impacts in the form of institutional change.

This 'transcendence' of impact assessment indicates a deliberate break with positivist approaches. It marks in particular a fundamental rejection of the 'predict-and-provide' paradigm that the positivist transport planning tradition has brought forth. Unsurprisingly, transport experts have warned however against a complete disconnect from transport planning practice: abstract systemic approaches lose their strategic value once they start neglecting the complexities of situated implementation and the needs for tailored solutions (Temenos et al., 2017:119). The systems-level insights on car dependency and induced travel may be crucial underpinnings for transport policy, yet they fail to inform decisions on concrete issues of road bypasses or traffic diversions. Regarding this troublesome policy–science interface, transition researchers have started considering how they could stay more in line with established modes of evidence-based policy-making (Köhler et al., 2019; Turnheim and Sovacool, 2020). The preoccupation with transformative impacts does not necessarily imply a disregard of quantitative impact assessments, it needs to be said: the MLP framework takes a long-term, systems-evolutionary perspective that as such tends to abstract from any concrete intervention and evaluation, yet its key concern remains how a socio-technical regime can shift towards a more sustainable organization. Most importantly, MLP insights are often combined with the innovation management frameworks of SNM, TM and TIS. The latter transitions-theoretical frameworks are focusing much more on concrete processes of knowledge development and evaluation, which often do involve impact assessment. This will become clearer in the following discussion of experimentation.

3.   GOVERNANCE PHILOSOPHY: FROM EVIDENCE-BASED IMPROVEMENT TO EXPERIMENTALIST GOVERNANCE

Innovation tends to involve experiments. Transport innovation relies on test circuits and crash test dummies, laboratories, and simulations. Beyond secluded experiments in laboratory settings, it also typically relies on real-world experiments and user testing – the key stage in the development of automated cars (Pel et al., 2020) – but also in the fine-tuning of mobility management arrangements (Gorris and van den Bosch, 2016). Like much innovation in large technical systems, transport innovation tends to involve high risks and pressures towards reliable infrastructure management. Experiments are therefore often undertaken as means towards systematic improvement, along the classical model of the controlled experiment – whether through testing circuits or through policy programmes like the famous Connecticut crackdown on speeding (Campbell and Ross, 1968). By contrast, transitions theory envisions experiments as means towards more fundamental system innovation. It approaches experiments therefore rather as procedures of experimentalist, uncertainty-embracing governance.

From Testing to Learning

Transport innovation can generally rely on abundant expertise and evidence. Whether through the R&D programmes of the automotive, rail and aviation sectors or through the extensive knowledge infrastructures available for transport engineering and transport planning, there is considerable capacity towards stepwise optimization. In the traditional transport engineering and planning approaches, experimentation is thus typically undertaken along the lines of the scientific experiment, under controlled conditions. The leading rationale is one of testing and falsification: How many miles on public roads can the driverless Waymo cars make faultlessly? Will the lowered speed limit lead to improved air quality conditions without negative side-effects on traffic flow and safety?

In contrast with this focus on testing, transitions theory rather considers experiments primarily as vehicles for social learning. This emphasis on learning reflects a constructivist critique of the positivist 'social engineering' approach to experimentation (Vergragt and Brown, 2007). Rather than organizing Popperian 'critical tests' on the feasibility of certain technologies, the aim is rather to explore how alternative socio-technical arrangements could, after some adjustments and some shifts in attitudes on the side of users, be eventually fitted in with regular 'regime' structures. The key idea is that experiments can be instruments towards 'double-loop learning', that is, towards the reframing of received ideas about performance, comfort, design and user practices. Literature on TM (Gorris and van den Bosch, 2016), 'bounded socio-technical experiments' (Brown et al., 2003) and SNM (Kemp et al., 1998) has thus demonstrated through various case studies how experiments have welded supportive actor networks around 'niche' innovations like electric vehicles, off-peak commuting or park-and-ride facilities. Rather than seeking to free the testing from interferences, these approaches typically embrace the complexity of experimentation under 'real-world' conditions.

From Pilots to Experimentation Cycles

In conventional approaches to innovation, experiments are often treated as bounded and temporary pilots. Experiments form part of implementation trajectories, serving to discard insufficient options at a low cost and to identify promising options for further diffusion and upscaling. Focusing on the efficient filtering out of the better options, experiments are ultimately considered as development costs – in the ideal case, one could do without these costly activities.

This project management logic of controlled innovation trajectories is considered less conducive to systemic innovation and transitions. The TM approaches typically aim for 'double-loop' learning, reframing of societal challenges, and the exploration of alternative futures. They seek to open up processes of political decision-making, rather than closing down prematurely on apparently safe options. This translates into innovation management that is geared towards sustained experimentation trajectories and portfolios, extended learning cycles, and broadly carried innovation ecosystems. Beyond the pilot projects, this development of innovation systems is crucial. SNM underlines the importance of developing alignments between diverse actors, as innovation ecosystems that support evolutionary 'niches' along their struggle for survival (Kemp et al., 1998). This programmatic view on experiments also follows from the transitions-theoretical idea that system innovation tends to rely on broader 'cascades' of innovations (Rotmans, 2005). System innovation is considered to result from co-evolution (Nykvist and Whitmarsh, 2008) and intersections between multiple innovation trajectories – for example those between information and communication technology (ICT)-induced dynamic traffic management and the organizational shifts towards joined-up mobility policy (Pel, 2014). Reaching for broader systemic impacts, a key challenge for TM is to secure the consolidation of learning processes. It revolves around the institutional anchorage of otherwise ephemeral projects (van Buuren and Loorbach, 2009).

From Failure Avoidance to High-Risk Experiments

The preceding considerations of purposes and management logics already indicate how transitions theory sheds a different light on experimental success and failure. Transitions theory, especially through its evolutionary economics background, can explain well why much transport innovation takes the form of cautious experimentation, controlled testing of marginal improvements, and reliance on certified procedures to produce evidence (Kemp et al., 1998). Whilst acknowledging the incentives towards such risk-avoiding approaches, it also underlines their shadow sides: they tend to favour system improvement, and fail to generate evidence on (and interest in) emergent 'niche' innovations. The activist model of TM stresses therefore that high-risk experimentation is needed (Rotmans, 2005). Emblematic examples are the experiments with off-beat, counterintuitive and controversial innovations like driverless transportation pods, rush hour avoidance (Gorris and van den Bosch, 2016), or the anarchist traffic management approach of Shared Space (Pel, 2016).

The ideals of learning-from-failure have proven difficult to live up to, however. As pointed out by de Bruijne et al. (2010), even transition experiments are eventually subjected to the project management logic of milestones, budgetary constraints and pressures towards positive outcomes. Project leaders, participants and stakeholders have expectations and requirements to deliver results. It is striking in this regard how accounts of experimental failure – by whichever standard – remain very sparse in transitions literature. Identifying several often-cumulating biases towards success stories, Turnheim and Sovacool (2020) observe how transitions research and practice remain pervaded by improvement-oriented innovation paradigms. Despite the emphasis on learning and acceptance of risk, the shift towards experimentalist governance is often counterbalanced by the desire to deliver concrete results and solutions.

4.   RELEVANT DIMENSIONS: FROM SUSTAINABLE TECHNOLOGIES TO SOCIAL-INSTITUTIONAL TRANSFORMATION

The transitions-theoretical emphasis on system innovation comes with somewhat alternative understandings of the appropriate scope and purposes of transport innovation and experimentation. The system innovation angle also stands out through its empirical focus on particular kinds and dimensions of innovation: transport innovation is sought less in efficiency enhancements, sustainable technologies and spatial adaptations; instead, it is sought in the transformation of the social-institutional structures – the 'regimes' – that the aforementioned transport innovations are embedded in.

From Technological Fix to Socio-Technical Transformation

As indicated earlier, the arguments towards system innovation take particular issue with technological fixes (Rotmans, 2005). The infatuation with technological advances is diagnosed to form an inherent part of the systemic problems: new combustion engines and automotive designs come and go, but cars and the car system essentially remain the same (Wells and Nieuwenhuis, 2012). Infrastructure expansions, safety measures and 'smart' traffic management solutions absorb and accommodate steadily increasing traffic volumes, but they leave the structural dynamics underlying car dependency and fossil fuel dependency intact (Geels et al., 2012). Careless and technology-pushed introduction of 'smart' intelligent transportation systems (ITS) solutions is therefore considered not intelligent (de Haan et al., 2011). In line with transportation planning concepts of transit-oriented development, park-and-ride facilities, modal shift and mobility management, transitions theory is inclined instead towards multi-dimensional strategies that target travel behaviour more comprehensively. The notion of the socio-technical 'regime' translates this knowledge on transport systems into strategies of transport system *innovation*: transitions theory essentially follows Latour (1992) in seeking innovations that change the socio-technical webs that technologies are simultaneously holding together and constituted by – not only the company-sponsored bicycle but also the showering facilities, not only the charging stations but also the fiscal infrastructure to support electric mobility. The historical studies of transition processes have thus elicited how the innovation of highway infrastructures relied upon the development of codified knowledge and institutionalized expertise (Geels, 2007), how underground parking became embedded in social practices

(Stanković et al., 2021), and how the removal of traffic lights can be undertaken as a social innovation initiative towards 'democracy on the streets' (Pel, 2016). The SNM repertoire is exemplary for this drive towards socio-technical innovation: it revolves around the transformation of the rules, conventions, beliefs, social relations and routines that technological innovations like electric cars are shaped by (Kemp et al., 1998).

From 'Human Factors' to Cultural-Institutional Shifts

Whilst still alive and kicking in contemporary ITS visions, many beliefs in technological fixes have already been broadened to include elements of mobility management. More generally, behaviour change and sensitivity to diversity in lifestyles have become key elements in transport policy and transport innovation. Transport economics and traffic psychology have developed a vast stock of knowledge on so-called Human Factors: the price elasticities of road pricing schemes, driver responses to road infrastructure extensions, the acceptance of speed limit restrictions, and the attention in vehicle automation to the resulting complexity of driving tasks (Horrey and Lee, 2020) – the behavioural dimensions of transportation systems have become key concerns of transportation research.

With regard to transport innovation, much of this Human Factors research remains limited however to issues of societal acceptance and user convenience. As social innovation scholars have pointed out, such innovation remains an extension of technology development – failing to trigger changes in social relations, power structures or institutions (Moulaert and MacCallum, 2019). By contrast, the transitions-theoretical aim for system innovation does aim for such social innovation, and for innovation that changes society's 'rules of the game'. With regard to transport innovation this leads to an approach in the spirit of Urry (2004): the innovation targets the social dynamics, the 'reconfigurations of time and space' that drive the self-perpetuating automobility system. Transitions theory indeed displays a clear convergence with approaches from mobilities research and practice theory (Temenos et al., 2017), seeking innovations in the form of reconfigured routines, slow urbanism and shared mobility, and more generally in the form of alternative mobility cultures (Freudendal-Pedersen et al., 2020). The latter approaches could be considered the somewhat more fine-grained and sociologically refined approaches: they highlight how everyday mobilities of individuals are shaped by mobility cultures, routines, activity patterns, spatial structures and technological landscapes, that is, through the various 'codes', 'scripts' and rules that make up society. In line with Latour (1992), much attention goes to the social practices that form around technologies like safety belts. By contrast, transitions theory is rather oriented towards more encompassing institutional change. In line with the idea of system innovation as organization-transcending innovation (Rotmans, 2005), innovation is often sought in structural couplings between policy arenas (van Brussel et al., 2016), in integrative planning models (Switzer et al., 2013), or in the formation of hybrid institutional arrangements like cooperative ITS (Pel et al., 2020). Following Geels (2007), the quests for system innovation tend to target the key institutions, the spiders in the social web that keep the mobility system on its unsustainable course.

From Sustainability Solutions to Ideological Lock-Out

Transitions theory strongly diverges from the technological fixes and associated 'Human Factors' approaches of conventional transport innovation. Scholars in mobilities studies and practice theory have rightly pointed out however that the professed shift to cultural-institutional innovation remains rather half-hearted (Temenos et al., 2017). Much transitions research remains focused on sustainability-enhancing technologies. Despite the systemic and institutionalist perspective on sustainable mobility, technologies and artefacts remain the central focus of empirical studies (Geels, 2005; Lin et al., 2018). This focus on 'sustainable technologies' is particularly strong in SNM and TIS approaches. Yet even the relatively more comprehensive strategies along the MLP or TM frameworks tend to aim for somewhat one-dimensional forms of 'regime' change: as strategies revolve around sustainability solutions and 'low-carbon' measures, systemic mobility problems are reduced to matters of clean energy.

The prevalence of this environmental rationale marks how transitions theory has retained certain 'ecological modernization' elements (cf. section 2). As a result it tends to neglect a range of other societal impacts and system pathologies that have been addressed in mobilities scholarship and related work on mobility politics (Cresswell, 2010): this comprises issues of exclusion and social inequality, segregated space and 'right to the street', surveillance and data ownership, cybersecurity, healthy lifestyles, and the various societal problems associated with hypermobility and social acceleration. To be sure, transitions-theoretical analyses have addressed ideological 'regime' elements such as the individualistic tendencies towards automobile 'cocooning' (Wells and Xenias, 2015). Also noteworthy are the accounts of the various cultural 'niches' (Sheller, 2012) and counter-cultural movements that wage resistance to the oppressive, alienating effects of a car-dominated society (Zijlstra and Avelino, 2012). The latter examples indicate how transitions theory has started to invoke mobilities scholarship to refine its critical analysis of contemporary mobility regimes. This also reveals however that not all of these reflections on 'mobile subjectivities' can be easily translated into practice and innovation management (Pel, 2016). This difficulty in moving from 'sustainable solutions' towards broader emancipation programmes also reflects the more fundamental difference that exists between social critique on the one hand and innovation theory on the other.

5.   SCOPE: FROM INNOVATION TO EXNOVATION

As indicated in the introduction, transport innovation remains surrounded with modernist beliefs in 'innovating our way out'. Strongly rooted in ecological modernization thinking, transitions theory is far from immune to this optimism. This is counterweighted however by its long-term perspective, which provides a confronting look in the rear-view mirror: What has all the past innovation delivered us? The transitions-theoretical perspective reminds us that – however path-breaking some current developments may seem – much transport innovation is already behind us. The analyses of technological path dependency demonstrate that the direction and scope of further innovation have largely been set already. The once so innovative development of underground parking facilities has gained considerable social and technological obduracy by now, for example. For current system innovation it is therefore vital to anticipate the expiry of the crucial long-term concessions to parking companies (Stanković et al., 2021). Transitions research is thus increasingly concerned with the counterparts of innovation. These phenomena of rewinding, destabilization, decay and phase-out are often referred to as 'exnovation'.

From 'Niche' Cultivation to Phase-Out

Transitions theory essentially conceptualizes societal transformation as a process of cumulating, 'cascading', innovations. Accordingly, frameworks like SNM, TM and TIS provide management repertoires for the cultivation of radical niche innovations. The key underlying idea is that emergent technologies need some time to become compatible with social practices and institutional structures: after gaining consistency, social support and economic interest, niches can then be scaled up, further refined, mainstreamed, and eventually become the standard (Kemp et al., 1998). Emblematic examples of such niche cultivation are last decade's sweeping changes in renewable energy. Past and present examples of transport 'niches' are safety belts, catalytic converters and automated driver assistance systems. In recent years it has been becoming increasingly apparent however that the cultivation of 'sustainable niches' is not enough for 'regime' transformation, and probably insufficiently committal to ensure that 'decarbonization' is achieved in a sufficiently timely way (Rosenbloom and Rinscheid, 2020). Research on transitions governance has therefore started to explore broader 'policy mixes': the cultivation strategies are complemented with broader policy packages that typically contain strategies towards the discontinuation, phase-out and decline of undesirable system elements (Kivimaa and Kern, 2016). Well-known examples in the transportation sector are the locally imposed Low Emission Zones, and the implied phasing out of polluting cars. Less evident examples are the Shared Space schemes; these initiatives towards self-organizing traffic are essentially rewinding a wave of earlier innovations through which traffic coordination became 'delegated' to traffic engineers, fences and traffic lights (Pel, 2016).

From Substitution to Layering

Strategies of 'niche cultivation', and conventional innovation understandings more generally, are working under strong assumptions of substitution. Transport innovation is then understood along the following common – yet idealized – patterns: successive automotive designs conforming with steadily stepped up vehicle type requirements, steady learning curves towards responsible and civilized traffic behaviour, gradual modal shifts, progressively updated generations of taxation schemes, and waves of transport measures that are being 'rolled out' over areas and jurisdictions. Through its evolutionary economics background, transitions theory is similarly guided by ideas of successive technological paradigms. On the other hand, transitions theory also draws on insights from institutional theory that paint a more variegated picture. These insights remind that institutional change – and system innovation – only seldom occurs through neat substitution, and only too often yields a rather messy layering of old and new arrangements. The Low Emission Zones may thus be ambitious and forceful impulses towards accelerated substitution of unsustainable technologies/practices, yet still these measures are inserted into a layered geology of taxation arrangements, spatial policies, mobility cultures and infrastructures (Callorda Fossati et al., under review). Analysing evidence of apparent shifts towards post-car lifestyles, Hopkins (2017) similarly questioned whether this could be taken as signs of regime 'destabilization' – the observed changes in travel behaviour could also reflect only transient phases in individual life courses. Actually, it is precisely in the transport domain that this institutional layering is particularly manifest: whether it concerns selective downgrading of infrastructure, local vehicle access restrictions, dedicated zones of reduced speed limits or the expansion of 'home-zone' regimes, all of these interventions evoke the typical debates about the fairness and the institutional consistency of the emerging 'patchwork' of local arrangements.

From Innovation-as-Progress to Innovation as Double-Edged Sword

The reorientation towards exnovation makes for a more historically informed and analytically balanced view on transport innovation. This attention to the destructive side of Schumpeter's 'creative destruction' is arguably inherent to the transitions perspective: innovation may be at the focus of many analyses, but it is not a goal in itself. In the ultimate instance, innovation is only secondary to the key concerns of system transformation and system *sustainability*. In other words, the 'exnovation turn' in transitions theory also has normative significance: it urges to reconsider what constitutes innovation success. With regard to driverless cars, arguably the frontier of transport innovation, it is common for example to take the stages of full automation and wide employment as natural end points of innovation trajectories (Pel et al., 2020). From the viewpoint of mobility transition this is not the obvious future to aspire to, however – it makes sense therefore to analyse activists' campaigns towards deceleration as efforts towards cautious and *prudent* transition (Hess, 2020). This exemplifies the broader trend that conventional innovation approaches, with their modernistic ideas of innovation-as-progress, are under increasing pressure (Godin and Vinck, 2017). This pressure is especially mounting with regard to sustainability-oriented innovation: as the gap between agreed climate targets and achievements 'on the ground' is widening, the time for non-committal and innovation-embracing governance strategies seems to be running out. The very term of 'decarbonization' implies restrictive strategies instead (Rosenbloom and Rinscheid, 2020). Equally significant is the rise of discourses on 'just transition', degrowth, societal deceleration and social-ecological collapse. It could be said that some of these principled rejections of innovation have as such little insight to offer for transport innovation. Concepts like 'responsible stagnation' (de Saille and Medvecky, 2016) and 'exnovation' do bring something to the table, however: they strengthen the transitions-theoretical insight that innovation is a double-edged sword.

6.   CONCLUSION: TRANSITIONS THEORY AS THEORY OF INNOVATION JOURNEYS

This book defines transport innovations as 'new elements of the transport system that are implemented in the real world'. Transitions theory is a strand of research that takes this systemic understanding very seriously. Other than seeking to improve or complement established repertoires of innovation management, it introduces a different innovation logic and a different governance philosophy that puts innovation in the service of structural societal transformation. Relentlessly asking what innovations bring in terms of a broader mobility 'regime shift', transitions theory proposes a different understanding of the purposes of innovation. The particular implications of this system innovation angle have been explored along four basic innovation aspects: the overarching rationale that guides it (section 2), the underlying governance philosophy (section 3), the relevant dimensions (section 4), and its scope

Table 2.1   *Conventional and transitions-theoretical views on transport innovation*

Innovation aspect	'Conventional' views	Transition perspective
Rationale: From ecological modernization to 'system innovation'	• Reducing externalities • System improvement • Impact assessment	• Regime shift • Systemic transformation • Institutional change
Governance philosophy: From evidence-based improvement to experimentalist governance	• Experimentation as testing • Experiments as pilots • Avoiding failure	• Experimentation as learning • Experiments as cycles • High-risk experiments
Relevant dimensions: From sustainable technologies to social-institutional transformation	• Technology fix • Human factors • Sustainability impacts	• Socio-technical change • Cultural-institutional shifts • Ideological lock-out
Scope: From innovation to exnovation	• Niche cultivation • Innovation as substitution • Innovation as progress	• Phase-out • Innovation as layering • Ambiguous innovation

(section 5). Providing a summary of observations, Table 2.1 helps in seeing how these four discussions hang together: the *rationale* of system innovation implies a broadening of perspective, which is consistent with arguments for complexity-embracing and experimentalist *governance*. This holistic system innovation rationale also corresponds with innovation strategies that cut across the technological, social and institutional *dimensions* of transport innovation. Finally, it can be said that the system innovation angle is relatively well equipped to deal with the recent calls to deepen the *scope* of innovation management – the rationale of system innovation also implies attentiveness to the various exnovation challenges that emerge in the process. Meanwhile, it is also possible to read the table as a compilation of 12 insights, each advancing from relatively reductionist ('conventional') approaches towards the more comprehensive innovation approaches of transitions theory.

Table 2.1 details the point made in the introduction: transitions theory challenges common understandings of transport innovation *purposes*. Hence the following research question:

How does the transitions theory perspective enrich existing insights on the success, failure and societal impacts of innovations in transportation?

A first answer is then that the transitions-theoretical perspective opens up the understanding of societal impacts. It takes a particularly critical, political angle on them: the notorious problems of rebound effects, displacement of problems and incremental improvements are confronted head-on through a principled focus on systemic impacts. Stressing the need for shifts in socio-technical 'regimes' and specifying the technologies, cultures, institutions, business models and infrastructures involved, it provides a comprehensive perspective. This meets the often-expressed need in transportation research for integrative insights. Reaching beyond one-dimensional solutions and antagonistic debates about the right approach to take, transitions-theoretical approaches seek innovation in both market and state logics, in cultures and in technologies, in community action and in R&D.

A second conclusion is however that the systemic overview and long-term orientation do not make for particularly *accurate* accounts of societal impacts. Transitions theory provides only quite rough diagnoses. The systems-theoretical disdain for quantified impact assessment, whether environmental, cost–benefit, spatial or socio-economic, is not without theoretical motivation – assessment models are often reductionist and static, thereby silently reproducing the dominant 'regime' logic. Still, it remains essential in transport innovation processes to develop tailored solutions, based on detailed analysis of local conditions and on evidence that carries political weight. It is thus not coincidental that much transitions research into transport innovation leans on a degree of engineering knowledge, economic analysis and methods of environmental assessment. Meanwhile, one could maintain that the transitions-theoretical focus on system-transformative impacts is just assessing societal impacts in a fundamentally different way. After all, the notion of the socio-technical 'regime' does allow one to register a wide range of social and technological developments that somehow impact the mobility system. Yet as discussed, transitions theory also remains a bit limited in these sociological, constructivist terms: some analyses are attentive to the mobility-related social pathologies as highlighted in mobilities research and critical geography, yet much transitions work remains focused on matters of clean energy, the stakes, politics and ethics of mobility innovation often remaining covered under a blanket of general 'sustainability' discourse. Regarding the societal impacts of transport innovation one can thus conclude that transitions theory holds a certain middle ground between the positivist-quantitative and the constructivist-qualitative approaches available – yet without being particularly strong in either of them.

This leads towards the third conclusion: transitions theory has the most added value regarding its nuanced and strategically sophisticated understanding of innovation success and failure. Various conventional views on transportation and mobility provide more accurate and refined understandings of the societal impacts at issue, it is true. They often remain stuck however in either idealistic accounts of envisioned 'solutions' (e.g. the societally optimal road charging scheme) or critical societal diagnoses that provide no leads towards possible ways out (e.g. taking unsustainable mobility patterns as derivatives of 'neoliberal' power structures). By contrast, transitions theory is strongly informed by the knowledge on innovation dynamics as gathered in Science & Technology Studies, sociology, institutional theory, reflexive governance, and innovation studies. As became especially clear in the discussion of the underlying governance philosophy (section 3), all of the transitions-theoretical innovation frameworks are frontally addressing the basic characteristic of innovation – *uncertainty*. Transitions literature, following van de Ven et al. (2000), commonly speaks of (sustainable) innovation *journeys*. It is along this same sensitivity to innovation dynamics and uncertainty that transitions theory seems well equipped to deal with the ambiguous transport innovation challenges of 'exnovation' and 'responsible stagnation' (section 5).

REFERENCES

Banister, D. (2008). The sustainable mobility paradigm. *Transport Policy*, 15(2), 73–80.

Berger, G., Feindt, P.H., Holden, E., and Rubik, F. (2014). Sustainable mobility – challenges for a complex transition. *Journal of Environmental Policy and Planning*, 16(3), 303–320.

Brown, H.S., Vergragt, P., Green, K., and Berchicci, L. (2003). Learning for sustainability transition through bounded socio-technical experiments in personal mobility. *Technology Analysis and Strategic Management*, 15(3), 291–315.

Callorda Fossati, E. et al. (under review). Exnovation: décloisonner les imaginaires de transition vers une économie durable à Bruxelles.

Campbell, D.T., and Ross, H.L. (1968). The Connecticut crackdown on speeding: Time-series data in quasi-experimental analysis. *Law and Society Review*, 3(1), 33–53.

Cohen, M. (2010). Destination unknown: Pursuing sustainable mobility in the face of rival societal aspirations. *Research Policy*, 39(4), 459–470.

Cresswell, T. (2010). Towards a politics of mobility. *Environment and Planning D: Society and Space*, 28(1), 17–31.

De Bruijne, M., van de Riet, O., de Haan, A., and Koppenjan, J. (2010). Dealing with dilemma's: How can experiments contribute to a more sustainable mobility system? *European Journal of Transport and Infrastructure Research*, 10(3), 274–289.

De Haan, J., Vrancken, J.L.M., and Lukszo, Z. (2011). Why is intelligent technology alone not an intelligent solution? *Futures*, 43(9), 970–978.

De Saille, S., and Medvecky, F. (2016). Innovation for a steady state: A case for responsible stagnation. *Economy and Society*, 45(1), 1–23.

Freudendal-Pedersen, M., Hartmann-Petersen, K., Friis, F., Rudolf Lindberg, M., and Grindsted, T.S. (2020). Sustainable mobility in the mobile risk society – designing innovative mobility solutions in Copenhagen. *Sustainability*, 12(17), 7218.

Geels, F.W. (2005). Processes and patterns in transitions and system innovations: Refining the co-evolutionary multi-level perspective. *Technological Forecasting and Social Change*, 72(6), 681–696.

Geels, F.W. (2007). Transformations of large technical systems: A multilevel analysis of the Dutch highway system (1950–2000). *Science, Technology, and Human Values*, 32(2), 123–149.

Geels, F.W., Kemp, R., Dudley, G., and Lyons, G. (eds) (2012). *Automobility in Transition? A Socio-technical Analysis of Sustainable Transport*. London: Routledge.

Godin, B., and Vinck, D. (eds) (2017). *Critical Studies of Innovation: Alternative Approaches to the Pro-innovation Bias*. Cheltenham, UK and Northampton, MA, USA: Edward Elgar Publishing.

Gorris, T., and van den Bosch, S.J.M. (2016). Applying transition management in ongoing programmes and projects in the Netherlands: The case of Transumo and rush hour avoidance. In Geerlings, H., Shiftan, Y., and Stead, D. (eds), *Transition towards Sustainable Mobility: The Role of Instruments, Individuals and Institutions* (pp. 71–94). London: Routledge.

Hajer, M.A. (1995). *The Politics of Environmental Discourse: Ecological Modernization and the Policy Process*. Oxford: Oxford University Press.

Hess, D.J. (2020). Incumbent-led transitions and civil society: Autonomous vehicle policy and consumer organizations in the United States. *Technological Forecasting and Social Change*, 151, 119825.

Hopkins, D. (2017). Destabilising automobility? The emergent mobilities of generation Y. *Ambio*, 46(3), 371–383.

Horrey, W.J., and Lee, J.D. (2020). Preface to the special issue on human factors and advanced vehicle automation: Of benefits, barriers, and bridges to safe and effective implementation. *Human Factors*, 62(2), 189–193.

Kemp, R., Schot, J., and Hoogma, R. (1998). Regime shifts to sustainability through processes of niche formation: The approach of strategic niche management. *Technology Analysis and Strategic Management*, 10(2), 175–198.

Kivimaa, P., and Kern, F. (2016). Creative destruction or mere niche support? Innovation policy mixes for sustainability transitions. *Research Policy*, 45(1), 205–217.

Köhler, J., Geels, F., Kern, F., Onsongo, E., Wieczorek, A.J., Alkemade, F., Avelino, F., Bergek, A., Boons, F., Fuenfschilling, L., Hess, D., Holtz, G., Hyysalo, S., Jenkins, K., Kivimaa, P., Markard, J., Martiskainen, M., McMeekin, A., Mühlemeier, M.S., Nykvist, B., Pel, B., Raven, R., Rohracher, H., Sanden, B., Schot, J., Sovacool, B., Turnheim, B., Welch, D., and Wells, P. (2019). An agenda for sustainability transitions research: State of the art and future directions. *Environmental Innovation and Societal Transitions*, 31, 1–32.

Latour, B. (1992). Where are the missing masses? The sociology of a few mundane artifacts. In Bijker, W. and Law, J. (eds), *Shaping Technology, Building Society: Studies in Sociotechnological Change* (pp. 225–258). Cambridge, MA: MIT Press.

Lin, X., Wells, P., and Sovacool, B.K. (2018). The death of a transport regime? The future of electric bicycles and transportation pathways for sustainable mobility in China. *Technological Forecasting and Social Change*, 132, 255–267.

Markard, J., Raven, R., and Truffer, B. (2012). Sustainability transitions: An emerging field of research and its prospects. *Research Policy*, 41(6), 955–967.

Meyer, M.D., and Miller, E.J. (2001). *Urban Transportation Planning: A Decision-Oriented Approach*. 2nd edition. Singapore: McGraw-Hill.

Moulaert, F., and MacCallum, D. (2019). *Advanced Introduction to Social Innovation*. Cheltenham, UK and Northampton, MA, USA: Edward Elgar Publishing.

Newman, P.G., and Kenworthy, J.R. (1989). *Cities and Automobile Dependence: An International Sourcebook*. Brookfield, VT: Gower Publishing.

Nykvist, B., and Whitmarsh, L. (2008). A multi-level analysis of sustainable mobility transitions: Niche development in the UK and Sweden. *Technological Forecasting and Social Change*, 75(9), 1373–1387.

Pel, B. (2014). Intersections in system innovation: A nested-case methodology to study co-evolving innovation journeys. *Technology Analysis and Strategic Management*, 26(3), 307–320.

Pel, B. (2016). Interactive metal fatigue: A conceptual contribution to social critique in mobilities research. *Mobilities*, 11(5), 662–680.

Pel, B., and Boons, F.A. (2010). Transition through subsystem innovation? The case of traffic management. *Technological Forecasting and Social Change*, 77(8), 1249–1259.

Pel, B., Raven, R.P.J.M., and van Est, Q. (2020). Transitions governance with a sense of direction: Synchronization challenges in the case of the Dutch 'driverless car' transition. *Technological Forecasting and Social Change*, 160, 120244.

Rosenbloom, D., and Rinscheid, A. (2020). Deliberate decline: An emerging frontier for the study and practice of decarbonization. *Wiley Interdisciplinary Reviews: Climate Change*, 11(6), e669.

Rotmans, J. (2005). Societal innovation: Between dream and reality lies complexity. Inaugural address, Erasmus Universiteit Rotterdam.

Sheller, M. (2012). The emergence of new cultures of mobility: Stability, openings and prospects. In Geels, F.W., Kemp, R., Dudley, G., and Lyons, G. (eds), *Automobility in Transition? A Socio-technical Analysis of Sustainable Transport* (pp. 180–202). New York: Routledge.

Smith, A., Voß, J.P., and Grin, J. (2010). Innovation studies and sustainability transitions: The allure of the multi-level perspective and its challenges. *Research Policy*, 39(4), 435–448.

Stanković, J., Dijk, M., and Hommels, A. (2021). Upscaling, obduracy, and underground parking in Maastricht (1965–present): Is there a way out? *Journal of Urban History*, 47(6), 1225–1250.

Switzer, A., Bertolini, L., and Grin, J. (2013). Transitions of mobility systems in urban regions: A heuristic framework. *Journal of Environmental Policy and Planning*, 15(2), 141–160.

Temenos, C., Nikolaeva, A., Schwanen, T., Cresswell, T., Sengers, F., Watson, M., and Sheller, M. (2017). Theorizing mobility transitions: An interdisciplinary conversation. *Transfers*, 7(1), 113–129.

Turnheim, B., and Sovacool, B.K. (2020). Exploring the role of failure in socio-technical transitions research. *Environmental Innovation and Societal Transitions*, 37, 267–289.

Urry, J. (2004). The 'system' of automobility. *Theory, Culture and Society*, 21(4–5), 25–39.

Urry, J. (2012). *Sociology beyond Societies: Mobilities for the Twenty-First Century*. London: Routledge.

Van Brussel, S., Boelens, L., and Lauwers, D. (2016). Unravelling the Flemish mobility orgware: The transition towards a sustainable mobility from an actor-network perspective. *European Planning Studies*, 24(7), 1336–1356.

Van Buuren, A., and Loorbach, D. (2009). Policy innovation in isolation? Conditions for policy renewal by transition arenas and pilot projects. *Public Management Review*, 11(3), 375–392.

Van de Ven, A.H., Angle, H.L., and Poole, M.S. (eds) (2000). *Research on the Management of Innovation: The Minnesota Studies*. Oxford University Press on Demand.

Van Wee, B. (2011). *Transport and Ethics: Ethics and the Evaluation of Transport Policies and Projects*. Cheltenham, UK and Northampton, MA, USA: Edward Elgar Publishing.

Vergragt, P.J., and Brown, H.S. (2007). Sustainable mobility: From technological innovation to societal learning. *Journal of Cleaner Production*, 15(11–12), 1104–1115.

Wells, P., and Nieuwenhuis, P. (2012). Transition failure: Understanding continuity in the automotive industry. *Technological Forecasting and Social Change*, 79(9), 1681–1692.

Wells, P., and Xenias, D. (2015). From 'freedom of the open road' to 'cocooning': Understanding resistance to change in personal private automobility. *Environmental Innovation and Societal Transitions*, 16, 106–119.

Zijlstra, T., and Avelino, F. (2012). A socio-spatial perspective on the car regime. In Geels, F.W., Kemp, R., Dudley, G., and Lyons, G. (eds), *Automobility in Transition? A Socio-technical Analysis of Sustainable Transport* (pp. 160–179). New York: Routledge.

3. Modelling innovations in freight transport: a business ecosystem perspective

Giovanni Zenezini and Lóránt A. Tavasszy

1. INTRODUCTION

Innovations in logistics services revolve around the fundamental challenge of making a step change in service quality towards the customer and reducing the costs to deliver these services. As service improvements involve higher costs, the two challenges are often tackled together. Changes in logistics service quality have mainly been driven by the digitalization of services and the servitization of product offerings (i.e. the addition of service elements to a product). Consumers have gotten used to the possibility to choose among different distribution options, including highly responsive services, like home delivery within the day, or even within hours. From a company perspective, omnichannel distribution – a separate channel for each customer segment – has become standard practice (Buldeo Rai et al., 2019; Taylor et al., 2019). As, by definition, customized services serve smaller segments of consumers, service providers miss out on the earlier benefits of economies of scale, which makes logistics more expensive. In addition, competition between service providers has put prices under pressure – home deliveries as well as returns are still done at low prices, or even for free. This pressure is absorbed by companies through innovations in logistics processes, either within or outside the company, by one or more of the options below:

- new logistics technology and organization (e.g. autonomous warehouses or delivery robots);
- improved yield management (i.e. higher prices for consumers willing to pay more);
- horizontal or vertical collaboration across the supply chain (e.g. co-procurement of services between competing firms; mergers and acquisitions of firms situated in different echelons of the supply chain);
- internalizing external costs of services (e.g. pricing environmental impacts).

The aim of this chapter is to discuss the modelling of transport innovations, with the purpose to predict impacts of innovations and thus support decision making. Our focus is on the multi-company city logistics environment that comprises several practical examples of these innovations such as:

- service providers or manufacturers pooling their transport orders to reduce costs;
- two manufacturers that source goods and services together, producing a similar effect;
- a shared warehouse available for multiple firms as opposed to a single-client warehouse;
- running an urban consolidation center (UCC) with public subsidy, justified by environmental impacts;
- price premiums for environmentally friendly services (e.g. zero emission vehicles).

These examples illustrate how changes can affect multiple actors in the system, and cross the boundaries of several institutions, with regulatory, legal, and even political challenges. Decision making around such innovations can be long-cycled (years or decades) for large-scale innovations, and short-cycled (weeks or months) for smaller, incremental innovations. In these situations, the relationships between stakeholders with different motives and business models are a critical aspect for understanding how to make effective decisions (Anand et al., 2012; Cagliano et al., 2017). Furthermore, decision making can transcend the concerns of private markets if public subsidization, investment, or regulation is involved. Models will help to predict the impacts for all the stakeholders of the city logistics system, public and private, and thereby aid the design and implementation of policies. In the context of co-creation of innovations by different stakeholders of city logistics, the role of modelling is changing – from supporting long-term cycles of policy making and implementation, to supporting short cycles of incremental innovation. These cycles are similar to the policy cycle but faster paced and shared between stakeholders. They include ex post analysis, predictions of upcoming states of the system, and optimization of control and implementation measures. The question for this chapter, then, is how models can help to assess economic, social, and environmental impacts of logistics innovations in the complex multi-actor systems called City Logistics (CL).

Descriptive and predictive freight transport models have come a long way, from the earliest econometric transport system equations to the current transparent agent-based simulation models, which aim to mimic everyday logistics decisions. The first approaches for freight modelling consolidated all logistics decisions into aggregate structures, describing freight production and attraction, trade, mode choice, and routing for an entire city, region, country, or even the world. Throughout the decades these models have evolved by providing further detailing of these structures in behavioral terms (Comi et al., 2014). More and more, logistics decisions were being considered in descriptive models, including decisions on distribution structures, multimodal chains, vehicle types, and routing and scheduling of trips. Also, disaggregate approaches provided empirically valid models at the level of the individual firm. Recent reviews of freight modelling emphasize the need to continue in the direction of a more realistic representation of actual logistics business processes (see e.g. Anand et al., 2015; Meersman and Van de Voorde, 2019; Tavasszy, 2020; Tavasszy et al., 2012). Also, the nature of modelling to support innovations is changing from an arm's length reflective role towards one similar to action research, where the modelling becomes part of the innovation cycle (OECD, 2020). Simulation models and agent-based models help to progress in this direction as they show how individual firm behavior and interactions between firms lead to an aggregate outcome, which is of interest for the policy maker who oversees innovation processes and might decide to intervene if negative externalities ensue. Moreover, these models differ from traditional models because they include many, heterogeneous agents and these agents receive feedbacks from other agents and are therefore better equipped to model the non-linear behaviors of complex innovative ecosystems.

We argue that models of firms and their interactions should preferably be built on a conceptual framework that recognizes the main interests of the model users. In this chapter we propose a framework for analysis, based on business ecosystems, that formally identifies the different actors in the system and their business interrelations, including private and public stakeholders. The main premise is that innovations will affect these actors through their relations and that innovations thus do not affect only one actor, but multiple or all actors. We illustrate how innovations propagate through the system of actors in city logistics through several examples of new public and private initiatives. Also, we explain how these ideas can be operationalized in empirically grounded agent-based models of cities.

The chapter is built up as follows. First, section 2 explores the theoretical background underlying the foundations of the business ecosystem agent-based modelling. Then, the general operationalization of the business ecosystems perspective on transport innovation is outlined in section 3, together with a practical implementation example for urban freight ecosystems in section 4. In section 5 we discuss theoretical and practical implications of this work, and finally we draw the conclusions in section 6.

2.   BUSINESS ECOSYSTEM AS LENS: THEORETICAL BACKGROUND

Business Ecosystems Theory

Theoretical and practical frameworks for designing and assessing business models and decisions "assume that the strategic outcome can be defined independently of the reactions of other players" (Tian et al., 2008, p. 102). However, a critical challenge that is not entirely dealt with by the business model concept lies in characterizing the relationships among business entities and understanding how decisions taken by one entity affect other interrelated entities (Tian et al., 2008). In some sectors, companies combine to provide services, thus taking the form of a business ecosystem (or network).

A business ecosystem is defined as a network of interrelated business entities, characterized by value transfer and value co-creation mechanisms (Wang et al., 2015), operational transactions, and interdependencies between business entities (Solaimani et al., 2015). This definition of a network of interrelated companies as a business ecosystem stems from the ecology research arena, whereby biological ecosystems are depicted as complex systems of organisms and relationships among them (Battistella et al., 2012). Likewise, within business ecosystems, "firms interact in complex ways, and the health and performance of each firm are dependent on the health and performance of the whole. Firms ... are therefore simultaneously influenced by their internal capabilities and by their complex interactions with the rest of the ecosystem" (Iansiti and Levien, 2002, p. 8). Business entities composing a business ecosystem can at the same time cooperate, to improve the growth of the business ecosystem, and compete for market shares (Battistella et al., 2012).

The business ecosystems literature recognizes the existence of roles and actors along the value chain, and draws attention to the necessity of making a clear distinction between roles due to the presence of different functions performed by the ecosystem companies (Pohlen and Farris, 1992). In fact, roles are defined in the pertinent literature as an aggregation of activities performed, as well as of the resources necessary to perform them. In this sense, roles serve as the basic element of a business ecosystem, whereby actors perform specific roles to achieve the overarching objectives of the ecosystem (Story et al., 2011). As a matter of fact, the profitability of the ecosystem is affected by the organizational structure underlying the assignment of actors (i.e. firms) to the role played, taking into consideration that different firms are able to take on the same role. Regarding this notion, most authors argue that, to some extent, it is possible to single out the most efficient firm–role assignment, through either qualitative inquiry or mathematical estimation (Savaskan et al.,

2004). However, in order to achieve and maintain the network structure at the efficient frontier it is necessary to understand and develop role-specific competences (Harland and Knight, 2001). Harland and Knight (2001) also argue that organizations can adjust the role played in the network, and thus respond to factors that have an impact on their performance by taking on different roles. Network management is also a very relevant role in a business ecosystem, and covers a wide range of activities, including collating and analyzing information and disseminating it to other actors so as to coordinate physical and information flows and facilitate communication and innovation (Harland and Knight, 2001).

In essence, by assessing through the theoretical lens of the business ecosystem framework how innovations affect changes inside a network of firms it is possible to achieve several objectives. On the one hand, this framework brings forward a perspective shift from the focal firm, typical of the traditional business model concept, to the whole ecosystem of firms. On the other hand, the framework still allows us to highlight all the individual business models of which the ecosystem is composed. Moreover, the business ecosystem framework acknowledges that when innovations are introduced the roles played by the firms change dynamically. Finally, business ecosystem theory provides more leeway for opening up the analysis towards all relevant actors in the ecosystem. In the context of transport innovation this means that public stakeholders, who ought to be included in the assessment as previously mentioned, are also given different roles and an actionable business model to drive their decisions.

These considerations make the business ecosystem framework well suited in our view to study not only business model changes through innovation but also technology transition regimes.

Transition Management in Business Ecosystems

Innovations in sustainable transport are wicked problems: they concern many actors and groups of actors with vested interests, who are not easily amenable to fundamental change (Kemp et al., 2007). Therefore, our view of organizations should consider more factors than those that cause short-term inertia in the system. Theories about change management, system transitions, and institutional economics have created the discipline of transition management to support the realization of major societal, or landscape, innovations (Geels, 2002). Here, the so-called "regimes" or robust structures of institutions prevent individual technological or organizational innovations – however radical they may be – from changing the system landscape. Therefore, the institutional economics of systems (see e.g. Williamson (2000) for a systematic description) – including the institutions themselves, their governance arrangements, and their management practices – should be understood and operationalized. This could provide an understanding of the detailed agenda of measures needed for change, which is directed at the system actors, their powers, and the value systems by which they are driven. We argue that the ecosystem lens is instrumental in this respect, as it recognizes the motivation and capability of actors to identify and create new inter-organizational business arrangements. Simply put, if we can predict how patchworks of regimes change, we may be able to predict system transitions.

The ecosystem approach is particularly useful in innovation contexts focused on both value creation and value capture, because it allows analysis to explicitly tackle not only the challenges faced by the focal firm but also those of the external partners and stakeholders (Adner and Kapoor, 2010). Therefore, the processes of technology substitution or business model innovation are in fact driven by the competition between "old" and "new" business ecosystems, and hindered by bottlenecks somewhere in the ecosystem that constrain the full realization of the new technology's (or business model innovation's) potential performance (Adner and Kapoor, 2016). Rong et al. (2015) argue that the process of new supply chain emergence cannot be explained using traditional supply chain theories. Instead, interoperability between different levels of organizations is necessary to cope with the uncertainties embedded in transition processes. Moreover, during the co-evolution of business ecosystems we see a process of emergence of dominant supply chains.

3.   A BUSINESS ECOSYSTEM PERSPECTIVE FOR TRANSPORT INNOVATION: GENERAL OPERATIONALIZATION

In the literature, several tools are available for modelling business ecosystems and analyzing the impacts of different business decisions taken by the business entities operating within the business ecosystem. A suitable implementation of agent-based modelling (ABM) to business ecosystem design and analysis is provided by the role-based modelling approach (Ok et al., 2013; Tian et al., 2008). In this approach, business entities can play multiple roles and make decisions reacting to the changes in the ecosystem over time, and based on their objectives, information, and constraints.

Modelling Business Ecosystems with ABM

As previously mentioned, traditional transport modelling approaches fall short of grasping the complex dynamics of multi-actor economic processes which determine the adoption of innovations. The proposed business ecosystem lens enables on the other hand the capturing of interdependencies and interrelations among firms in dynamic ecosystems where reconfigurations of roles and functions emerge continuously, aiming to create and capture value and generating patterns of competition or cooperation. A good fit for modelling business ecosystems is agent-based modelling, insofar as it is able to model organizational complexities and the interdependencies among organizational design elements and decision making (Rivkin and Siggelkow, 2003) better than other modelling approaches. Moreover, the processes of emergence and self-organization are very important features of agent-based models, and they imply that some properties belong only to the system as a whole and not to its individual components (Grimm et al., 2005).

In agent-based models, a bottom-up approach is adopted to define and represent a complex system, rather than identifying global variables ruling the system as a whole. Hence, there are three basic elements in each agent-based model:

- a set of agents, together with their attributes and behaviors;
- a set of relationships and rules that drives agents' interaction;
- the agents' environment.

General Theory

The main pillars of this framework are roles and business entities, representing the most important agents in the business ecosystem. These two types of agents operate differently, whereby business entities represent the firms operating in the ecosystem that enter into contractual relationships with each other, and roles are the functional agents of the system carrying out operational activities.

The first pillar of the framework requires a working representation of how to define a role. The definitions available in the literature are however very context-specific and, while pointing to the notion that multiple companies can play the same role, they do not indicate specifically what categories and variables can be used in order to separate roles and companies. To solve this dilemma and achieve more precision, a role is here defined as a bundle of different functions and activities, but since companies can perform similar functions the distinction between the roles can be somewhat blurred, and this could generate problems and conflicts between actors. Hence, a specific role k can be defined as:

$$R_k = \{A_k, D_k, M_k\} \tag{3.1}$$

where A_k, D_k, and M_k are sub-sets of activities, decisions, and metrics available in the ecosystem.

Table 3.1   *Elements of the framework*

Component	Definition	Properties
Role	A role is a bundle of different activities, decisions, and metrics available in the ecosystem.	Activity(s) Decision(s) Metric(s)
Business Entity	A business entity is an actor of the business ecosystem. A business entity can be associated with a particular type depending on the ecosystem context.	Type Role(s)
Resource	A resource can be a physical (e.g. a vehicle, a warehouse), intangible (e.g. knowledge, intellectual property), or financial asset. Resources are owned by the business entities and are necessary for the roles to be performed.	Owner Unit cost Operational characteristics
Activity	An activity is performed by a business entity while playing a specific role, in order to offer a service. Activities consume resources.	Resource usage
Metric	A metric is a key performance indicator (KPI) measuring a certain business object, namely activities, resources, value proposition exchange, business entity, ecosystem.	Business object Value
Decision	Business entities make operative and economic decisions in the fulfilment of their roles, based on a set of constraints, variables, decision parameters.	Objective Decision variable set Constraint set
Service	A service is an aggregation of activities that use resources and are characteristics of a role.	Service attributes Activity set
Value Proposition	A value proposition is a set of service offerings characterized by different gained benefits that are valued by users.	Provider and user Services Evaluation method

The value proposition represents the component of the system which dictates if a certain role will be taken by a business entity, thus driving a contractual relationship with another business entity. A value proposition has been defined as a bundle of products and services which represents a value for a specific customer (Osterwalder, 2004).

In a business ecosystem, the interrelations between resources, activities, value propositions exchanged, and decisions are fundamental. As anticipated, a business entity performs activities and requires investment in resources to build a sustainable business model. Then, the value proposition exchanged lies at the core of a specific business model configuration, which in turn determines which business entity takes certain decisions as well as the partnership model. These decisions have an impact on activity execution, and metrics are used to assess quantitatively the outcome of activity execution so as to evaluate the role-playing performance (Table 3.1).

Business entities must choose which roles to play in the business ecosystem, thus deciding which of the roles' specific activities, decisions, value propositions, and metrics to inherit. Business entities also have entity-specific attributes and relationships. The most important attribute possessed by a business entity regardless of the roles played is represented by the resources (human, financial, physical etc.) owned. As a matter of fact, the availability of resources has a significant influence on the types of roles a business entity can play in the ecosystem. A depiction of the general workings of the role-based business ecosystem is given in Figure 3.1. It centers around the assignment of roles to business entities. The physical flow of goods relates to the roles in the system, independent of their business owner, as these are the agents executing the physical process. Due to this property, next to the physical flows, the roles can also give rise, together with the contractually determined service and payment agreements that flow between business entities, to other intangible benefits (e.g. process status information, or social involvement). The execution process also provides information and feedback to the business entities that own the role.

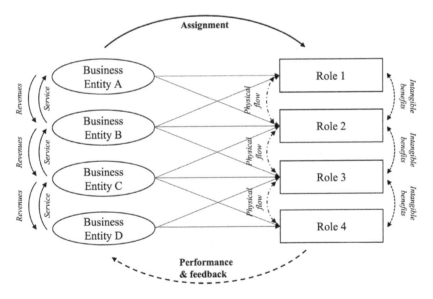

Figure 3.1   Relationships between roles, business entities and their roles

In other words, goods flow between roles and services flow between business entities in return for the exchange of revenues and intangible benefits. Business entities own monetary resources and thus are able to enter into logistics contracts and acquire services from other business entities. In essence, the value exchanges of money, goods, and services, as well as the intangible benefits (e.g. value proposition), are dependent on the role assignment, and are thus created (or co-created) and exchanged during the actual execution of the roles. For this reason, the boundaries between the roles have to be defined in a clear-cut way so as to identify the most basic elements of a business ecosystem that are still capable of providing value to the ecosystem and entice business entities to develop a sustainable business model around them.

The business model of a business entity is thus identified with the set of roles the business entity is playing and its relationship with other business entities, which are substantiated through formal or informal contractual obligations. This will lead to the coexistence of different business models in the system, such as the case of global players (e.g. Logistics Service Providers (LSPs)) offering a wide array of services for different market segments. Hence, each business ecosystem consists of a set of business entities and roles, together with the assignment of business entities to the roles. A business ecosystem then represents just one of the possible configurations of the system stakeholders and interactions.

Innovation in transport ecosystems brings forward a transition, either radical or incremental, from one configuration to another. For example, new business entities enter the ecosystem to provide value added services to other business entities and can enhance the overall profitability of the ecosystem in two ways. First, they can marginally improve the performance of the status quo role assignment through technological advancements that increase operational efficiency, without changing the underlying structure of the system. Second, they can create new logistics value and business relationships by either aggregating or separating the existing roles, thus contributing to a potential shift from one regime to another. In such a way, existing business entities are able to change some of the roles they play, moving towards a specialization (i.e. playing fewer roles) or a vertical integration (i.e. aggregating roles). The former case may be exemplified by a business entity outsourcing a purely operational role to a more specialized business entity, such as is the case with freight transportation tasks, which are usually carried out by haulers on behalf of large LSP organizations. The latter case instead involves business entities deciding to internalize more roles if synergies arise from the aggregation and bundling of services and products.

By the same token, changes in the role configuration of a business ecosystem may be fostered by the repositioning of existing business entities not necessarily driven by the entrance of new players. One could think for example of the breadth of roles being played by the online retailing giant Amazon, which goes beyond the traditional role of retailer to include those of logistics service provider (i.e. through the separate entity Amazon Logistics) and cloud computing platform offering Internet-as-a-service (Iaas) services to small and large businesses. These role changes were enabled by the availability of resources and by the fact that other business entities evaluated positively the benefits being generated through the service delivery.

The general description of the theoretical framework presented in this section is expanded upon in the next section via an application to urban freight transportation (UFT) systems.

4.   AN APPLICATION TO URBAN FREIGHT TRANSPORTATION[1] SYSTEMS

UFT systems are characterized by a multitude of stakeholders with different and often conflicting objectives (Anand et al., 2014; Macharis et al., 2014). Moreover, urban freight has been center stage for the introduction of several logistics and transport innovations that cooperate or compete with incumbent players, such as cargo-bike delivery (Arnold et al., 2018; Gruber et al., 2014; Melo and Baptista, 2017), delivery through crowd-sourcing (Buldeo Rai et al., 2018; Devari et al., 2017; Guo et al., 2019; Le et al., 2019), or urban consolidation centers (Browne et al., 2005; Johansson and Björklund, 2017; Morganti and Gonzalez-Feliu, 2015; Paddeu et al., 2017).

The general role-based business ecosystem theoretical framework has been applied to UFT systems by Zenezini (2018; Zenezini et al., 2017, 2019). First of all, a practical implementation of the general theory must begin with the identification of the ecosystem boundaries. For urban freight systems, these are represented by the logistics process entailed in the last mile of freight transportation from a local source to the final recipients of the goods, which comprise retailers and final customers. For instance, this could be represented by the last leg of the physical distribution journey from the reception of goods at the distribution center located in the outskirts of an urban area to the final customer.

Then, agents must be defined, including business entities and roles. In particular, the roles are identified and classified:

1. Receiver. This role generates the demand for freight but is not in charge of any decision regarding the delivery process and only acts as recipient of the goods. This role is usually covered by final customers and local retailers.
2. User of logistics and city delivery services. These two roles also generate the demand for freight but actively decide to use one or more logistics or transportation services. Users of logistics service providers often require a wider array of services including warehousing and cross-docking, while users of city delivery services only need to outsource the transportation

of goods in the last mile. Usually shippers take on the role of users of logistics service providers, while express couriers often use local freight carriers for the city delivery.

3. Logistics service provider and city delivery operator. These two roles are taken by business entities that are appointed by the two previous roles to deliver parcels and other goods. These roles comprise the functions of goods consolidation as well as last-mile planning and delivery. City delivery operators offer only the transportation service. Usually express couriers such as DHL or small city transport companies take on these roles.

4. Network coordinator. While the previous roles are centered on the physical flow of goods downstream along the last-mile chain, this role covers those necessary activities and competences required for a smooth and transparent flow of information between users and providers. In other words, they provide the interface between the service providers and the users. Usually transport providers provide coordination services but, in some cases, intermediary platforms or public authorities can take on this role.

5. Logistics space planner and policy maker. These roles comprise the functions of land-use planning, in both public and private areas – for instance, facility managers who decide to offer a logistics concierge service for their employees or a public authority that wishes to add more loading/unloading bays for transport companies. The major interest of policy making is to evaluate the aggregate outcome of the ecosystem and intervene when necessary to steer it towards more sustainable goals (e.g. by limiting fossil fuel vehicles).

In addition to the UFT roles, nine business entity types are identified ranging from large global players such as express couriers to facility managers and local freight transportation companies. A more thorough and encompassing definition of CL business entities, roles, activities, resources, decisions, and value propositions is available in Zenezini (2018, pp. 20, 42).

The next step of an ABM implementation is the assignment of roles to business entities. In this regard, entities can only perform a handful of roles due to their inherent constraints or internal objectives. Nevertheless, most entities have significant leeway to change their status quo situation and move towards new roles, thus triggering the value creation process and ultimately the generation of a new business ecosystem. In any case, CL systems need to comprise all the roles identified in the matrix (Table 3.2), but, since business entities can take up more than one role, they can consist of only a sub-set of business entities.

Table 3.2   Role assignment matrix

Business entity	Role							
	Receiver	User of logistics services	User of city delivery services	Logistics service provider	City delivery operator	Network coordinator	Logistics space planner	Policy maker
Express courier		X	X	X	X	X	X	
City freight carrier					X		X	
Last-mile operator				X	X	X	X	
UCC operator	X			X	X	X	X	
Parcel locker operator	X		X	X	X	X	X	
Shipper		X	X		X			
Large retailer	X	X	X		X		X	
Local retailer	X	X	X				X	
Local authority	X	X				X	X	X
ICT-platform operator						X		
Facility manager	X	X				X	X	
Final customer	X	X	X					

Note: X marks a potential role–entity assignment.

Table 3.3   Business and operative decisions of CL roles

Role	Strategic decisions	Operational decisions
Receiver	Choice of logistics services Evaluation of level of service Evaluation of intangible benefits	Decide stock levels Inventory policy: economic order quantity (EOQ), frequency of delivery, time of delivery
User of logistics services	Choice of logistics services Demand allocation (long term) Evaluation of level of service Evaluation of intangible benefits	Demand allocation (short term)
User of city delivery services	Suppliers selection Evaluation of level of service Evaluation of intangible benefits	Demand allocation (short term)
Logistics service provider	Value proposition setting Level of service provided	Fleet allocation Vehicle routing Demand allocation
City delivery operator	Pricing scheme Budget allocation	Fleet allocation Vehicle routing
Network coordinator	Resource acquisition	Data quality control Computational capacity allocation

Each role–entity assignment configuration implies an allocation of decisions to business entities. Therefore, a business entity makes different decisions based on the roles played, and thus adopts different decision-making attributes. In the CL business ecosystem, decisions are related to both business and operational aspects of role execution (Table 3.3).

In the next two sub-sections we show a working example of how the theoretical framework is used by comparing and contrasting a traditional business ecosystem with an innovative one.

Traditional Urban Freight

A traditional urban freight business ecosystem focused on home delivery is usually composed of four entities taking on eight different roles, as shown in Table 3.4. Generally speaking, online retailers outsource the physical distribution to express couriers, who in turn consolidate different flows at their cross-docking centers as well as sorting the final delivery to delivery vans which are mostly operated by small local carriers. Final customers pay for the delivery but usually do not get to choose the LSP in charge of the delivery. Finally, local authorities are responsible for setting local regulations for freight vehicles.

Table 3.4   *Traditional last-mile delivery role–business entity assignment*

Business entity	Role						
	Receiver	User of logistics services	User of city delivery services	Logistics service provider	City delivery operator	Network coordinator	Logistics space planner	Policy maker
Express courier			X	X	X	X	X	
City freight carrier					X			
Online retailer		X						
Local authority								X
Final customer	X	X						

The value propositions offered by providers to users are often centered on speed, reliability, flexibility, visibility, and total cost of ownership (Ghodsypour and O'Brien, 2001; Dulmin and Mininno, 2003; Awasthi et al., 2016; Hwang et al., 2016), as shown in Figure 3.2. Local authorities are seemingly outside of the picture in traditional city logistics ecosystems because they do not offer logistics services directly. However, their actions, aimed at increasing the sustainability of transport operations, have an impact on providers. For instance, restrictions on polluting vehicles reduce the overall emissions level but increase the cost of transport providers (Broaddus et al., 2015; Dablanc and Montenon, 2015). On the other hand, a congestion charge might reduce the number of vehicles and thus increase the commercial speed.

By comparing Tables 3.2 and 3.4, we see that in Table 3.4 (i.e. the traditional UFT business ecosystem) fewer Xs are marked and thus there is some untapped potential for innovation due to several missing assignments between business entities and roles.

Innovative Urban Freight

As mentioned above, previous literature has explored a variety of innovative urban freight innovations that have attempted to alter the ecosystem by changing the traditional assignments of business entities to roles.

In this sub-section, we will focus on a sub-set of innovations which relate to the concept of the UCC. A UCC is a logistics facility that bundles consignments coming from multiple carriers and aims to consolidate deliveries to local retailers and final customers in order to reduce the number of vehicles required, the distance travelled, and the CO_2 emissions (Browne et al., 2011; Heeswijk et al., 2017; Johansson and Björklund, 2017). In particular, the four cases of UCC-based UFT business ecosystems presented here show that very similar innovations can shape the ecosystem in radically different ways.

Case 1: Targeting new customer segments and consolidating last-mile deliveries

The first case of innovation depicts a new business entity operating a last-mile delivery service through a distribution center and a network of parcel lockers located inside the office buildings of large employers (Table 3.5). This new company is opening up a new market in the traditional urban freight ecosystem by offering a dual value proposition: for employers the value proposition consists in the fact that the additional workload at the reception desk of the employer will be relieved if employees ship their items to an unmanned automated locker; for employees the service reduces the risk of missed deliveries without bearing additional cost.

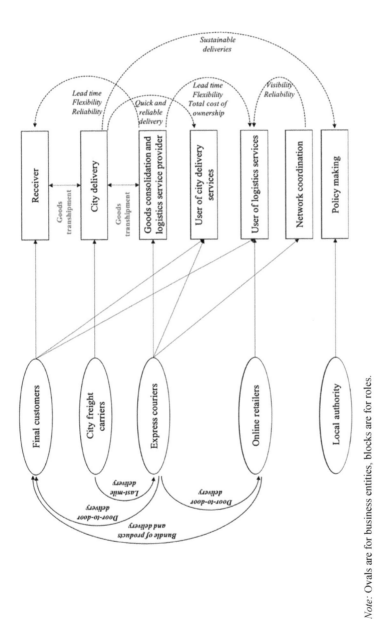

Note: Ovals are for business entities, blocks are for roles.

Figure 3.2   Traditional last-mile delivery business ecosystem

Table 3.5   First case of urban freight innovation – business ecosystem configuration*

Business entity		Role						
	Receiver	User of logistics services	User of city delivery services	Logistics service provider	City delivery operator	Network coordinator	Logistics space planner
Express courier						X	X
City freight carrier				X	X		
UCC operator	X		X	X	X	X	X
Online retailer		X					
Facility manager		X					X
Final customer		X					

Note: * Local authorities play the role of policy maker.

After signing up to the service, employees make their online purchase and enter the company's distribution center as a delivery address while receiving a code to open the parcel locker which will contain their parcel. Express couriers then deliver goods to the company's distribution center on behalf of the shippers. Finally, the company receives the parcels from the couriers and sorts them onto the delivery vans operated by a city freight carrier for the final leg. The value proposition then rests upon good coordination between the company and the express couriers, who are required to deliver at the distribution center early in the day in order to comply with the delivery service levels.

Case 2: A UCC subsidized by a local authority

The second case shows a typical example of a UCC implemented by a local administration and operated by an LSP (Table 3.6). The value proposition for this UCC operator is again dual. First, express couriers outsource the city delivery to a city freight carrier, a situation akin to a business-as-usual configuration. Second, local retailers pay the last-mile delivery service in a bundle with the extra storage service provided by the UCC, which in turn increases delivery service flexibility and speed (i.e. local retailers can have their parcels delivered on very short notice from the close-by UCC). Local retailers are thus asked to be more proactive in their logistics choices in comparison to the traditional urban freight.

The UCC consolidates goods destined to retailers in the Central Business District (CBD) of the city, and then operates a fleet of electric vehicles for the delivery. Besides subsidies provided by the local city council, which account for 45 percent of operation costs, the revenue streams derive from usage fees paid by both local retailers and express couriers. The value proposition in this case is sustained largely with very low and competitive fees, which could put the UCC's financial stability in jeopardy once subsidies are terminated.

Case 3: A privately owned UCC

This case represents a company operating a network of urban consolidation centers in Dutch cities (Table 3.7). It focuses on offering goods consolidation and other logistics services (e.g. delayed cross-docking, home deliveries, waste returns) to small local retailers. The major value proposition for the company is aimed at local retailers, who can take advantage of a decreased number of deliveries and a lower inventory, which are typical benefits of a receiver. Hence, the UCC operator receives monetary remuneration from local retailers, who need to be proactive and shift towards the role of logistics services users.

The UCC operator acts as a logistics service provider and organizes the last-mile delivery process, as in the previous cases. Moreover, as in the previous cases, there is an overlapping of logistics service provider and network coordinator roles between the new business entity and incumbents (i.e. express

Innovations in transport

Table 3.6   Second case of urban freight innovation – business ecosystem configuration

Business entity	Role							
	Receiver	User of logistics services	User of city delivery services	Logistics service provider	City delivery operator	Network coordinator	Logistics space planner	Policy maker
Express courier			X	X		X	X	
UCC operator				X	X	X	X	
Local retailer	X	X	X					
Shipper		X						
Local authority		X					X	X

Table 3.7   *Third case of urban freight innovation – business ecosystem configuration*

Business entity	Role						
	Receiver	User of logistics services	User of city delivery services	Logistics service provider	City delivery operator	Network coordinator	Logistics space planner
Express courier				X		X	X
City freight carrier					X		
UCC operator			X	X		X	X
Local retailer	X	X					
Shipper		X					

couriers). Retailers in fact pay a monthly membership fee plus an additional fee for the extra logistics services. The last-mile delivery is outsourced to city freight carriers. Contrary to the previous UCC case, this UCC operator does not target express couriers specifically and instead hopes to target shippers by offering them an ICT system integration package that provides a single interface to receive real time Proof of Delivery (POD) for all their shipments and enables them to combine shipments for geographical areas. As a consequence, network coordinator is a role where the UCC operator is putting in considerable effort in order to offer a valuable service and provide intangible benefits to shippers.

Case 4: A pickup point for employees

This case hinges on an internal pickup point and consolidation center located within a university (Table 3.8). The main value proposition in this case is to provide a service to employees. Moreover, intangible benefits are also reaped by the express couriers, who can be certain that their deliveries will not fail and can optimize their routing by consolidating deliveries in a single stop. In some regards, this case study is akin to Case 1. However, in this particular case the delivery process is not automated as in Case 1, where parcel lockers were installed and no interaction between the driver and the personnel occurs.

Daily operations include receiving deliveries for all employees (about 2000 people) and sorting them by university department, and are subcontracted to a third-party company by the university. Thus, for the employees this center operates as a pickup point for their online purchases, whereby delivery receipt is notified via electronic exchange and employees can pick their purchases up within office hours. Express couriers retain their business-as-usual business and operational model. Again, network coordinator is a role of paramount importance for the success of the service, even though the pickup point operator does not guarantee any level of service for the delivery.

From the cases presented in this section we can draw some implications for CL business ecosystems, as well as make insightful linkages between CL practice and the business entity theoretical lens and its application to transport innovations.

Competition between old and new ecosystems, and related challenges

The new company entering the market in Case 1 becomes a logistics service provider, thus competing with larger firms. The decisive success factor for the new player here is to improve the goods consolidation and logistics service provider role performance, and find a coordination mechanism with the express couriers in the absence of a contractual agreement. Challenges arise when competition ensues between ecosystems. The UCC operator of Case 2 for instance acts as an additional decoupling point, bearing operational costs

Table 3.8   *Fourth case of urban freight innovation – business ecosystem configuration*

Business entity	Role						
	Receiver	User of logistics services	User of city delivery services	Logistics service provider	City delivery operator	Network coordinator	Logistics space planner
Express courier			X	X		X	X
City freight carrier					X		
Pickup point operator	X			X		X	X
Online retailer		X					
Facility manager		X					X
Final customer		X					

without creating additional value to exchange for higher revenues. Moreover, the UCC operator performs the role of city delivery operator and offers the service to the local retailers, who have already paid for a part of the delivery process and are not always able to negotiate a reduction of delivery fees with shippers and couriers. Hence, acting as both logistics service provider and city delivery operator might not yield economic and financial sustainability for the business entity aggregating those roles. Finally, a very important role that each of the previous new business entities had to perform is that of network coordinator. To perform such a role, the business entities had to develop skills and acquire resources. As previously mentioned, when the complexity and number of linkages among business entities and roles increases, the network coordinator ensures that the delivery goes as smoothly as possible and different supply chains integrate seamlessly. On the operational side, it is often required that new business entities develop an integrated ICT platform from scratch. As a matter of fact, an ICT platform is a required asset for the network coordinator role, which can be performed by new business entities in a more effective and efficient way than other business entities.

Value creation mechanisms

Creating and providing value for existing and new customer segments is key for ecosystem innovation. For instance, the network coordinator does not only help stakeholders switch to the new business model, but could also provide additional value and constitute a profitable service, as in Case 3 for shippers. The new company in Case 1 must compete in performing the same role as the express couriers but by adding an additional consolidation point hopes to gain revenues by providing value to a new customer in the network, namely the employer. In turn, express couriers might benefit from disengaging from the last leg of the delivery process, which accounts for a large share of the total logistics cost. Similar intangible benefits are achieved by express couriers in the case of the university pickup point operator in Case 4. Ideally, monetary flows should be generated in exchange from all stakeholders who benefit from value creation. Unfortunately, this is not always possible due to the ecosystem's inertia and the bargaining power of large incumbent players. Gaining a critical mass of users must then be achieved in order to shift part of the power from incumbents to entrants.

Role improvement

The new business entity in Case 1 takes advantage of the fact that it is not profitable for employers to act as receiver, since it is not rewarding for them and it generates hidden costs of inbound operations. The key to becoming profitable and attractive to employers is to evaluate correctly the value of the solution from the employers' point of view and propose a service fee lower than that value. Reducing the cost entailed in playing a certain role in the ecosystem is thus an efficient way to improve the overall profitability and the ecosystem and to thrive in it.

5. IMPLICATIONS FOR RESEARCH AND PRACTICE

This work generates several research and practical implications for researchers, practitioners, public bodies, and transport innovators.

Implications for Research

This work opens up a variety of potential implications for business ecosystem modelling, by pivoting on the linkages between the strategic decisions taken at the firm level in the face of innovation and the intrinsic operational processes of a transport ecosystem. These linkages work both ways, since the decision from a firm to take on a role and enter into contractual relationships is ultimately driven by the operational aspects entailed by that specific role. Hence, more strategic decisions should be added at the role level to investigate endogeneity in the model. For example, the decision to change a role might be triggered by the failure of an entity to make profit, or by other conditions such as an entity not maximizing other objectives.

In order to release the full capability of the business ecosystem framework and turn its underlying tenets into actionable and useful bottom-up simulation models, it is however necessary to gain more understanding of the behaviors of firms when innovation occurs, bearing implications at different levels of decision making. Bottom-up modelling requires a lot of trial-and-error due to the fact that acquiring behavioral data is a complicated enterprise as it requires abstracting complex behaviors from real life. In this context, the rules that govern decision-making processes can be set up in various ways. Strictly logical, deterministic rules would assign only one possible behavior to an individual in a particular circumstance (Grimm and Railsback, 2005). Alternatively, a rule may be probabilistic, with a different probability for each choice in an array of possible actions in response to some stimulus. Rules may furthermore be a combination of probabilistic and deterministic.

We thus point out some of the thornier issues that researchers need to address while implementing the business ecosystem perspective in a full-fledged agent-based model.

First, while the identification of roles metrics is quite straightforward when they are concerned with tangible objects such as services and resources, it is much more complex when intangible benefits are exchanged between roles and business entities. Second, the decision to take a certain role is binary, meaning that behaviors change abruptly after certain thresholds are achieved. Are these thresholds only represented by the effectiveness of a value proposition offering or by a better cost–benefit analysis, or else are there other aspects to be considered, such as conforming with others or long-term goal-seeking behaviors? Firms for instance might look beyond their immediate payoff and maximize their long-term profits. It is worthwhile in this sense to explore reinforcement learning mechanisms for firm agents (Kara and Dogan, 2018; Teo et al., 2012).

Third, innovation in the ecosystem may trigger various reactions from incumbents. Incumbents can either:

• keep on playing the same roles and cooperate by complementing some innovators' activities or markets;
• imitate the business model, if the innovation is incremental and requires only minor changes in competences and resources (Casadesus-Masanell and Zhu, 2013);
• create new market needs, leveraging the innovation to change their products and service offering (Bucherer et al., 2012).

The agent-based model could then incorporate separate behavioral foundations in the agents' code in order to simulate the effects of these different strategic decisions taken by incumbents of the ecosystem. In this regard, the ABM implementation is not the focus but rather the means through which researchers are able to address the multi-faceted complexities of transport innovation business ecosystems and achieve quantitative evaluations for all stakeholders involved. Hence, the ABM implementation would involve using the most consolidated software available (see Borshchev and Filippov, 2004; Macal and North, 2010; for a more comprehensive review of ABM implementation tools and softwares). Nevertheless, researchers should deal with more manageable applications of the theoretical framework by focusing on a sub-set of activities, decisions, and metrics among the ones included in the business ecosystem roles. The goal of the ABM implementation is thus to describe the business entities as agents that are able to adapt by taking proactive or reactive decisions based on the level of metrics. Zenezini (2018, p. 42) provides an example of such a line of thinking based on Case 1 outlined in section 4.

Implications for Practice

The business ecosystem approach marks a major change in modelling from the current methods, which rarely describe the business models of actors, let alone the distinction between the institutional and the business levels. It introduces several innovations in the way we describe actors and processes, for example:

- From a practical standpoint, the business ecosystem agent-based framework enables the assessment of the operational and economic feasibility of innovative solutions in the transport sector, by assigning a business model to all the stakeholders involved. Hence, it allows us to pinpoint the benefits, the responsibilities, and the related challenges for each actor of the ecosystem. As a by-product of this capability, scholars and practitioners may use the framework to identify "winners" and "losers" of a transport innovation.
- Then, several outcomes can unfold. Should transport innovation convey the benefits for other entities in the ecosystem effectively, we would see the positive effects on the ecosystem as a whole already in the short term. Otherwise, other actors might fail to recognize the value of the innovation, thus hindering the long-term sustainability of the innovation itself. The framework in this context may be used to highlight where the discrepancies between the global potential benefits and such barriers to the diffusion of innovation reside.
- Moreover, it is possible to evaluate winners and losers using different scales of evaluation. Innovation in the transport sector can be implemented as a means to achieve environmental sustainability rather than pure economic sustainability. Hence, the business ecosystem agent-based framework already envisions that public authorities become part of the business cycle of the transport innovation. Therefore, when tensions arise between environmental sustainability and financial remuneration of investments by private operators, devising a business model for the local authority supports its entry in the ecosystem as a proactive agent able to smooth those tensions with incentives or regulation.

In the past decades, the history of innovations in city logistics has shown that step-by-step innovations, based on a minimum viable product perspective, are often more effective than radical innovations and large-scale investments. Radical innovations may fail because they only focus on long-term impacts while underestimating the strains and barriers inherent to the change of roles required from the actors in the short term. This is the case of innovation processes, including those in transport, that are positioned in a context of Living Labs (this can be a factory, a consumer group, or an entire city) (Quak et al., 2016). Here, innovations are not presented as big-bang scenarios but in an incremental fashion, where the lab context provides feedback about what works and what does not and allows incremental design and implementation of change. Models in this context have also been named "digital twins" of cities, and form the instrumentation of these labs, where they are fed by sensors that measure all activities, and supply decision makers with scenarios for the future. The business ecosystem lens and the ABM approach allow a dynamic simulation of the interactions among the stakeholders, and allow the evaluation of the implications of multiple, simultaneous or sequential decisions by actors in the system. They could be the backbone for a digital twin of a city's logistics community, which helps to predict and visualize how existing regimes will respond to changes. The effects of radical versus incremental innovations can also be explored with this approach.

6.   CONCLUSIONS

Market pressures from customers and competition forces companies to innovate their logistics processes, either within or outside the company. Innovations in the transport and logistics sectors can affect multiple actors in the system, and entail various decision-making processes ranging from long-cycled (years or decades) for large-scale innovations, and short-cycled (weeks or months) for smaller, incremental innovations.

In these contexts, transport innovation modelling should consider the relationships that take place between stakeholders with different motives and business models. In this chapter, we aimed to fill this gap by introducing a new modelling paradigm that depicts transport systems as business ecosystems. To this end, we explained the antecedents of the paradigm, operationalized its theoretical concepts with a practical application in the area of urban freight transport, and proposed several implications for practice and theory.

Four cases of application to the UFT context of our framework were presented. These cases are based on the innovative concept of a UCC, but differ significantly in terms of business entities involved and reconfiguration of the ecosystem. In Case 1 a new business entity enters the market, aggregating the roles being traditionally played by larger incumbents, and thus hopes to improve the performance of those roles as well as involve more business entities in the ecosystem in order to be successful. In Case 2 the innovator provides the same service without a reconfiguration of the system, and thus simply replicates the same business relationships without providing added value. Case 3 is very similar to Case 2 but aims at providing added value to another business entity, hoping therefore to compete with the larger incumbents. Case 4, finally, is focused on including more business entities in the ecosystem by specializing in a specific role and not by overlapping with the roles being played by the traditional business entities.

The proposed framework may be used in other transport ecosystem contexts where innovations occur, beyond the geographical scope underlying the cases presented in this chapter. The business ecosystem framework may be used for instance to evaluate the transition towards the Physical Internet (PI). PI is a revolutionary concept that aims to coordinate different actors situated in different geographical areas and at different functional levels for a more transparent, smooth, efficient, and sustainable supply chain (Pan et al., 2017). This new paradigm forces us to rethink the traditional roles of global supply chain actors and add new ones such as open warehouses and distribution centers (Crainic and Montreuil, 2016; Oktaei et al., 2014).

This work unveils several opportunities for further research. In fact, researchers can make use of the proposed framework to model the uptake process of transport innovations. In order to do so, however, it is necessary to delve into the links between operational and strategic decisions of the role-based framework. For instance, it could be possible to investigate endogeneity in the model by integrating strategic decisions at the role level. Further exploration is also required in terms of understanding the behaviors of the ecosystem firms in the face of incremental or radical innovation.

This work also engenders several implications for practice. First, the business ecosystem agent-based framework allows pinpointing the benefits, the responsibilities, and the related challenges for each actor of the ecosystem, including the public authorities (e.g. by including environmental benefits as well). Second, it allows shedding a light on barriers that exist in the ecosystem and hinder the success of transport innovations, potentially preventing global benefits being achieved in the long term.

Finally, the business ecosystem lens integrated with the ABM approach could provide a backbone for a digital twin of a transport and logistics environment, providing feedback to ecosystem actors about what works and what does not, thus allowing for an incremental design of innovation and further implementation of change.

NOTE

1. As a synonym for City Logistics.

REFERENCES

Adner, R. and Kapoor, R. (2010), "Value creation in innovation ecosystems: How the structure of technological interdependence affects firm performance in new technology generations", *Strategic Management Journal*, 31(3), 306–333.
Adner, R. and Kapoor, R. (2016), "Innovation ecosystems and the pace of substitution: Re-examining technology S-curves", *Strategic Management Journal*, 37(4), 625–648.
Anand, N., van Duin, R., Quak, H. and Tavasszy, L. (2015), "Relevance of city logistics modelling efforts: A review", *Transport Reviews*, 35(6), 701–719.
Anand, N., van Duin, R. and Tavasszy, L. (2014), "Ontology-based multi-agent system for urban freight transportation", *International Journal of Urban Sciences*, 18(2), 133–153.
Anand, N., Yang, M., van Duin, J.H.R. and Tavasszy, L. (2012), "GenCLOn: An ontology for city logistics", *Expert Systems with Applications*, 39(15), 11944–11960.

Arnold, F., Cardenas, I., Sörensen, K. and Dewulf, W. (2018), "Simulation of B2C e-commerce distribution in Antwerp using cargo bikes and delivery points", *European Transport Research Review*, 10, 2, available at: https://doi.org/10.1007/s12544-017-0272-6.

Awasthi, A., Adetiloye, T. and Crainic, T.G. (2016), "Collaboration partner selection for city logistics planning under municipal freight regulations", *Applied Mathematical Modelling*, 40(1), 510–525.

Battistella, C., Colucci, K. and Nonino, F. (2012), "Methodology of business ecosystems network analysis: A field study in Telecom Italia Future Centre", in De Marco, M., Te'eni, D., Albano, V. and Za, S. (eds.), *Information Systems: Crossroads for Organization, Management, Accounting and Engineering: ItAIS: The Italian Association for Information Systems*, Berlin and Heidelberg: Spinger-Verlag, pp. 239–249.

Borshchev, A. and Filippov, A. (2004), "From system dynamics and discrete event to practical agent based modeling: Reasons, techniques, tools", in *Proceedings of the 22nd International Conference of the System Dynamics Society*, Vol. 22, Citeseer.

Broaddus, A., Browne, M. and Allen, J. (2015), "Sustainable freight: Impacts of the London congestion charge and low emissions zones", *Transportation Research Record*, 2478(1), 1–11.

Browne, M., Allen, J. and Leonardi, J. (2011), "Evaluating the use of an urban consolidation centre and electric vehicles in central London", *IATSS Research*, 35(1), 1–6.

Browne, M., Sweet, M., Woodburn, A. and Allen, J. (2005), *Urban Freight Consolidation Centres Final Report*, available at: http://ukerc.rl.ac.uk/pdf/RR3_Urban_Freight_Consolidation_Centre_Report.pdf.

Bucherer, E., Eisert, U. and Gassmann, O. (2012), "Towards systematic business model innovation: Lessons from product innovation management", *Creativity and Innovation Management*, 21(2), 183–198.

Buldeo Rai, H., Verlinde, S. and Macharis, C. (2018), "Shipping outside the box. Environmental impact and stakeholder analysis of a crowd logistics platform in Belgium", *Journal of Cleaner Production*, 202, 806–816, available at: https://doi.org/10.1016/j.jclepro.2018.08.210.

Buldeo Rai, H., Verlinde, S., Macharis, C., Schoutteet, P. and Vanhaverbeke, L. (2019), "Logistics outsourcing in omnichannel retail", *International Journal of Physical Distribution and Logistics Management*, 49(3), 267–286.

Cagliano, A.C., Carlin, A., Mangano, G. and Rafele, C. (2017), "Analyzing the diffusion of eco-friendly vans for urban freight distribution", *International Journal of Logistics Management*, 28(4), 1218–1242.

Casadesus-Masanell, R. and Zhu, F. (2013), "Business model innovation and competitive imitation: The case of sponsor-based business models", *Strategic Management Journal*, 34(4), 464–482.

Comi, A., Donnelly, R. and Russo, F. (2014), "Urban freight models", in Tavasszy, L. and de Jong, G. (eds.), *Modelling Freight Transport*, Oxford: Elsevier, pp. 163–200.

Crainic, T.G. and Montreuil, B. (2016), "Physical Internet enabled hyperconnected city logistics", *Transportation Research Procedia*, 12, 383–398.

Dablanc, L. and Montenon, A. (2015), "Impacts of environmental access restrictions on freight delivery activities: Example of low emissions zones in Europe", *Transportation Research Record*, 2478(1), 12–18.

Devari, A., Nikolaev, A.G. and He, Q. (2017), "Crowdsourcing the last mile delivery of online orders by exploiting the social networks of retail store customers",

Transportation Research Part E: Logistics and Transportation Review, 105, 105–122, available at: https://doi.org/10.1016/j.tre.2017.06.011.

Dulmin, R. and Mininno, V. (2003), "Supplier selection using a multi-criteria decision aid method", *Journal of Purchasing and Supply Management*, 9(4), 177–187.

Geels, F.W. (2002), "Technological transitions as evolutionary reconfiguration processes: A multi-level perspective and a case-study", *Research Policy*, 31(8–9), 1257–1274.

Ghodsypour, S.H. and O'Brien, C. (2001), "The total cost of logistics in supplier selection, under conditions of multiple sourcing, multiple criteria and capacity constraint", *International Journal of Production Economics*, 73(1), 15–27.

Grimm, V. and Railsback, S.F. (2005), *Individual-Based Modeling and Ecology*, Vol. 8, Princeton, NJ: Princeton University Press.

Grimm, V., Revilla, E., Berger, U., Jeltsch, F., Mooij, W.M., Railsback, S.F., Thulke, H.H., et al. (2005), "Pattern-oriented modeling of agent-based complex systems: Lessons from ecology", *Science*, 310(5750), 987–991.

Gruber, J., Kihm, A. and Lenz, B. (2014), "A new vehicle for urban freight? An ex-ante evaluation of electric cargo bikes in courier services", *Research in Transportation Business and Management*, 11, 53–62.

Guo, X., Lujan Jaramillo, Y.J., Bloemhof-Ruwaard, J. and Claassen, G.D.H. (2019), "On integrating crowdsourced delivery in last-mile logistics: A simulation study to quantify its feasibility", *Journal of Cleaner Production*, 241, 118365, available at: https://doi.org/10.1016/j.jclepro.2019.118365.

Harland, C.M. and Knight, L.A. (2001), "Supply network strategy: Role and competence requirements", *International Journal of Operations and Production Management*, 21(4), 476–489.

Heeswijk, W. van, Larsen, R. and Larsen, A. (2017), "An urban consolidation center in the city of Copenhagen: A simulation study", Beta Working Paper series 523, TU Eindhoven, Research School for Operations Management and Logistics (BETA), Eindhoven, the Netherlands, available at: http://onderzoeksschool-beta.nl/wp-content/uploads/wp_523.pdf.

Hwang, B.-N., Chen, T.-T. and Lin, J.T. (2016), "3PL selection criteria in integrated circuit manufacturing industry in Taiwan", *Supply Chain Management: An International Journal*, 21(1), 103–124.

Iansiti, M. and Levien, R. (2002), "Keystones and dominators: Framing operating and technology strategy in a business ecosystem", Harvard Business School Working Paper, No. 03-061.

Johansson, H. and Björklund, M. (2017), "Urban consolidation centres: Retail stores' demands for UCC services", *International Journal of Physical Distribution and Logistics Management*, 47(7), 646–662.

Kara, A. and Dogan, I. (2018), "Reinforcement learning approaches for specifying ordering policies of perishable inventory systems", *Expert Systems with Applications*, 91, 150–158.

Kemp, R., Loorbach, D. and Rotmans, J. (2007), "Transition management as a model for managing processes of co-evolution towards sustainable development", *International Journal of Sustainable Development and World Ecology*, 14(1), 78–91.

Le, T.V., Stathopoulos, A., Van Woensel, T. and Ukkusuri, S.V. (2019), "Supply, demand, operations, and management of crowd-shipping services: A review and empirical evidence", *Transportation Research Part C: Emerging Technologies*, 103, 83–103.

Macal, C.M. and North, M.J. (2010), "Tutorial on agent-based modelling and simulation", *Journal of Simulation*, 4(3), 151–162.

Macharis, C., Milan, L. and Verlinde, S. (2014), "A stakeholder-based multicriteria evaluation framework for city distribution", *Research in Transportation Business and Management*, 11, 75–84.

Meersman, H. and Van de Voorde, E. (2019), "Freight transport models: Ready to support transport policy of the future?", *Transport Policy*, 83, 97–101.

Melo, S. and Baptista, P. (2017), "Evaluating the impacts of using cargo cycles on urban logistics: Integrating traffic, environmental and operational boundaries", *European Transport Research Review*, 9, 30, available at: https://doi.org/10.1007/s12544-017-0246-8.

Morganti, E. and Gonzalez-Feliu, J. (2015), "City logistics for perishable products. The case of the Parma's Food Hub", *Case Studies on Transport Policy*, 3(2), 120–128.

OECD (2020), "Addressing societal challenges using transdisciplinary research". OECD Science, Technology and Industry Policy Papers No. 88, Paris: OECD Publishing.

Ok, K., Coskun, V., Ozdenizci, B. and Aydin, M.N. (2013), "A role-based service level NFC ecosystem model", *Wireless Personal Communications*, 68(3), 811–841.

Oktaei, P., Lehoux, N. and Montreuil, B. (2014), "Designing business models for Physical Internet transit centers", in *Proceedings of 1st International Physical Internet Conference*, Québec City, Canada.

Osterwalder, A. (2004), "The business model ontology: a proposition in a design science approach", Doctoral Thesis, University of Lausanne.

Paddeu, D., Fancello, G. and Fadda, P. (2017), "An experimental customer satisfaction index to evaluate the performance of city logistics services", *Transport*, 32(3), 262–271.

Pan, S., Ballot, E., Huang, G.Q. and Montreuil, B. (2017), "Physical Internet and interconnected logistics services: Research and applications", *International Journal of Production Research*, 55(9), 2603–2609.

Pohlen, T. and Farris, T. (1992), "Reverse logistics in plastics recycling", *International Journal of Physical Distribution and Logistics Management*, 12(7), 35–47.

Quak, H., Lindholm, M., Tavasszy, L. and Browne, M. (2016), "From freight partnerships to city logistics living labs – Giving meaning to the elusive concept of living labs", *Transportation Research Procedia*, 12, 461–473.

Rivkin, J.W. and Siggelkow, N. (2003), "Balancing search and stability: Interdependencies among elements of organizational design", *Management Science*, 49(3), 290–311.

Rong, K., Hu, G., Lin, Y., Shi, Y. and Guo, L. (2015), "Understanding business ecosystem using a 6C framework in Internet-of-Things-based sectors", *International Journal of Production Economics*, 159, 41–55.

Savaskan, R.C., Bhattacharya, S. and Van Wassenhove, L.N. (2004), "Closed-loop supply chain models with product remanufacturing", *Management Science*, 50(2), 239–252.

Solaimani, S., Bouwman, H. and Itälä, T. (2015), "Networked enterprise business model alignment: A case study on smart living", *Information Systems Frontiers*, 17(4), 871–887.

Story, V., O'Malley, L. and Hart, S. (2011), "Roles, role performance, and radical innovation competences", *Industrial Marketing Management*, 40(6), 952–966.

Tavasszy, L.A. (2020), "Predicting the effects of logistics innovations on freight systems: Directions for research", *Transport Policy*, 86, A1–A6.

Tavasszy, L.A., Ruijgrok, K. and Davydenko, I. (2012), "Incorporating logistics in freight transport demand models: State-of-the-art and research opportunities", *Transport Reviews*, 32(2), 203–219.

Taylor, D., Brockhaus, S., Knemeyer, A.M. and Murphy, P. (2019), "Omnichannel fulfillment strategies: Defining the concept and building an agenda for future inquiry", *International Journal of Logistics Management*, 30(3), 863–891.

Teo, J.S.E., Taniguchi, E. and Qureshi, A.G. (2012), "Evaluating city logistics measure in e-commerce with multiagent systems", *Procedia – Social and Behavioral Sciences*, 39, 349–359.

Tian, C.H., Ray, B.K., Lee, J., Cao, R. and Ding, W. (2008), "BEAM: A framework for business ecosystem analysis and modeling", *IBM Systems Journal*, 47(1), 101–114.

Wang, J., Lai, J.-Y. and Hsiao, L.-C. (2015), "Value network analysis for complex service systems: A case study on Taiwan's mobile application services", *Service Business*, 9(3), 381–407.

Williamson, O.E. (2000), "The new institutional economics: Taking stock, looking ahead", *Journal of Economic Literature*, 38(3), 595–613.

Zenezini, G. (2018), "A new evaluation approach to city logistics projects: a business-oriented agent-based model", PhD Thesis, Politecnico di Torino.

Zenezini, G., van Duin, R., Tavasszy, L. and De Marco, A. (2017), "Stakeholders' roles for business modelling in a city logistics ecosystem: Towards a conceptual model", paper presented at the 10th International City Logistics Conference, Institute for City Logistics, 14–16 June, Phuket, Thailand.

Zenezini, G., Gonzalez-Feliu, J., Mangano, G. and Palacios-Arguello, L. (2019), "A business model assessment and evaluation framework for city logistics collaborative strategic decision support", in Camarinha-Matos, L., Afsarmanesh, H. and Antonelli, D. (eds.) *Collaborative Networks and Digital Transformation. PRO-VE 2019. IFIP Advances in Information and Communication Technology*, Vol. 568. Springer, Cham, available at: https://doi.org/10.1007/978-3-030-28464-0_48.

4. Understanding mobility biographies: conceptual and empirical advancements and practical innovation

Henrike Rau and Joachim Scheiner

1. INTRODUCTION

How people travel in everyday life has been the subject of a burgeoning body of transport research in various disciplines, often in interdisciplinary projects. Much of this research has focused on the empirical investigation of the links between individuals' travel behaviour and wider (infra)structural and urban conditions. Cross-sectional research designs for collecting large-scale quantitative data on topics such as modal choice, route selection and spatio-temporal commuting patterns have dominated the research landscape to date. How and why travel patterns change across the life course has commanded much less attention from transport researchers, resulting in major knowledge gaps. Mobility biographies research (MBR), which has emerged as a field in its own right, has tried to close some of these gaps through innovative longitudinal research approaches that capture mobility practices across the life course (cf. Scheiner and Rau, 2020 for a collection of contributions by prominent international scholars that captures past and current developments in MBR). MBR has also opened up new avenues for concrete change in the transport sector, for example through the promotion of behaviour change initiatives that treat critical life events as windows of opportunity to transform resource-intensive mobility practices (e.g. Schäfer et al., 2012; Rau and Manton, 2016). Tracing the development of MBR thus presents interesting opportunities for understanding the evolution of transport research more generally, and related shifts in transport policy and practices towards demand-side measures that seek to change how (much) people travel. In this chapter we present and discuss some recent trends in researching mobility across the life course. We focus in particular on three main areas: (1) theoretical and conceptual innovations in understanding how and why people travel, including ideas that move well beyond the previously dominant focus on individual behaviour, (2) methodological advancements that push the field beyond its current emphasis on quantitative methods and (3) groundbreaking empirical findings that are likely to send future MBR onto new trajectories. At various points we refer to the implications of MBR for ongoing and future innovations in the transport sector, for example with respect to innovative initiatives such as 'critical mass' that work with the interlinkages between mobility biographies, or school-based opportunities to learn about cycling, public transport, multimodality and related topics.

The remainder of this chapter is divided into five principal sections. Section 2 sketches the main developments of MBR to date, with a view to identifying strengths and weaknesses. Section 3 covers novel theoretical approaches to mobility biographies. Particular emphasis is placed on approaches that cover stability and change from different angles and that seek to identify reasons why changes in daily travel occur. Section 4 examines recent conceptual and methodological innovations that aim to advance the empirical investigation of mobility biographies, including retrospective forms of inquiry. Moreover, it presents and discusses novel themes and trends that have emerged from empirical studies of mobility biographies in the last few years. Their potential to initiate new strands of inquiry in the field as well as a new generation of demand-focused change initiatives in the transport sector is also discussed. Finally, Section 5 offers some concluding remarks and reflections on possible future trends in mobility research, policy and practice, with a focus on the potential role of MBR for transport innovation.

2.   KEY CONCEPTS AND METHODOLOGIES IN MBR

2.1   State of the Art – Theories and Concepts

MBR represents a framework for understanding the dynamics of travel behaviour across the life course, paying particular attention to incisive life events and their impact on how people travel, and why. Although focused primarily on mobility and travel behaviour, it has been influenced by life course and biography research in other fields and disciplines (see Mortimer and Shanahan, 2003 for life course studies; Chamberlayne et al., 2000 and Roberts, 2002 for biography studies). Perhaps most importantly, the work of developmental psychologist Glen Elder and associates has provided important impulses for the conceptual development of MBR (see also Viry, 2020 for a summary of key concepts and ideas arising from this strand of inquiry).

Time geography, a distinct temporal-spatial perspective developed by Swedish geographer Torsten Hägerstrand, has provided another key point of departure for MBR. Hägerstrand's (1970) concept of space–time paths inspired early seminal studies of travel behaviour by life-cycle stages (Kostyniuk and

Kitamura, 1982), with references to life course dynamics (Clarke et al., 1982). The life-cycle perspective was later applied in dynamic transport models such as MIDAS (Goulias and Kitamura, 1997) or ALBATROSS (Timmermans and Arentze, 2011). The term mobility biographies itself was introduced by Lanzendorf (2003), with similar ideas appearing simultaneously in other studies (see overview in Scheiner, 2017).

A number of key concepts in life course studies have made their way into MBR. The central idea is that lives can be understood as temporal paths or trajectories, and any point on such a path captures and reflects past experiences and decisions and future expectations, anticipations and aspirations (Giele and Elder, 1998). These paths are, in turn, structured by life stages (or phases) that reflect social norms, expectations and prescriptions regarding roles and statuses, and combinations thereof (e.g. employed father, student mother). Changes in social role or status are called transitions, and frequently coincide with incisive life events (e.g. entry into school, marriage) (Chatterjee and Scheiner, 2015; Sharmeen et al., 2014; Müggenburg et al., 2015). Moreover, life stages and life events can initiate distinct travel and mobility patterns, including particular modal shifts, changes in distances travelled, or diverse meanings attached to different forms of mobility. This also points to the significance of MBR for the design and implementation of a new generation of change initiatives aimed at transforming how (much) people travel, promising progress in an area that has been very slow to change.

Five major theoretical elements can be identified as playing a key role in MBR.

1. Though individual in nature, life courses develop within a societal aggregate. Hence, mobility biographies need to be understood in a wider context, that is, in historical circumstances and processes in time and space. This is why they may be cohort specific, rather than universal, and may even be specific for certain sub-groups within a cohort, for example for men or women. This suggests that different levels of change (society, interpersonal relations, individual, etc.) need to be distinguished (see Section 3.1), with significant consequences for real-world change initiatives.
2. The habitual character of daily travel results in strong behavioural routines and stability over time. Habits are a powerful, but not the only, factor that counters change. This and other factors that serve resistance to change are discussed in Sections 3.2 and 3.4.
3. There are close relationships between mobility and other domains of the life course, as highlighted in Zhang's (2017) life-oriented approach to travel. These relationships point towards the links between the 'dividuals' of an 'in-dividual' (Hägerstrand, 1970), but they do not in themselves

provide theoretical mechanisms to explain why changes in a domain result in changes in another domain. Similarly, one may ask why events and processes that occur outside the life course of an individual (e.g. in the transport and land-use system) trigger change in travel (see Sections 3.3 and 3.4 for discussion).

4. Significant changes in mobility are motivated by transitions, events and learning processes over an individual's biography, as well as intended and accidental ruptures in routines. A large part of empirical MBR focuses on the impact of key life events and life stage transitions on mobility, in particular mode choice and car ownership (Zhang et al., 2014). These events have been classified by Müggenburg et al. (2015) into (1) life events (or life-cycle events, life course events) that directly relate to the private or professional career, (2) adaptations in long-term mobility decisions (similar to mobility milestones, as defined by Rau and Manton, 2016) and (3) exogenous interventions (e.g. intentional measures or incidents) that change travel behaviour. The last includes disruptive events (Marsden and Docherty, 2013) and critical incidents. It should be noted that such a classification certainly deserves more attention in the future. For instance, critical incidents such as an accident or a flood disaster are not interventions, as the very notion of an intervention presupposes intention.

5. People's life courses are linked to the life courses of others in their social environment. These links may be studied using terms such as 'linked lives' (Elder et al., 2003), socialisation or peer groups. As the MBR has been developed as an individualist approach, this point has attracted attention only in recent years, creating greater awareness of the inherently social nature of individuals' mobility biographies (Sattlegger and Rau, 2016). This said, a more sociological reading of this idea of 'linked lives' remains under-developed in many psychological studies of the life course, which continue to treat individuals' lives as distinct trajectories that are intertwined with those of other significant individuals (e.g. family and household members, neighbours, colleagues), potentially overemphasising individuals' ability to make choices in the process. This point suggests strong links to recent research on the role of personal social networks for travel including both intra-household (or intra-family) (Scheiner, 2020) and extra-household (Lin and Wang, 2014; Sharmeen et al., 2014) interactions, and also indicates the importance of socialisation (Section 3.4) that links the individual to his/her wider social environment and is expressed in different levels of change (Section 3.1). These insights also point to the need to fundamentally rethink policies and change initiatives that focus more or less exclusively on the transformation of individuals' behaviour, with a view to developing innovative initiatives that work with

the interlinkages between mobility biographies (see also Sections 3.3 and 3.4 below).

MBR has closer links to life course research than to biography research. As outlined above, the life course is typically conceived as a sequence of events and role transitions that a person lives through from birth to death (Elder et al., 2003). In contrast, a biography is understood as a subject's self-reflective, meaningful action within the temporal structure of his/her own life (Sackmann, 2007, 50). Accordingly, biography studies tend to reconstruct the subjective meanings someone associates with his/her own life (Antikainen and Komonen, 2003) while life course studies attempt to objectively measure sequences and structures in people's lives, for example by asking for pre-defined stations, events or sequences. With these debates in mind, we nevertheless decided to use the term mobility biography as it has been applied widely for related research in the past decade.

2.2   State of the Art – Epistemologies and Methodologies

In the past few years, the reliance on statistical significance of cause–impact relationships has raised criticism of MBR. Sattlegger and Rau (2016) argue for the adoption of a reconstructive-interpretative approach[1] to MBR that views people's memories as oral history that shapes present action. This approach aims to discover latent, less conscious structures of meaning in mobility whereas more conventional mobility biographies studies typically examine realised behaviour and the impact of measurable variables or, in the case of qualitative studies (e.g. Jones et al., 2014; Bonham and Wilson, 2012), look for more overt meanings. At the same time, biographies are understood as social entities that do not necessarily follow linear temporal and cause–impact structures, and that combine and reflect social reality as well as subjective experiences (Sattlegger and Rau, 2016).

Overall, MBR covers a wide spectrum of epistemological positions, with a positivist-structuralist perspective and a hermeneutic-interpretive-reconstructive position at each end. The former approaches the study of mobility biographies with a set of conceptual and methodological tools that aim at objectivity, validity and reliability and that mirror approaches to research that dominate in the natural sciences and engineering. The last incorporates a range of perspectives and methodologies that emphasise the role of meaning in understanding how and why people act in certain ways and that reflect strong traditions in the social sciences and humanities. Attempts have been made to reconcile these divergent positions, at least to some extent (e.g. Scheiner, 2005; Goetz et al., 2009; Schwanen, 2011; Rau et al., 2020; Chatterjee and Clark, 2020). At the same time, there are arguments for maintaining these epistemological divergences as a source of innovation in MBR, and in transport studies more generally. It is beyond the scope of this chapter to capture the full spectrum of views and issues related to this matter, but see Scheiner and Rau (2020) for an extensive treatment of epistemological orientations in MBR.

Since the inception of MBR as an independent field of research, the MBR landscape has been predominantly shaped by empirical studies and their outcomes, contrasting with a relative lack of theoretically grounded explanations of findings. Moreover, the persistent dominance of quantitative methods for collecting and analysing mobility biography data has contributed to the marginalisation of potentially promising qualitative and mixed methods alternatives, at least until recently (Rau et al., 2020). This said, these obvious limitations have provided impetus for further innovation in the field, the benefits of which are starting to emerge.

3.   UNDERSTANDING STABILITY AND CHANGE IN MOBILITY BIOGRAPHIES

People constantly develop their travel behaviour while interacting with each other on various levels – in their partnerships, families, personal networks, in the transport system, in organisations and communities, and so on. At the same time there are powerful forces that serve to prevent change on the individual (e.g. habits, heuristics) and collective (e.g. social norms, socialisation, regimes) levels. Change requires motivation, self-efficacy and social support on the individual level, and – on a societal level – strong innovations, niches, ruptures to established practices and/or existential threats. Transport innovations need to take into account the conditions for change. This section shows how MBR makes valuable contributions to the study of stability and change in mobility over the life course and its context.

MBR has been criticised for its lack of theoretical backbone (Chatterjee and Scheiner, 2015; Scheiner and Rau, 2020). At the same time, empirical studies in the field tend to limit their focus to behaviour change as a function of events, thereby ignoring (1) behavioural stability and (2) behavioural change that is not linked to discrete events. What is more, MBR tends to see individuals as being disconnected from their social environment, a view that has also dominated many demand-oriented change initiatives in the transport sector to date. For example, a plethora of information campaigns have sought to encourage individuals to change their travel mode to more sustainable options, without due regard to their life-stage-specific social circumstances (the presence of children in the household who need to be transported, care responsibilities for older parents who live far away, etc.). This section offers a nuanced critique of such views and initiatives, with a view to identifying credible and practicable alternatives. We start by distinguishing various levels of change to appropriately consider the wider societal context (Section 3.1). Secondly, we discuss factors on different levels that serve stability (Section 3.2) and change (Section 3.3). Lastly, the term socialisation helps to further understanding of how an individual's life course unfolds in a wider social context (Section 3.4).

3.1   Sorting Levels of Change and Stability

The literature on theories of change identifies distinct levels of change, with authors focusing on individual (behaviour, attitude) change and systems change (system transformation, macro social change) respectively. Theories of organisational change are located in between these two extremes. Looking at the spectrum between individual and organisational levels, Ampt and Engwicht (2007) provide a helpful classification following Halpern et al. (2004):

1. Individual-level theories include 'classical' psychological theories that help explain behaviour stability and change, such as instrumental and classical conditioning theory (Pavlov, 1927; Skinner, 1953), cognitive dissonance (Festinger, 1957), and the consumer information-processing model in economics with its idea of heuristics (Tversky and Kahneman, 1974). While this basic literature has clearly remained extremely valuable, it is very much limited to the individual.
2. Interpersonal behavioural theories focus on the role of interactions, social embedding, role models or mentoring. They include social cognitive theory with its key concept of self-efficacy (Bandura, 1977, see below), social networks and support theory (House, 1981), and social influence and interpersonal communication approaches (Kelley and Thibaut, 1978). The theory of interpersonal behaviour highlights the role of affects, attitudes, social norms and roles for the formation of behaviour while at the same time recognising the importance of past behaviour and the formation of habits and facilitating conditions for behaviour (Triandis, 1977).
3. Community theories of behaviour focus explicitly on groups, organisations, social institutions and communities. They include, firstly, social capital theory (Bourdieu, 1985; Coleman, 1988; Putnam, 1995), which highlights the interactions and cooperation between people, typically in a neighbourhood or community. Secondly, innovation diffusion theory seeks to understand how new practices, ideas or goods spread in society and space over time (Rogers, 1962). Thirdly, tipping point theory highlights the role of thresholds or critical mass in the process of change (Gladwell, 2000). Here, the role of 'mass' points to system change theories that focus on an aggregate 'system' level. This includes the interweaving and coordination of activities between various societal subsystems (economic, social, cultural, infrastructural, regulative) (Köhler

et al., 2009) or the cognitive, normative and regulative institutions that have been called 'sociotechnical regimes' (Geels, 2012). Interestingly, grassroots movements and bottom-up change initiatives such as 'critical mass', an international social movement that engages in pro-cycling campaigning and direct action, rely on this notion of tipping points. Finally, all theories of human behaviour discussed here emphasise the importance of interpersonal interaction, either in a setting with more or less clearly defined spatio-temporal boundaries (social capital theory, tipping point theory) or across space and time (innovation diffusion theory).

Undoubtedly, different levels of social organisation must be fruitfully linked, given that communities and organisations are made up of individuals and individuals' actions cannot be properly understood without reference to the social, economic, administrative and political settings within which they occur. Taken together, they shape networks of actors who negotiate stability and change of mobility practices on the individual and collective level. It is this networked, interconnected character that fosters inertia but also presents chances for fast and wide-ranging innovation in the transport sector whenever the opportunity arises. For example, the rise in power and popularity of cycling activist groups during the Covid-19 pandemic in 2020 coincided with pre-existing sympathies among some urban transport planners who used the growth in cycling (and other forms of individualised mobility, at the expense of public transport) to realise 'pop-up bike lanes' and other measures.

3.2 Resistance to Change: Habits, Heuristics, Personalities and Regimes

Travel behaviour has frequently been conceptualised as relatively stable, habitual day-to-day behaviour that resists efforts to change it. Habits are based on internal scripts, and they work as behavioural scripts that can easily be applied in situations that are experienced as similar to previous ones (see Fujii and Gärling, 2003 for a transport context). They thus resemble the idea of heuristics (or shortcuts) (Tversky and Kahneman, 1974). Heuristics are simplifications of situations applied to handle complexity and uncertainty. They may be used in a conscious, 'controlled' way, rather than being necessarily linked to 'automatic' behaviour, such as habits. In any case, both habits and heuristics are expected to stabilise behaviour, making it highly resistant to deliberate efforts to reconfigure it.

Resistance to change may also be a personality trait that is specific to some people but not others. Risk aversion and aversion of regret that may occur due to choices made under conditions of uncertainty (e.g. one does not know whether an alternative mode or route is as beneficial as it appears) may lead to 'choice inertia', that is, resistance to change (Chorus, 2014; Ben-Elia and Avineri, 2015). Also, a perceived lack of self-efficacy can cause resistance to change when someone does not believe that he/she is able to achieve behavioural change or the desired outcomes (see below).

Persistence of behaviour can also be due to a higher-order systemic level that prevents change, as is argued by prominent advocates of multi-level transition theory (Geels, 2012). Here, the concept of regimes has been used to understand the forces that prevent change by focusing on system preservation and optimisation rather than system innovation. This is due to path dependencies caused by habits, existing competencies, past investment, regulation, and dominating norms and perspectives (Köhler et al., 2009).

Even though regimes tend to achieve a stable state, they develop over time, and they may experience gradual trends as well as disruption. Individual life courses are embedded in such historical changes. To ensure the adequate interpretation of mobility against the background of economic, social, technological and political conditions at a particular time and place, these conditions need to be taken into account. Their reconstruction is a considerable challenge for mobility biography studies, and requires considerable knowledge of historical context and great care.

3.3   Initiating Change

Explanations of stability, stagnation and inertia that have emerged in transport research more generally, and MBR in particular, have been contrasted with approaches that emphasise change, the dynamics of everyday practices, and opportunities for a more or less radical transformation of how people travel, and why. The last issue also relates very closely to the conception and implementation of practical changes that derive from both innovation (e.g. electric vehicles, autonomous driving) and exnovation (e.g. phasing out of the combustion engine) in the transport sector.

Regarding notions of change in MBR, the idea that changes in social role or status and associated events in various life domains trigger change in travel behaviour has been of central importance. This idea reflects the well-documented links between activity patterns and travel (e.g. Zhang, 2017), but does not suggest that a life course is a sequence of disjointed combinations of states in various domains. Rather, the temporalities of different domains intersect in multiple and complex ways, resulting in the opening and closing of options and various forms of path dependency. Long-term travel evolution may be the outcome of choices made in other life domains, for example car dependence based on residential choice or job choice. However, these considerations do not answer the question of *why* key events affect travel. Busch-Geertsema and Lanzendorf (2015, 36–37) argue that key events (or their anticipation) may change travel *r*equirements, *o*pportunities and/or *a*bilities ('ROA model', Harms, 2003), resulting in dysfunctional habits. The stress that emerges from such a mismatch thus works as a conceptual trigger to reconsider travel behaviour (Miller, 2005; Clark et al., 2016).

However, motivation to change may result from changing requirements. People develop, and they may change their goals, aspirations and, consequently, motivations for action. On the other hand, motivation only drives action if there is some expectation of success. Bandura (1977) used the term self-efficacy to describe whether someone expects to be able to successfully perform a specific action. In his framework, self-efficacy is influenced by vicarious experience, verbal persuasion, physiological feedback and, most importantly, performance accomplishment. Similar ideas can be found in the theory of planned behaviour (Ajzen, 1991), which includes perceived behavioural control as a determinant of behaviour.

Rogers's (1975) protection motivation theory provides a similar, perhaps more nuanced understanding. It claims that four conditions are needed for change: (1) high (perceived) risks of current behaviour, (2) severe behavioural consequences (threat appraisal), (3) high expectancy that behaviour change decreases the threat (coping appraisal) and (4) high self-efficacy to perform the new behaviour.

Transition theory aims to capture how change occurs within systems or societies[2] (see Temenos et al., 2017 for an application of transition theory to the topic of mobility), contrasting with the focus on the individual that characterises many of the theories discussed in this section. Transitions are understood here as radical, systemic changes (e.g. the transition to sustainable mobility) that require strong innovations. Here, niches are important 'as the locus for radical innovations' (Geels, 2012, 472), as exemplified by the 'critical mass' movement. Regimes and incumbents, on the other hand, tend to reinforce stability and prevent change.

3.4   The Role of Socialisation in Stability and Change

Individuals' life courses are not isolated from those of others. Instead, they are embedded in family, kin, friendship and neighbourhood networks as well as wider social, economic, political or administrative structures (e.g. schools, the media). These may serve as important sources of (mobility) socialisation, providing individuals with the communicative and behavioural tools to participate in society, including its transport and mobility system. Importantly, these processes are not confined to earlier phases in people's lives (e.g. childhood, young adulthood) but occur across the entire life course. Learning from significant others (e.g. parents, peers) represents a key mechanism of socialisation through which social norms, values, knowledge, prescriptions and behavioural scripts are transmitted (Haustein et al., 2009). Recent evidence for socialisation effects generally finds positive behavioural links, suggesting that conformity in behaviour dominates over non-conformity (Haustein et al., 2009; Kroesen, 2015; Scheiner, 2020; Sunitiyoso et al., 2013). In other words, socialisation processes establish what is deemed to be 'the normal' and what needs to be known or done to be part of society. This tends to perpetuate existing norms, values and social structures (as opposed to challenging or transforming them). Mobility socialisation helps to establish a 'habit on the aggregate (or system) level' that offers a degree of certainty and predictability regarding people's travel behaviour. For example, urban transport systems rely on a complex set of formal and informal rules (e.g. traffic laws, local bylaws regulating the use of public space for parking, 'typical' conduct regarding how close to drive to the person in front). People learn them through formal instruction (e.g. cycling training in primary schools, driving test) and informal learning opportunities (e.g. peers, 'learning by doing'). At the same time, these 'societal habits' regarding mobility can act as major obstacles to transformation and innovation, especially in cases where they coincide with particularly hard-to-change infrastructure (e.g. road and rail networks).

Interestingly, the idea that mobility socialisation is a process that extends across the entire life course and that requires repeated instruction has not yet been translated into innovative learning offers. For example, in Germany school-based opportunities to learn about traffic regulations, cycling safety, multimodality and other related topics remain scarce. Moreover, fewer people may avail of established learning options, or do so later in life than has previously been the case (e.g. evidence of young urban dwellers delaying getting their driver's licence or not getting it at all, cf. Kuhnimhof et al., 2012; Buehler et al., 2017).

4.   NEW DEPARTURES: CONCEPTUAL AND METHODOLOGICAL INNOVATIONS IN MBR

An increasing diversification of MBR is clearly discernible today, reflecting conceptual and methodological developments within the field as well as innovative impulses spilling over from cognate areas of socio-environmental inquiry. The latter includes the growing use of theories of practice in (environmental) sociology more generally and sustainable consumption research in particular. A trend towards mixing methods is also evident (see the collection by Scheiner and Rau, 2020).

4.1   Mobility Practices across the Life Course

The growing influence of theories of practice on different strands of sustainability research has also led to new conceptual foci and approaches in MBR. A range of studies has emerged that focus on socially negotiated and shared mobility practices (as opposed to individuals' travel behaviour), with a view to capturing both social and material aspects of daily mobility and its development across the life course. Drawing on major practice-theoretical contributions (e.g. Schatzki et al., 2001; Reckwitz, 2002; Shove, 2010), these practice-centred inquiries into mobility biographies treat the life course as a series of phases of stability and change in social and material conditions that manifest themselves in everyday mobility practices. Moving 'beyond the ABC' (Shove, 2010), that is, away from individuals' attitudes, behaviour and choice, practice-theoretical approaches to mobility biographies shift attention to the 'careers of practices' and those who 'carry' them (Shove and Pantzar, 2007). Greene and Rau (2018) connect existing mobility biographies work with practice-theoretical approaches to consumption research to develop a biographic, practice-centred approach for researching mobility practices. Using an illustrative case of a female participant's career in car driving, their work demonstrates the benefits of reconceptualising mobility as a set of highly dynamic practices that address key societal needs and (re)produce social structures in the process (see also Barr and Prillwitz, 2014). This framework thus reclaims some space for individual agency by focusing on people's practice careers. Importantly, it redefines mobility socialisation as a lifelong process that moves beyond an emphasis on how individuals learn from others to focus primarily on short- and long-term changes in the practice landscape regarding mobility.

Another benefit of a practice-centred approach to mobility biographies arises from its commitment to treating 'context' as an integral part of a practice (as opposed to something that 'sits' outside it, cf. Shove, 2010). This implies that wider material and social conditions (e.g. transport infrastructure, a dominant 'car culture') are not external to a practice but represent its constitutive elements. This departure from more conventional ideas of context (e.g. to individuals' behaviour) brings with it a firm commitment to integrated thinking and research, including with regard to deliberate efforts to transform mobility practices towards greater sustainability. At the same time, this highly integrated way of thinking about and investigating everyday life is not without its challenges. Efforts to date to clearly demarcate and subsequently 'operationalise' daily mobility practices have been shown to be highly complex and difficult (e.g. Cass and Faulconbridge, 2017).

This also opens up fruitful avenues for interdisciplinary inquiry into *why* people (do not) travel and how this changes across the life course. For example, recent research in Germany on reasons for people cycling (or not) reveals how cycling as a social-material practice is contingent upon favourable social conditions (e.g. parents and peers who cycle, cultural recognition of cycling as a viable mode of transport) as well as (infra)structural factors that encourage cycling (Mahne-Bieder et al., 2020; Driller et al., 2020). Few of these can be directly influenced by people making the 'right choices'. Instead, they reveal the ebbs and flows of historical developments in transport planning and policy, related shifts in mobility culture (as exemplified by the recent revival of cycling as an urban mobility practice) and exposure to certain life events and 'mobility milestones' (Rau and Manton, 2016), such as residential relocation, changes in the composition of the family or household, or the acquisition of new mobility skills (e.g. learning how to cycle).

Explanations of change rooted in practice theory differ quite significantly from behavioural change approaches. According to Spurling et al. (2013), change comes about as a result of recrafting or substituting established practices (e.g. driving a car) or changing how practices interlock (e.g. work and daily commuting, driving to the supermarket to do the weekly shopping). Instead of focusing on what individuals say and do, the emphasis is shifted towards the 'biographies of practices' and key events resulting in their (de)stabilisation. The recent revival of cycling as an urban mobility practice in many large European cities, which results largely from a shift in meaning combined with a more or less radical improvement in cycling infrastructure, serves as a prime example of a major turn in the biography of a mobility practice, which in turn shapes and reflects the practice careers of a large number of people.

A practice-oriented perspective also has significant implications for the design of practice-focused real-world experiments and change initiatives in the transport sector. Here, transport innovations may be understood as changes in practices that incorporate social, psychological, economic, technological, spatial and temporal conditions. Efforts to shift these practices towards greater sustainability thus require in-depth knowledge of the material and social conditions that lead to their stabilisation as well as the full range of options to disrupt, transform or replace them in ways that meet the needs of society and the environment.

Another example is the rapid rise in long-distance travel that may represent the 'next level' of travel demand beyond regional boundaries that follows the railway and the private car in past centuries (Scheiner and Holz-Rau, 2013). Here, practice-theoretical approaches to mobility biographies can shed light on the social causes of this rise in long-distance travel, including the development of geographically extensive social networks that require considerable synchronisation across time and space, (in)voluntary migration, tourism, and other 'mobility links' (Mattioli, 2020).

4.2   Novel Tools and Daring Designs in MBR: Qualitative and Mixed Methods Approaches and Big Data

The dominance of quantitative approaches to MBR has been well documented (e.g. Scheiner and Rau, 2020). However, there is evidence of qualitative and mixed methods approaches gaining traction. This partly relates to the emergence of novel questions regarding *why* people (do not) change their mode of transport, many of which lend themselves to qualitative inquiry (Sattlegger and Rau, 2016; Rau and Sattlegger, 2018; Viry, 2020).

At the same time, the growing significance of 'big data' in mobility research cannot be overestimated. For example, large data sets collected through location tracking via mobile phone and subsequently presented, have become an influential source of 'evidence' in the context of mobility debates. Their benefits and drawbacks have become particularly visible during the Corona pandemic that started in 2020 and that represented a period of heightened immobility in many countries across the world.

Ethnographic inquiries into mobilities and disruptions can yield fresh insights into the (in)stability of daily mobility practices. A UK-based study by Cass et al. (2015) reveals the regularity of disruption in everyday life, highlighting people's extensive capacities to change how they move. This leads these authors to argue that 'concepts of "normality", "routine" and "habit" should be discarded as the baselines for mobility. People are constantly negotiating disruptions to their everyday mobility, and this suggests that there is capacity for change that needs to be unlocked' (Cass et al., 2015, 6).

4.3   Fresh Insights into Mobility Biographies: Blind Spots and Future Innovations

Based on this review and our recent collection of state-of-the-art MBR (Scheiner and Rau, 2020), we identify a number of blind spots that may serve to instigate future work and deliver fresh insights.

The significance of social networks (especially those beyond the household) has already been recognised but they are not yet fully integrated into the mobility biographies framework. Within-family relationships as well as those within and between generations deserve much more attention than has hitherto been the case (e.g. Plyushteva and Schwanen, 2018).

Blind spots also remain in relation to specific life events, for example regarding the impact of health events, accidents and 'near misses' on mobility practices (cf. Rau et al., 2020; Mahne-Bieder et al., 2020), complementing the effects of more general changes in health status over the life course. Related to this, investigations are urgently needed into the 'sphere of change' that surrounds life events where changes in mobility practices may occur both prior to and well after the event, as well as the potential of these spheres of change for transport innovations.

Longer-term trajectories also remain seriously under-researched, partly because of the limited temporal range of much of the data collected. Travel habits tend to be largely organised in overlapping temporalities, in daily, weekly, annual and long-term cycles that connect with other domains of daily life (Freudendal-Pedersen, 2009; Sattlegger and Rau, 2016). At the same time, the temporal organisation of mobility reflects people's social and economic aspirations for both themselves and their families, opening up new avenues for future research on the linkages between physical and social mobility (see also Plyushteva and Schwanen, 2018 for reflections on this issue).

Regarding the spatial organisation of travel behaviour, there has been a vast array of research on the linkages between innovations in spatial planning and transport planning and shifts in how people travel. Rather than looking at isolated effects of various environmental variables on travel behaviour, understanding the intersectionality of physical, social, economic and cultural changes may be a promising line of inquiry worth pursuing in the future.

Prospective interviewing to capture future aspirations and expectations has hardly been used in mobility biography studies (exceptions are Delbosc and Nakanishi, 2017 and Delbosc and Naznin, 2019). Admittedly, asking people about their expectations regarding possible futures is not a reliable method to forecast what will likely happen, especially when the inquiries refer to a distant future. Still, these kinds of inquiries may yield fresh insights into people's hopes, aspirations, goals and expectations concerning their future mobility, with strong links to past and present circumstances. Moreover, a greater focus on qualitative and quantitative scenarios regarding the future of the transport sector could also help to support innovations in transport planning and research techniques such as practice-oriented participatory backcasting (Davies et al., 2014).

Finally, we would like to reiterate previous calls for bridging the gap between qualitative and quantitative research. Creating opportunities for interaction between researchers who follow different approaches, such as in the context of joint research and publications, can foster conceptual and methodological exchange and reflection. Recent attempts to add multiple layers of meaning to 'positivist' empiricist approaches to mobility biographies have provided a promising start in this direction.

5.   CONCLUSIONS

Investigations of mobility biographies in various disciplines and in interdisciplinary collaborations have cast new light on changes in how people travel that relate to specific life events and phases. Following its inception as a distinct field of research and its initial focus on quantitative empirical studies of individual behaviour, MBR is now in the process of diversification. Qualitative approaches informed by practice-theoretical considerations form an important new strand of work within MBR that treats mobility as an inherently social phenomenon and that brings to the fore the interplay of social, material and psychological aspects in the (re)production of mobility practices. Due to the dynamic nature of MBR and its increasing significance in transport research and beyond, important further conceptual and methodological innovations can be expected in the near future. These include further work on the linked mobility biographies of members of different social and cultural groups and a more detailed scientific engagement with different types of events and transitions, and their role in reshaping mobility practices.

Recent innovations in MBR outlined in Section 4 add that individuals cannot make autonomous choices but that their decisions are embedded in complex contextual conditions. Practice-theoretical approaches imply that wider material and social conditions (including transport innovations) represent constitutive elements of practices. The recent revival of cycling serves as a prime example of a major turn in the biography of a mobility practice. Social marketing innovations in transport provide examples of how to support novel practices, for example providing free monthly public transport tickets for those who already have a subscription that can be used to invite friends to join in on out-of-home activities linked to public transport use (Kasper et al., 2008).

To successfully initiate transport innovations also requires knowledge about the emergence of opportunities in space and time. It is also essential to understand how these innovations will affect people in different ways. Here, the segmentation of target groups according to the life course, social needs and restrictions (e.g. social roles, resources), and psychological state (e.g. with respect to self-efficacy and motivation) can open up new and promising pathways. Taking individuals' and policy stakeholders' subjective representations of transport issues seriously (as opposed to simply relying on 'matters of fact') is another issue that may help to identify opportunities for change. MBR contributes to creating such knowledge by studying stability and change in mobility over the life course.

NOTES

1. This approach advocates for a more inductive, exploratory examination of mobility biographies through a narrative-interpretative lens. Importantly, it promotes a more holistic treatment of mobility biographies that includes social influences such as shared mobility-related norms and that moves beyond a focus on single life events or stages.

2.   Note that the term transition theory is also sometimes used in the context of individual-level learning theories, but this is avoided in this chapter to prevent confusion.

REFERENCES

Ajzen, I. (1991). The theory of planned behavior. *Organizational Behavior and Human Decision Processes* 50, 179–211. https://doi.org/10.1016/0749-5978(91)90020-T.

Ampt, E. and Engwicht, D. (2007). A personal responsibility perspective to behaviour change. Paper presented at the 30th Australasian Transport Research Forum, Melbourne (www.atrf.info, 13.10.2016).

Antikainen, A. and Komonen, K. (2003). Biography, life course, and the sociology of education. In: Torres, C.A. and Antikainen, A. (eds), *The International Handbook on the Sociology of Education*. Lanham, MD: Rowman & Littlefield, pp. 143–159.

Bandura, A. (1977). Self-efficacy: toward a unifying theory of behavioral change. *Psychological Review* 84(2), 191–215. https://doi.org/10.1037/0033-295X.84.2.191.

Barr, S. and Prillwitz, J. (2014). A smarter choice? Exploring the behaviour change agenda for environmentally sustainable mobility. *Environment and Planning C: Government and Policy* 32(1), 1–19.

Ben-Elia, E. and Avineri, E. (2015). Response to travel information: a behavioural review. *Transport Reviews* 35(3), 352–377. https://doi.org/10.1080/01441647.2015 .1015471.

Bonham, J. and Wilson, A. (2012). Bicycling and the life course: the start-stop-start experiences of women cycling. *International Journal of Sustainable Transportation* 6, 195–213. https://doi.org/10.1080/15568318.2011.585219.

Bourdieu, P. (1985). *Sozialer Raum und 'Klassen'. Leçon sur la leçon. Zwei Vorlesungen*. Frankfurt am Main: Suhrkamp.

Buehler, R., Pucher, J., Gerike, R. and Götschi, T. (2017). Reducing car dependence in the heart of Europe: lessons from Germany, Austria, and Switzerland. *Transport Reviews* 37(1), 4–28.

Busch-Geertsema, A. and Lanzendorf, M. (2015). Mode decisions and context change – what about the attitudes? A conceptual framework. In: Attard, M. and Shiftan, Y. (eds), *Sustainable Urban Transport*. Bingley, UK: Emerald, pp. 23–42. https://doi .org/10.1108/S2044-994120150000007012.

Cass, N., Doughty, K., Faulconbridge, J. and Murray, L. (2015). *Ethnographies of Mobilities and Disruption*. https://eprints.lancs.ac.uk/id/eprint/84305/1/Cass_et_al _2015_Ethnographies_of_mobilities_and_disruptions.pdf.

Cass, N. and Faulconbridge, J. (2017). Satisfying everyday mobility. *Mobilities* 12(1), 97–115.

Chamberlayne, P., Bornat, J. and Wengraf, T. (2000). *The Turn to Biographical Methods in Social Science: Comparative Issues and Examples*. London: Routledge. https://doi.org/10.4324/9780203466049.

Chatterjee, K. and Clark, B. (2020). Turning points in car ownership over the life course: contributions from biographical interviews and panel data. In: Scheiner, J. and Rau, H. (eds), *Mobility and Travel Behaviour across the Life Course: Qualitative and Quantitative Approaches*. Cheltenham, UK and Northampton, MA, USA: Edward Elgar Publishing, pp. 17–32.

Chatterjee, K. and Scheiner, J. (2015). Understanding changing travel behaviour over the life course: contributions from biographical research. Resource paper presented at the 14th International Conference on Travel Behaviour Research, Windsor, UK, 19–23 July 2015.

Chorus, C.G. (2014). Risk aversion, regret aversion and travel choice inertia: an experimental study. *Transportation Planning and Technology* 37(4), 321–332. https://doi .org/10.1080/03081060.2014.899076.

Clark, B., Lyons, G. and Chatterjee, K. (2016). Understanding the process that gives rise to household car ownership level changes. *Journal of Transport Geography* 55, 110–120. https://doi.org/10.1016/j.jtrangeo.2016.07.009.

Clarke, M., Dix, M. and Goodwin, P. (1982). Some issues of dynamics in forecasting travel behaviour – a discussion paper. *Transportation* 11, 153–172. https://doi.org/ 10.1007/BF00167929.

Coleman, J.S. (1988). Social capital in the creation of human capital. *American Journal of Sociology* 94, Supplement, S95–S120. https://doi.org/10.1086/228943.

Davies, A.R., Fahy, F. and Rau, H. (2014). *Challenging Consumption*. Abingdon, UK: Routledge.

Delbosc, A. and Nakanishi, H. (2017). A life course perspective on the travel of Australian millennials. *Transportation Research Part A: Policy and Practice* 104, 319–336.

Delbosc, A. and Naznin, F. (2019). Future life course and mobility: a latent class analysis of young adults in Victoria, Australia. *Transport Policy* 77, 104–116.

Driller, B., Thigpen, C.G. and Handy, S. (2020). A qualitative exploration of children's attitudes toward bicycling in Davis, California. In: Scheiner, J. and Rau, H. (eds), *Mobility and Travel Behaviour across the Life Course*. Cheltenham, UK and Northampton, MA, USA: Edward Elgar Publishing, pp. 172–189.

Elder, G.H., Jr, Johnson, M.K. and Crosnoe, R. (2003). The emergence and development of life course theory. In: Mortimer, J. and Shanahan, M.J. (eds), *Handbook of the Life Course*. New York: Kluwer Academic Publishers, pp. 3–19.

Festinger, L. (1957). *A Theory of Cognitive Dissonance*. Stanford, CA: Stanford University Press.

Freudendal-Pedersen, M. (2009). *Mobility in Daily Life: Between Freedom and Unfreedom*. Farnham, UK: Ashgate Publishing.

Fujii, S. and Gärling, T. (2003). Development of script-based travel mode choice after forced change. *Transportation Research Part F: Traffic Psychology and Behaviour* 6(2), 117–124. https://doi.org/10.1016/S1369-8478(03)00019-6.

Geels, F.W. (2012). A socio-technical analysis of low-carbon transitions: introducing the multi-level perspective into transport studies. *Journal of Transport Geography* 24, 471–482.

Giele, J.Z. and Elder, G.H., Jr (eds) (1998). *Methods of Life Course Research: Qualitative and Quantitative Approaches*. Thousand Oaks, CA: SAGE. https://doi .org/10.4135/9781483348919.

Gladwell, M. (2000). *The Tipping Point: How Little Things Can Make a Big Difference*. Boston, MA: Little, Brown and Company.

Goetz, A., Vowles, T. and Tierney, S. (2009). Bridging the qualitative–quantitative divide in transport geography. *Professional Geographer* 61(3), 323–335. https://doi .org/10.1080/00330120902931960.

Goulias, K.G. and Kitamura, R. (1997). A dynamic microsimulation model system for regional travel demand forecasting. In: Golob, T.F., Kitamura, R. and Long, L. (eds), *Panels for Transportation Planning: Methods and Applications*. New York: Springer Science + Business Media, pp. 321–348. https://doi.org/10.1007/978-1 -4757-2642-8_13.

Greene, M. and Rau, H. (2018). Moving across the life course: the potential of a bio-graphic approach to researching dynamics of everyday mobility practices. *Journal of Consumer Culture* 18(1), 60–82.

Hägerstrand, T. (1970). What about people in regional science? *Papers in Regional Science* 24, 7–21. https://doi.org/10.1111/j.1435-5597.1970.tb01464.x.

Halpern, D., Bates, C., Mulgan, G. and Aldridge, S. (2004). *Personal Responsibility and Changing Behaviour*. London: Cabinet Office.

Harms, S. (2003). *Besitzen oder Teilen: Sozialwissenschaftliche Analyse des Carsharings*. Zurich: Rüegger.

Haustein, S., Klöckner, C.A. and Blöbaum, A. (2009). Car use of young adults: the role of travel socialization. *Transportation Research Part F: Traffic Psychology and Behaviour* 12, 168–178. https://doi.org/10.1016/j.trf.2008.10.003.

House, J.S. (1981). *Work Stress and Social Support*. Reading, MA: Addison-Wesley.

Jones, H., Chatterjee, K. and Gray, S. (2014). A biographical approach to studying individual change and continuity in walking and cycling over the life course. *Journal of Transport Health* 1, 182–189. https://doi.org/10.1016/j.jth.2014.07.004.

Kasper, B., Schubert, S. and Toepsch, J. (2008). Still young enough to learn – how to grow information and competence for multimodal mobility of old age people. In: Anderl, R., Arich-Gerz, B. and Schmiede, R. (eds), *Technologies of Globalization*. Darmstadt: Technische Universität Darmstad, pp. 366–387.

Kelley, H. and Thibaut, J.W. (1978). *Interpersonal Relations: A Theory of Interdependence*. New York: John Wiley & Sons.

Köhler, J., Whitmarsh, L., Nykvist, B., Schilperoord, M., Bergman, N. and Haxeltine, A. (2009). A transitions model for sustainable mobility. *Ecological Economics* 68(12), 2985–2995. https://doi.org/10.1016/j.ecolecon.2009.06.027.

Kostyniuk, L. and Kitamura, R. (1982). Life cycle and household time-space paths: empirical investigation. *Transportation Research Record* 879, 28–37.

Kroesen, M. (2015). Do partners influence each other's travel patterns? A new approach to study the role of social norms. *Transportation Research Part A: Policy and Practice* 78, 489–505. https://doi.org/10.1016/j.tra.2015.06.015.

Kuhnimhof, T., Buehler, R., Wirtz, M. and Kalinowska, D. (2012). Travel trends among young adults in Germany: increasing multimodality and declining car use for men. *Journal of Transport Geography* 24, 443–450.

Lanzendorf, M. (2003). Mobility biographies. A new perspective for understanding travel behaviour. Paper presented at the 10th International Conference on Travel Behaviour Research, Lucerne, 10–15 August 2003.

Lin, T. and Wang, D. (2014). Social networks and joint/solo activity-travel behavior. *Transportation Research Part A: Policy and Practice* 68, 18–31. https://doi.org/10.1016/j.tra.2014.04.011.

Mahne-Bieder, J., Popp, M. and Rau, H. (2020). Welche Barrieren und Hindernisse haben Nicht-Radfahrende in Deutschland? Eine vergleichende Betrachtung und Typisierung. In: Appel, A., Scheiner, J. and Wilde, M. (eds), *Mobilität, Erreichbarkeit, Raum*. Wiesbaden: Springer VS, pp. 83–98.

Marsden, G. and Docherty, I. (2013). Insights on disruptions as opportunities for trans-port policy change. *Transportation Research Part A: Policy and Practice* 51, 46–55. https://doi.org/10.1016/j.tra.2013.03.004.

Mattioli, G. (2020). Towards a mobility biography approach to long-distance travel and 'mobility links'. In: Scheiner, J. and Rau, H. (eds), *Mobility and Travel Behaviour across the Life Course*. Cheltenham, UK and Northampton, MA, USA: Edward Elgar Publishing, pp. 82–99.

Miller, E. (2005). An integrated framework for modelling short- and long-run household decision-making. In: Timmermans, H. (ed.), *Progress in Activity-Based Analysis*. Bingley, UK: Emerald, pp. 175–201.

Mortimer, J. and Shanahan, M.J. (eds) (2003). *Handbook of the Life Course*. New York: Kluwer Academic Publishers. https://doi.org/10.1007/b100507.

Müggenburg, H., Busch-Geertsema, A. and Lanzendorf, M. (2015). A review of achievements and challenges of the mobility biographies approach and a framework for further research. *Journal of Transport Geography* 46, 151–163. https://doi.org/10.1016/j.jtrangeo.2015.06.004.

Pavlov, I.P. (1927). *Conditioned Reflexes*. Oxford: Oxford University Press.

Plyushteva, A. and Schwanen, T. (2018). Care-related journeys over the life course: Thinking mobility biographies with gender, care and the household. *Geoforum* 97, 131–141.

Putnam, R. (1995). Bowling alone: America's declining social capital. *Journal of Democracy* 6(1), 65–78. https://doi.org/10.1353/jod.1995.0002.

Rau, H. and Manton, R. (2016). Life events and mobility milestones: advances in mobility biography theory and research. *Journal of Transport Geography* 52, 51–60. https://doi.org/10.1016/j.jtrangeo.2016.02.010.

Rau, H., Popp, M. and Mahne-Bieder, J. (2020). Quality and quantity in mobility biographies research: experiences from a mixed method study of non-cyclists in Germany. In: Scheiner, J. and Rau, H. (eds), *Mobility and Travel Behaviour across the Life Course: Qualitative and Quantitative Research Approaches*. Cheltenham, UK and Northampton, MA, USA: Edward Elgar Publishing, pp. 33–49.

Rau, H. and Sattlegger, L. (2018). Shared journeys, linked lives: a relational-biographical approach to mobility practices. *Mobilities* 13(1), 45–63.

Reckwitz, A. (2002). Toward a theory of social practices: a development in culturalist theorizing. *European Journal of Social Theory* 5(2), 243–263. https://doi.org/10.1177/13684310222225432.

Roberts, B. (2002). *Biographical Research*. Buckingham, UK: Open University Press.

Rogers, E.M. (1962). *Diffusion of Innovations*. New York: Free Press.

Rogers, R.W. (1975). A protection motivation theory of fear appeals and attitude change. *Journal of Psychology* 91, 93–114. https://doi.org/10.1080/00223980.1975.9915803.

Sackmann, R. (2007). *Lebenslaufanalyse und Biografieforschung*. Wiesbaden: Springer VS.

Sattlegger, L. and Rau, H. (2016). Carlessness in a car-centric world: a reconstructive approach to qualitative mobility biographies research. *Journal of Transport Geography* 53, 22–31. https://doi.org/10.1016/j.jtrangeo.2016.04.003.

Schäfer, M., Jaeger-Erben, M. and Bamberg, S. (2012). Life events as windows of opportunity for changing towards sustainable consumption patterns? *Journal of Consumer Policy* 35(1), 65–84. DOI 10.1007/s10603-011-9181-6.

Schatzki, T.R., Cetina, K.K. and von Savigny, E. (eds) (2001). *The Practice Turn in Contemporary Theory*. Abingdon, UK: Routledge.

Scheiner, J. (2005). Daily mobility in Berlin: on 'inner unity' and the explanation of travel behaviour. *European Journal of Transport and Infrastructure Research* 5(3), 159–186.

Scheiner, J. (2017). Mobility biographies and mobility socialisation – new approaches to an old research field. In: Zhang, J. (ed.), *Life-Oriented Behavioral Research for Urban Policy*. Tokyo: Springer Japan, pp. 385–401.

Scheiner, J. (2020). Changes in travel mode use over the life course with partner inter-actions in couple households. *Transportation Research Part A: Policy and Practice* 132, 791–807. https://doi.org/10.1016/j.tra.2019.12.031.

Scheiner, J. and Holz-Rau, C. (2013). A comprehensive study of life course, cohort, and period effects on changes in travel mode use. *Transportation Research Part A: Policy and Practice* 47, 167–181. https://doi.org/10.1016/j.tra.2012.10.019.

Scheiner, J. and Rau, H. (eds) (2020). *Mobility and Travel Behaviour across the Life Course: Qualitative and Quantitative Approaches*. Cheltenham, UK and Northampton, MA, USA: Edward Elgar Publishing.

Schwanen, T. (2011). Car use and gender: the case of dual-earner families in Utrecht, The Netherlands. In: Lucas, K., Blumenberg, E. and Weinberger, R. (eds), *Auto Motives: Understanding Car Use Behaviours*. Bingley, UK: Emerald, pp. 151–171. https://doi.org/10.1108/9780857242341-008.

Sharmeen, F., Arentze, T. and Timmermans, H. (2014). An analysis of the dynamics of activity and travel needs in response to social network evolution and life-cycle events: a structural equation model. *Transportation Research Part A: Policy and Practice* 59, 159–171. https://doi.org/10.1016/j.tra.2013.11.006.

Shove, E. (2010). Beyond the ABC: climate change policy and theories of social change. *Environment and Planning A: Economy and Space* 42(6), 1273–1285.

Shove, E. and Pantzar, M. (2007). Recruitment and reproduction: the careers and carri-ers of digital photography and floorball. *Human Affairs* 17(2), 154–167.

Skinner, B.F. (1953). *Science and Human Behavior*. New York: Free Press.

Spurling, N.J., McMeekin, A., Southerton, D., Shove, E.A. and Welch, D. (2013). *Interventions in Practice: Reframing Policy Approaches to Consumer Behaviour*. Sustainable Practices Research Group, Manchester. https://eprints.lancs.ac.uk/id/eprint/85608.

Sunitiyoso, Y., Avineri, E. and Chatterjee, K. (2013). Dynamic modelling of travel-lers' social interactions and social learning. *Journal of Transport Geography* 31, 258–266. https://doi.org/10.1016/j.jtrangeo.2013.05.012.

Temenos, C., Nikolaeva, A., Schwanen, T., Cresswell, T., Sengers, F., Watson, M. and Sheller, M. (2017). Theorizing mobility transitions: an interdisciplinary conversa-tion. *Transfers* 7(1), 113–129.

Timmermans, H. and Arentze, T.A. (2011). Transport models and urban planning practice: experiences with Albatross. *Transport Reviews* 31(2), 199–207. https://doi.org/10.1080/01441647.2010.518292.

Triandis, H.C. (1977). *Interpersonal Behaviour*. Monterey, CA: Brooks/Cole Publishing Company.

Tversky, A. and Kahneman, D. (1974). Judgment under uncertainty: heuristics and biases. *Science*, New Series 185(4157), 1124–1131.

Viry, G. (2020). Life course and mobility. In Jensen, O.B. et al. (eds), *Handbook of Urban Mobilities*. London: Routledge, pp. 127–136.

Zhang, J. (ed.) (2017). *Life-Oriented Behavioral Research for Urban Policy*. Tokyo: Springer Japan. https://doi.org/10.1007/978-4-431-56472-0.

Zhang, J., Yu, B. and Chikaraishi, M. (2014). Interdependences between household residential and car ownership behavior: a life history analysis. *Journal of Transport Geography* 34, 165–174. https://doi.org/10.1016/j.jtrangeo.2013.12.008.

5. Behavioral economics and social nudges in sustainable travel

William Riggs

INTRODUCTION

In recent years travel behavior has become increasingly complex as new forms of mobility have emerged to expand the palette of options available to many consumers. Ridesharing and bike/scooter services have emerged to reveal small trips that are just simply segments: one part of a larger trip comprising multiple modes that often include public transportation and walking (Clewlow and Mishra, 2017; Rayle et al., 2016; Shaheen, 2018). Many times these "last mile" trips serve individuals that have been previously underserved by transit. They also supplant trips that were previously made by driving private vehicles, bicycling, or walking (Shaheen and Chan, 2016).

This new mobility-on-demand ecosystem has been built largely on the advent of mobile technology and brought up new dialogues of how planners, engineers, and policy makers need to consider mobile and transactive technology to nudge travel behavior (Riggs, 2016; Riggs and Gordon, 2015). It brings up the idea of the "quantified self", or the ability for individuals to know and disseminate their location-based information including trips, time traveling, money spent, activities conducted, and so on (Jariyasunant et al., 2015). These concepts relate to basic principles of behavioral economics which are increasingly used to shape human behavior across a wide range of disciplines (Thaler and Sunstein, 2008).

Traditional travel and commute programs have primarily relied on more standard economic measures. For example, price and supply restrictions on parking have been used to deter driving as well as to reduce auto use and ownership (Guo, 2013; Shoup, 2005; Weinberger et al., 2008). Likewise, similar strategies have been leveraged in many cities to reduce congestion, improve air quality, lower energy use and greenhouse gas emissions, and protect residential quality of life (Shoup, 2005). Transit passes and other financial incentives have been used to encourage driving alternatives (Riggs, 2014; Senft and Calgary, 2005).

In sum, literature has shown that transportation choice is tied to both social and economic factors/levers and is strongly influenced by public policy (Brock and Durlauf, 2003; Dugundji and Walker, 2005; Marchal and Nagel, 2005), but more recent work has explored how some of these social and economic levers may be different than previously expected. Research has found that searching for parking accounts for only a small percentage of induced traffic and congestion, because individuals self-regulate and choose to accept available parking spaces before they reach their destination (Millard-Ball et al., 2020; Weinberger et al., 2020). These choice-responses illustrate that how consumers face constrained choices (such as parking scarcity) and when given options may tend to use more sustainable travel choices like walking, cycling, or transit. This presents an opportunity to explore how focusing on how behavioral nudges and better/optimized choice sets can improve transportation sustainability.

This chapter first focuses on the basic principles of behavioral economics and how small nudges and variations in how individuals look at complex information can change the decisions they make. It then translates this to transportation and travel incentives. It goes on to discuss a series of case studies on how behavioral economics are being applied in transportation to promote innovations in travel. Finally, the chapter concludes with a discussion of two broad themes of environmental and fiscal sustainability that offer opportunities for the future of behavioral nudges in transport.

BACKGROUND ON THE BEHAVIORAL ECONOMY OF NUDGING

Nudges are non-invasive actions that lead people toward some alternative/ desired policy action. Cass Sunstein and Richard Thaler, in their 2008 book *Nudge* (Thaler and Sunstein, 2008), explain nudges as aspects of choice architecture that can change the behavior of people. Individuals face choices that are complex, and directive information or more clear incentives can help them better understand their options and the related implications. Many individuals also have planned behaviors that may be based out of habit as opposed to being grounded in traditional rational economic theory.

In this light many individuals make "predictably irrational" decisions when information is complex (Ariely, 2008), meaning that they will regularly make decisions against their best interest in these situations. Psychological studies have shown that people can become cognitively anchored to numbers or information that then frame any subsequent interaction (Thaler and Sunstein, 2003). Many times, individuals choose less than optimal outcomes when they are forced to make "snap judgments" with complicated information or numbers (Kahneman and Tversky, 1972; Tversky, 1972; Tversky and Kahneman,

1973). Referred to sometimes as "prospect theory", or more commonly as "availability bias", they tend to make subjectively probabilistic decisions that tend to be overly optimistic and overestimate favorable outcomes for themselves in situations that are in fact probabilistically unfavorable. In other words, individuals tend to draw on numbers, information, and experiences available to them in making decisions. Even if they are provided new information, they are anchored to this past information.

For example, studies have shown that when physicians look at small snippets of data on an illness they make judgments that are more subjective (Redelmeier et al., 1995; Redelmeier and Tversky, 1992). Likewise, research shows that individuals will misjudge the price of parking when presented with limited information, and underestimate the price of parking on a daily basis when presented with a monthly price (Riggs and Yudowitz, 2020). Further, when individuals are shown a number, for example at the front of a store or at the top of a form, the higher the number, the more likely that any subsequent numbers that person comes up with will be high (Ariely, 2008, 2016).

This idea of price anchoring also has a relationship with social norms, and relates to many types of decisions. As shown in a 2004 experiment by Heyman and Ariely, financial incentives or nudges are effective at shifting behavior but can also be at odds with equally effective social nudges (Heyman and Ariely, 2004). In their experiments, individuals were asked to do certain tasks, for example moving a couch or solving a puzzle. Some were offered monetary incentives, and some were offered a gift of candy. Both were effective at incentivizing behavior. Yet when the candy was assigned a dollar value it was perceived as less valuable and was not as effective as an incentive (e.g. fewer individuals elected to accept it as payment for their effort). These types of psychological decisions have direct applicability to transportation decisions.

Sustainable Mobility through Nudging and Behavioral Insights

Translating this to travel behavior brings forth social as well as psychological factors that influence travel choices, such as loss aversion, travel habits, and personal views (Riggs, 2017a, 2017c, 2019b, 2020a). These can lead to individuals exhibiting "predictably irrational" behaviors while choosing their mode of transportation. For example, the structure of travel choices or "choice architecture" may on the surface make habitual behavior appear very rational; choices may post too much time cost each day to check which mode works best for a commute trip. Yet if these choices were more clear or readily available, they might be different. Research shows that there are ample possibilities for cities to change the way commute choices are structured or prioritized and subsequently nudge people toward sustainable modes of transit, whether by providing incentives, social nudges, or games (Weber et al., 2018).

For example, when framing travel choices, studies have shown that social nudges such as doing something good for the environment have a significant impact on travel behavior. These incentives can include things like bike vouchers or freebies for individuals who commit to not drive and do not accept a parking pass or opt into a parking program, but they also can include more altruistic requests—for example to do something good for the environment (Riggs et al., 2019a; Riggs and Ross, 2016). In some cases, research has shown social nudges can be even more effective than financial incentives (Heyman and Ariely, 2004; Riggs, 2017b). Furthermore, research by those such as Heyman, Ariely, and Riggs confirms behavioral psychology research finding that mixing financial and social messages causes both to be less effective. Using this lens, various examples and case studies are worth exploring.

APPLYING BEHAVIORAL ECONOMICS TO TRANSPORT—A CASE STUDY APPROACH

Research has shown that nudge programs can focus on reduced congestion, more comfortable commutes, and less pollution (FHWA, 2019). Applications can be used by transit groups, employers, and users to help acquaint themselves with means to find parking, change travel time, obtain carpool rides, and even locate alternate commute options like bicycles, walking routes, and public transit. Highlights of examples of case study projects in which cities have mutually brought together technology and behavioral insights to increase the use of green transit are presented next. They include case studies in a variety of areas as shown in Table 5.1.

Case Studies on Combining Behavioral Nudges with Technology

Technology has provided a new way to frame incentive programs that motivate travelers to switch their travel mode. Many times this is done through the commuting platforms or new forms of software. Much more than a marketing trend or passing transportation fad, this trend of using geospatial data, technology, and games to influence behavior has the propensity for great social impact: Shape Up, a social exercise game, claims to have helped 700,000 people lose weight; PayOff, a debt-management game, has documented that $41 million of credit debt has been paid off; and OPower, an energy conservation tool, has worked with power providers to reduce energy consumption by 1.6 billion kilowatt hours.

Research has indicated there are over 30 computer and video games designed to improve health and physical education, each found to positively influence young people's knowledge, skills, attitudes, and behaviors in relation to health and physical exercise (Papastergiou, 2009). Many of these

Table 5.1   *Case studies by topic*

Topic	Location	Attributes	Source
Technology	Vancouver	Technology Service Reliability Adjustments Targeted Marketing Gamification	(Alta, 2019)
Technology	Durham	Technology Targeted Marketing Customized Mapping Dynamic Nudges/Incentives	(Alta, 2019; Bliss, 2018; FHWA, 2019)
Technology	Hong Kong	Technology Service Reliability Adjustments Peak Pricing	(Alta, 2019; FHWA, 2019)
Technology	London	Technology Gamification	(Weber et al., 2018)
Sustainable Mode Choice, Congestion and Parking	Seattle	Targeted Marketing Participant Journaling/Trip Diaries	(Benson et al., 2008; FHWA, 2019; McCoy et al., 2016; Moore, 2004)
Sustainable Mode Choice, Congestion and Parking	Stanford	Targeted Marketing Peak Pricing Gamification	(Green, 2007; Hamilton, 2008; Mandayam and Prabhakar, 2014; Zhu et al., 2015)
Sustainable Mode Choice, Congestion and Parking	Berkeley	Targeted Marketing Peak Pricing Gamification	(Deakin et al., 2004; Riggs et al., 2011; Riggs, 2014; Riggs and Kuo, 2015; Rivadeneyra et al., 2017)
New Mobility, Carsharing and Ridesharing	Monrovia	Technology Service Reliability Adjustments Peak Pricing	(APTA, 2018; Metro, 2018; Perk and Hinebaugh, 2016; Riggs and Beiker, 2020; Walmsley, 2019)
New Mobility, Carsharing and Ridesharing	Multiple (Car2Go)	Technology Service Reliability Adjustments Targeted Marketing	(Namazu et al., 2018)
New Mobility, Carsharing and Ridesharing	San Francisco Bay Area	Technology Service Reliability Adjustments Targeted Marketing	(Harb et al., 2018, 2021)

harness smartphones and use large volumes of accurate data on behavior, rider demographics, motivations, barriers, and more, which then can become action-able for smarter nudges and better roadway and route planning with targeted education and marketing (Broach et al., 2012; Heinen and Maat, 2012; Riggs, 2015; Riggs and Kuo, 2015; Weber et al., 2018).

Travel-based programs and initiatives can use web or smartphone apps, making it easy for travelers to figure out alternate means of transportation. Encouragement or nudges can be used to frame travel via walking, cycling or scooting, via carpool or transit, or even transition to cleaner vehicles powered by electricity or hydrogen rather than fossil fuels. In sum these programs can harness innovation, reduce single-occupancy trips, and encourage reduced emissions travel that meets lifestyle needs (Riggs and Sethi, 2020).

Vancouver

Metro Vancouver/Translink joined with the Behavioral Insights Team (BIT) and Alta Planning and Design (Alta) to develop a behavioral strategy to increase the use of public transit (Alta Planning & Design, 2019). For context, Alta found that less than about a quarter of the population in the region used public transit to get to work, despite having access to it. The goal of the behavioral program was to nudge users to increase their use of transit, but also to become more social in doing so—creating a community of transit riders and attempting to socialize them further through conversations and incentive gamification to encourage interaction.

The approach had three basic strategies for travel behavior. Technology was used to send out messages about routes and to encourage an increase in the amount of ridership, similar to other smaller-scale efforts that illustrated these kinds of targeted efforts can influence ridership by as much as 30 percent (Riggs, 2015; Riggs and Kuo, 2015). Low-frequency transit users were requested to "Try it Again", while moderate-frequency users were encouraged to "Make it a Habit". High-intensity users were asked to "Use it Well".

In the case of the "Try it Again" group, psychological barriers like the perception of vehicles, and biases against transit were detected which served a cognitive bias against the frequent use of public transportation (Alta, 2019). Researchers developed multiple other interventions that addressed these biases, working on platform reliability and quality of transit service to mitigate the cognitive barriers. In addition to service improvements, a variety of gam-ification techniques—like competitions, prizes, lotteries, and other low-cost social nudges—were used as the program evolved. This increase in trip certainty and quality was focused on making individuals less anxious while on public transit. As a result of the program, 11 percent of the low-use group became moderate users and jumped into the "Make it a Habit" group. The program illustrates and underscores the power of low- and no-cost strategies to increase sustainable travel.

Durham

In Durham, North Carolina, planners and engineers teamed up with behavioral scientists to nudge daily travelers out of single-occupancy cars and shift them to other sustainable travel modes (FHWA, 2019). In 2018 Durham implemented nudge strategies for six months to encourage approximately 15,000 auto users to leave their cars at home. The first strategy used targeted marketing, where participants received personalized maps with walking, bus, and bike routes. This built on work from Riggs and Kuo (2015) done at UC Berkeley where customized mapping and personal outreach resulted in a 30 percent mode shift from driving to walking, cycling, or transit. In the case of Durham, these maps were sent to commuters via email with messages about the benefits of the increased physical activity and financial savings from not driving. As shown in Figure 5.1, the maps "included trip time comparisons and listed the potential benefits of alternatives to solo driving, including the weight loss potential, the savings in gas money, and the time commuters could reclaim from the city's infamous traffic … [alongside the slogan] 'Driving downtown is so 2017'" (Bliss, 2018).

A second strategy targeted city employees, bundling this marketing effort with a financial incentive for taking transit. Called the GoDurham lottery, the program allowed people to participate in a bus lottery. Based on pre/post surveys the program not only incentivized ridership but it also increased levels of satisfaction among commuters. In all the program resulted in a 12 percent decrease in driving and a 16 percent decrease in driving alone (Webber, 2018). The program subsequently won the city a $1 million award as a part of the Bloomberg Philanthropies Mayors Challenge.

Hong Kong

Similarly, in Hong Kong financial behavioral nudges have been studied to shift ridership of transit from peak to off-peak in order for the system to handle more users (Halvorsen et al., 2016; Koutsopoulos et al., 2016). Using a mobile phone application and texts, users were sent messages and asked to opt for rides during off-peak hours. For context, Hong Kong is one of the world's mostly densely populated areas, with a robust public transit system that serves over 12 million passenger transits every day. This messaging paralleled a fare discount which offered a 25 percent reduction if passengers rode to any one of 29 heavy-use stations prior to 7:15 a.m. The nine-month-long experiment succeeded in bringing down ridership by 3 percent in the peak morning hours (Halvorsen et al., 2016; Koutsopoulos et al., 2016).

London

Likewise, in London, as a part of the UK National Cycle Challenge, the Love to Ride mobile application was used to gamify active commuting via cycling.

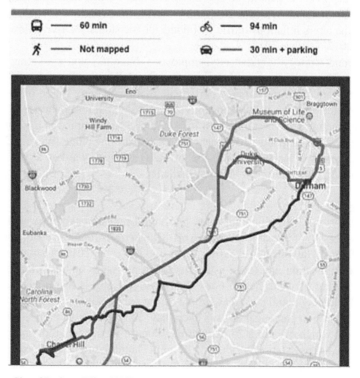

Source: City of Durham (Bliss, 2018).

*Figure 5.1   City of Durham Transportation Demand Management (TDM)
tool screenshot*

Table 5.2   Trips logged by year

	Mean	Standard Deviation	Observations	Compared to Prior Year
Trips Logged				
2015	8.0	8.3	18,613	N/A
2016	12.5	13.4	17,071	+**
Trips Logged: App				
2015	14.1	10.8	2,974	N/A
2016	18.0	15.3	6,822	+**
Trips Logged: No App				
2015	6.86	7.14	15,639	N/A
2016	8.8	10.56	10,249	+**

Note: ** Significant at the 0.01 level.

Love to Ride is one among a number of private firms providing contract services for government agencies and assisting behavior change programs to increase cycling, leveraging web and smartphone connectivity. As detailed by Weber et al. (2018), the mobile application allowed for teams of people to compete virtually if they signed up and rode a bike for at least ten minutes. Points were awarded based on distance travelled but also on how participants interacted and nudged one another through social connections.

As a result of the competitions the number of cycling trips for the over 40,000 participants increased from 149,238 in 2015 to 213,070 in 2016. As shown in Table 5.2, those with app engagement logged many more trips than those without the app. This is consistent with other work that suggests that gamification but also cultural creation through personal connection can be effective social nudges even in the absence of financial incentives (Riggs et al., 2019a, 2019b).

Case Studies on Sustainable Mode Choice, Congestion and Parking

Seattle
In considering cases that were structured to facilitate sustainable mode choices, the Transportation Operations Program in Seattle, Washington, provides a good example (FHWA, 2019). The program promoted the incorporation of walking, biking, and transit into the lives of citizens as opposed to driving. Initiated in 2000, the program, called "Way to Go", brought about a marketing challenge to have users engage in sustainable travel and therefore help by having "One Less Car" on the road. In this light the main aim of the

program was to reduce driving and auto ownership alongside promoting more eco-friendly ways of traveling.

To be eligible as a part of the study, participants needed fewer cars than drivers in their household and to live within city limits. A total of 86 participants participated in the nine-week-long study, with participants being varied to represent a large variety of commute patterns. Participants kept a journal detailing their transportation and submitted the data every week. After a baseline period second vehicles were required to be parked and participants were offered a stipend of $80 each week as an incentive if they did not drive.

By the end of the study there was a reduction in the total miles traveled of over 27 percent, which was coupled with a 38 percent increase in bicycling, a 25 percent increase in transit, a 23 percent increase in carpooling, and a 30 percent increase in overall walking. In addition, families discovered alternate means of meeting mobility without using a second car. Participants found that they could decrease their car costs by taking advantage of other travel options. The study showed that individuals became aware of the additional money a second car required each year ($4,200) while alternate modes of transportation like biking and buses would only cost $1,300. By the end of the study, 26 percent of participant households had sold their second cars. A survey conducted after six months revealed that over 80 percent of households had decreased their additional car usage (Moore, 2004).

Stanford

In 2012 Stanford University began the Federal Highway Administration funded Congestion and Parking Relief Incentives (CAPRI) program. The program was designed to use behavioral economics and nudges to reduce peak hour congestion around the university. For context, Stanford University is one of the largest employers in the San Francisco region, and as a part of its development permits was required by local government to generate "no net new commute trips" during peak hours.

The CAPRI program looked at rewarding travelers with points to play games. When travelers made trips outside of peak hours, they got more points than others (Mandayam and Prabhakar, 2014; Zhu et al., 2015). They could redeem the points in two ways: by trading 100 points to get $1 or by playing a game (chutes and ladders) on the CAPRI website. For those who played the game, cash rewards were given, ranging from $1 to $150. The random rewards/game option was particularly successful, with 87.3 percent of participants choosing to play games of chance for higher rewards and 13.2 percent of participants switching from fixed incentives to randomized higher rewards. Research has shown that these gamification strategies have the capacity to drive participants toward particular behaviors that are sustainable, such as riding a bike or walking to work (Riggs, 2017b; Millonig et al., 2016; Hamari et al., 2014).

Berkeley

In 2015 UC Berkeley began a program aimed at using parking pricing as an incentive to drive sustainable transportation behavior. UC parking rates had traditionally been lower than those in city lots because of the university's permit structure (Riggs et al., 2011; Riggs, 2014; Tang et al., 2016). The university allowed faculty and staff to pay transportation costs, including parking fees, through pre-tax salary deductions. As a result, there was an economic incentive for those who park three days or more per week to purchase monthly parking. Only a limited number (44) of daily permits per person, per year could be purchased and a person who drove two–three days per week did not have a viable option to purchase daily parking as opposed to monthly parking—they were purchased at one time in a bundle and not in an ad hoc/situational way. This made it difficult for people to choose driving along with another mode. Once a person had a permit of any type, they became drivers because of the sunk costs. There was no incentive to take another mode.

Source: Tang et al. (2016).

Figure 5.2   Screenshots from the Flex Pass interface

To rectify this imbalance in the parking system the campus established a nudge program called the Flex Pass. The option allowed for daily variable pricing delivered via mobile application as shown in Figure 5.2. Those who opted-in to the program could build a schedule and commit not to drive, selecting another mode. In exchange they would get variable rewards for their behavior and their permit would not be active on campus that day. By pricing this permit on a daily basis, the university patrons had to choose to pay a market-based, daily rate rather than investing in the full cost of a monthly permit.

Pricing for the pay-per-use option was higher than the monthly permit on a daily, per-use basis—somewhere between $8 and $15 per day—and users would alternatively earn that value by not using spaces. Patrons were incentivized by potential earnings/savings and by the convenience of being able to use the permit on-demand—as it suited their daily needs/schedule. For example, if a professor only came to campus four days a week but wanted to walk, ride a bike, or take transit one or more of those days, they would save $30 or more per month by using a Flex Pass as opposed to buying a monthly permit.

Seventy-seven percent of the campus population ended up participating in the program and of that group the treatment had a 6 percent effect on reducing parking demand. It offered an incremental approach to shifting away from the monthly permit that incentivized driving alternatives. To this effect it focused on incremental decisions and habit, reinforcing the idea of daily trip-making via auto, transit, or non-motorized means (Riggs and Yudowitz, 2020).

Case Studies on New Mobility, Carsharing and Ridesharing

Ridesharing and carsharing has grown in recent years as a method of reducing auto ownership and providing more access to travel (Deakin et al., 2010). Ridesharing usually involves a reservable ride that is experienced in a passenger/chauffeur relationship. Conversely, carsharing usually involves vehicles that are owned by individuals and leased via a digital interface, or institutionally owned vehicles which are reservable through a digital interface. In either situation the idea of reservable, sharable rides provides many opportunities for thinking differently about using price or other incentives.

The topic is particularly relevant in light of new and emerging forms of transportation such as automated and autonomous vehicles (AVs) (Guerra, 2015b; Riggs and Boswell, 2016). AVs present new opportunities to connect individuals to jobs and change the way cities organize space and optimize trips (Fagnant and Kockelman, 2014; Guerra, 2015a). While much of this is predicated on how much car manufacturers are able to transition, or perhaps wean, customers from the idea of owning a car, many car companies are working on this and the idea of challenging the centrality of the automobile in our public realm (Larco, 2017). Cities are already seeing increases in driving and changes to how people travel because of ridesharing and e-commerce (Clark and Larco, 2018; Clewlow and Mishra, 2017).

Because on-demand transportation services allow for convenient point-to-point mobility, they have the potential to reduce automobile ownership in urban areas, with the potential to facilitate first and last mile connections (Rayle et al., 2016; Shaheen and Chan, 2016) and complement transit (Rassman, 2014). Other work has suggested that this smart and connected form of mobility actually reduces public transit ridership and increases vehicle miles traveled (VMT; Clewlow and Mishra, 2017), and creates complicated pick up and drop off issues (Riggs et al., 2017). With regard to carsharing services specifically, in 2013 Shaheen and Cohen estimate over 1,600,000 users sharing more than 24,000 vehicles in the United States and Canada alone (Shaheen and Cohen, 2013).

Monrovia

The Go Monrovia Incentive Partnership carsharing and ridesharing ecosystem has been shown to supplement as well as to replace transit, walking, and biking trips (Gehrke et al., 2019; Young et al., 2020). Marketing of reliability, convenience, and safety benefits has been shown to generate increased use of platforms like Uber and Lyft (Dong, 2015). In this context, many partnership programs have emerged that harness these attributes as public transportation authorities in the United States have been exploring these kinds of flexible route/on-demand services (Metro, 2018; Perk and Hinebaugh, 2016; Riggs and Beiker, 2020; Walmsley, 2019). These partnerships extend to a variety of platforms to help address gaps in the transportation network (McMahon, 2018; Riggs et al., 2011; Riggs, 2014). They also map to a broader trend toward multimodal travel behavior and smaller platforms, from jitneys to robo-taxis, that can make mobility more efficient and equitable (Cervero, 2017).

In the example of Monrovia, California, a partnership between Lyft, Lime, and the city provided reduced fare incentives for individuals to use rideshare services for carpooling (City of Monrovia, 2020). The city had not expanded transportation infrastructure in a way that supports driving alternatives. Through a public–private transportation partnership the city came up with a program to reduce traffic and congestion. The program offered people access to a Lyft ride within the city limits for $0.50 and dockless bike rides for $1 (APTA, 2018). Bundling this price schema with a targeted marketing campaign the city was able to provide vastly more rides at the same cost as their traditional on-demand ride service, which cost about $1 million annually. The program allowed for 30,000 rides a month whereas this had previously been the capacity on an annual basis, showing the efficacy with which price, convenience, and marketing can nudge consumers to shared vehicles.

Multiple (Car2Go)

Extending the idea of nudges to more traditional carshare services, experiments with the Car2Go rental platform provide a good example of how behavioral economics can be used to nudge other aspects of transportation markets (Namazu et al., 2018). Focusing on the idea of vehicle inspections, a group of researchers designed an informational notecard to encourage users to "Please inspect the car while waiting" to get into the vehicle (Figure 5.3). The idea behind the nudge experiment was to encourage appropriate car-use etiquette and ensure that users looked at the vehicle for damage before they got in. The nudge tactic used psychological triggers, such as the word "waiting" to give individuals the idea that they had time, and a smiley face to encourage people to turn the notecard over to participate in a survey and drawing. The color red was chosen to contrast with the white and blue Car2Go vehicles.

Source: Namazu et al. (2018).

Figure 5.3   Placard design for Car2Go nudge program

Based on this intervention the program showed that consistent with other work on targeting marketing and information, individuals did behave as the nudge encouraged. Most took the "waiting time" to inspect the vehicle and fill out the evaluation card. Over the 927 trips observed, individuals spent more time inspecting vehicles before driving them when given the nudge from the notecard. As a result, the total number of inspections jumped to 40–50 percent from a previous high of 24 percent. While there were variations in behavior based on weather and time of day, the program showed that simple reminders, colors, and notes can be effective at nudging behaviors.

These results will become increasingly important in the context of automation and smart and connected mobility. Connected vehicles and connected nomadic devices like smartphones present a powerful new opportunity to directly connect vehicles and humans but also to encourage shared automotive travel. And as vehicles become smarter, these kinds of programs and partnerships that facilitate decisions will be increasingly important. Future vehicles will require human decisions and interfaces that necessitate responsible behavior, and nudge programs have the potential to guide not only how patrons use the vehicles but also how often they are used. This will be particularly important as riding in automated vehicles becomes easier, less costly, and more frictionless. There may be new ways of providing affordable or free transit, capitalizing on social norms and gifts, making greater use of targeted marketing data, information, or emotional pleas to nudge sustainable travel behavior.

San Francisco Bay Area

Extending how transportation innovations can relate to social nudges, experimental work led by researchers at UC Davis has simulated the nudge of increased mobility from an easily accessible vehicle driven by someone else. Using a naturalistic experiment by providing a chauffeur, researchers have been able to simulate that the convenience and reliability of increased service that is enabled by cell phone technology increases driving (Harb et al., 2018, 2021). This corresponds to other evaluations of application-based travel via ridesharing tools, which show that their convenience can nudge or induce driving behavior, increasing total distance traveled by as much as 85 percent (Henao and Marshall, 2019; Schaller, 2017; Ward et al., 2021).

To conduct the nudge survey, a number of individuals were observed for three weeks. During the middle week they were given a driver or "chauffeur" to simulate the experience of riding in an autonomous vehicle. During this week researchers found that the availability of the car caused a sizable increase in distance traveled and number of trips. While the increase in accessibility brought about by the availability of a vehicle may be a net-positive impact (particularly for older adults no longer able to drive themselves), the ease of such "autonomous" services is worth considering in future travel nudge research.

CONCLUSIONS

In sum, financial and social nudges can be effectively used to bring about a change in behavior. These nudges can take many forms, from targeted marketing and information to the provision of incentives, and, as the cases indicate, have a variety of applications. Many behavioral nudge campaigns involve using technology to deliver targeted or curated information. Many reduce the number of choices—the choice set—that travelers have to allow for more environmentally or societally optimal outcomes. Key success points include the opportunity to provide custom mapping or gamify healthy behaviors and non-automotive travel. The best nudges target physiological heuristics that form the basis of prospect theory and disrupt some of the baseline availability biases that travelers have.

Moreover, what is clear is that technology has the capacity to enable nudge programs to become more effective. Planners and engineers have the power to directly and dynamically message consumers—what has been called transactive engagement. This has allowed for the creation of mobile applications that enable users to compete to walk or cycle against one another, or to receive real-time pricing information to help them make better decisions to walk, cycle, or share rides with others.

This trends toward using technology to frame travel decisions is likely to increase alongside transportation innovations, and in this light it is important to underscore how behavioral nudge campaigns relate to broader themes of environmental and fiscal sustainability. First, in the area of environmental sustainability, gaining a better understanding of how to nudge sustainable behaviors is essential as public and private organizations across the globe attempt to address climate change. If travel decisions can be better understood and cities can nudge consumers in ways that reduce complexities and irrationalities, then substantive gains could be made in encouraging more environmentally responsible travel.

Second, from a fiscal standpoint, nudges can save consumers money and help them make more astute financial decisions. As the cases herein illustrate, the innovative use of behavioral economics and nudge can incentivize other travel modes and encourage alternatives to driving—and in that light they can link back to environmental efforts and are likely be more cost-effective and equitable.

These two themes are very important in the present context of COVID-19, and particularly how nudges can be used in recovering from the pandemic. Like never before, we live in a time when technology exists that can revolutionize our transportation industry, and the pandemic shows that system disruptions can nudge travel behavior—both reducing trips and changing the way they are made or the location or distance of those trips (Riggs, 2020b). As the success of shared and on-demand mobility as well as many of the cases evaluated in this chapter illustrate, increasing service reliability, access, and convenience can encourage people to use new modes of travel and open up opportunities for supplemental trips made by walking, cycling, and transit (Meyer and Shaheen, 2017; Riggs, 2019a).

This acceleration of technology when people are experiencing new life-stage events presents a significant opportunity for nudges moving forward—as individuals reestablish travel habits and explore new ways of getting around as the pandemic subsides. Cities, towns, and regions can think about the future of nudges, information, and behavioral science in transport disciplines and develop new behavioral strategies to encourage sustainable travel behavior. They can use nudges and behavioral science to fundamentally change the way we think about transportation.

REFERENCES

Alta (2019). *Behavioral Insights to Transportation Demand Management*. Alta Planning + Design. https://altago.com/resources/behavioral-insights-transportation-demand-management/.

APTA (2018). *City of Monrovia: Leveraging Emerging Ridesharing Services to Expand Mobility Options*. APTA. https://www.apta.com/research-technical-resources/mobility-innovation-hub/case-studies/.

Ariely, D. (2008). *Predictably Irrational: The Hidden Forces that Shape Our Decisions*. Harper.

Ariely, D. (2016). *Payoff: The Hidden Logic that Shapes Our Motivations*. Simon and Schuster.

Benson, T., Cooper, C., and Knott, S. (2008). King County embraces social marketing to change travel behavior. *TDM Review*, No. 4, 15–18. http://trid.trb.org/view.aspx?id=886481.

Bliss, L. (2018, October 30). How Durham is using nudge theory to drive people out of their cars. *Bloomberg.com*. https://www.bloomberg.com/news/articles/2018-10-30/durham-s-1-million-plan-to-nudge-drivers-out-of-cars.

Broach, J., Dill, J., and Gliebe, J. (2012). Where do cyclists ride? A route choice model developed with revealed preference GPS data. *Transportation Research Part A: Policy and Practice*, 46(10), 1730–1740.

Brock, W., and Durlauf, S. (2003). A multinomial choice model with social interactions. In L. Blume and S. Durlauf (eds.), *The Economy as an Evolving Complex System*. Oxford University Press.

Cervero, R. (2017). Mobility niches: Jitneys to robo-taxis. *Journal of the American Planning Association*, 83(4), 404–412. https://doi.org/10.1080/01944363.2017.1353433.

City of Monrovia (2020). *GoMonrovia*. https://www.cityofmonrovia.org/your-government/public-works/transportation/gomonrovia.

Clark, B., and Larco, N. (2018). *The Impacts of Autonomous Vehicles and E-commerce on Local Government Budgeting and Finance*. UrbanismNext. https://urbanismnext.uoregon.edu/files/2017/07/Impacts-of-AV-Ecommerce-on-Local-Govt-Budget-and-Finance-SCI-08-2017-2n8wgfg.pdf.

Clewlow, R.R., and Mishra, G.S. (2017). *Disruptive Transportation: The Adoption, Utilization, and Impacts of Ride-Hailing in the United States*. Research Report UCD-ITS-RR-17-07, University of California, Davis, Institute of Transportation Studies. https://itspubs.ucdavis.edu/wp-content/themes/ucdavis/pubs/download_pdf.php?id=2752.

Deakin, E., Bechtel, A., Crabbe, A., Archer, M., Cairns, S., Kluter, A., Leung, K., and Ni, J. (2004). Parking management and downtown land development in Berkeley, California. *Transportation Research Record: Journal of the Transportation Research Board*, 1898(1), 124–129. https://doi.org/10.3141/1898-15.

Deakin, E., Frick, K., and Shively, K. (2010). Markets for dynamic ridesharing? *Transportation Research Record: Journal of the Transportation Research Board*, 2187(1), 131–137. https://doi.org/10.3141/2187-17.

Dong, H. (2015). Were home prices in new urbanist neighborhoods more resilient in the recent housing downturn? *Journal of Planning Education and Research*, 35(1), 5–18. https://doi.org/10.1177/0739456X14560769.

Dugundji, E., and Walker, J. (2005). Discrete choice with social and spatial network interdependencies: an empirical example using mixed GEVmodels with field and panel effects. *Transportation Research Record: Journal of the Transportation Research Board, 1921*(1), 70–78. https://doi.org/10.1177/0361198105192100109.

Fagnant, D.J., and Kockelman, K.M. (2014). The travel and environmental implications of shared autonomous vehicles, using agent-based model scenarios. *Transportation Research Part C: Emerging Technologies, 40*, 1–13.

FHWA (2019). Applying incentives to shift mode of travel. In *Expanding Traveler Choices through the Use of Incentives: A Compendium of Examples* (Chapter 4). FHWA Office of Operations. https://ops.fhwa.dot.gov/publications/fhwahop18071/ch4.htm.

Gehrke, S.R., Felix, A., and Reardon, T.G. (2019). Substitution of ride-hailing services for more sustainable travel options in the Greater Boston Region. *Transportation Research Record: Journal of the Transportation Research Board, 2673*(1), 438–446. https://doi.org/10.1177/0361198118821903.

Green, J. (2007). Stanford bets on new program to encourage walking, biking. *MercuryNews.com*. http://www.mercurynews.com/ci_23271742/stanford-bets-new-program-encourage-walking-biking.

Guerra, E. (2015a). When autonomous cars take to the road. *Planning, 81*(5), 36–38. https://trid.trb.org/view.aspx?id=1358466.

Guerra, E. (2015b). Planning for cars that drive themselves: Metropolitan planning organizations, regional transportation plans, and autonomous vehicles. *Journal of Planning Education and Research*, 0739456X15613591. https://doi.org/10.1177/0739456X15613591.

Guo, Z. (2013). Does residential parking supply affect household car ownership? The case of New York City. *Journal of Transport Geography, 26*, 18–28. https://doi.org/10.1016/j.jtrangeo.2012.08.006.

Halvorsen, A., Koutsopoulos, H.N., Lau, S., Au, T., and Zhao, J. (2016). Reducing subway crowding: Analysis of an off-peak discount experiment in Hong Kong. *Transportation Research Record: Journal of the Transportation Research Board, 2544*(1), 38–46. https://doi.org/10.3141/2544-05.

Hamari, J., Koivisto, J., and Sarsa, H. (2014). Does gamification work? A literature review of empirical studies on gamification. *47th Hawaii International Conference on System Sciences (HICSS)*, 3025–3034. http://ieeexplore.ieee.org/abstract/document/6758978/.

Hamilton, B. (2008). The transportation demand management experience at Stanford University. *TDM Review, 16*(2), 16–21. http://trid.trb.org/view.aspx?id=869540.

Harb, M., Stathopoulos, A., Shiftan, Y., and Walker, J.L. (2021). What do we (not) know about our future with automated vehicles? *Transportation Research Part C: Emerging Technologies, 123*, 102948. https://doi.org/10.1016/j.trc.2020.102948.

Harb, M., Xiao, Y., Circella, G., Mokhtarian, P.L., and Walker, J.L. (2018). Projecting travelers into a world of self-driving vehicles: Estimating travel behavior implications via a naturalistic experiment. *Transportation, 45*(6), 1671–1685. https://doi.org/10.1007/s11116-018-9937-9.

Heinen, E., and Maat, K. (2012). Are longitudinal data unavoidable? Measuring variation in bicycle mode choice. *Transportation Research Record: Journal of the Transportation Research Board, 2314*(1), 72–80. https://doi.org/10.3141/2314-10.

Henao, A., and Marshall, W.E. (2019). The impact of ride-hailing on vehicle miles traveled. *Transportation, 46*(6), 2173–2194. https://doi.org/10.1007/s11116-018-9923-2.

Heyman, J., and Ariely, D. (2004). Effort for payment: A tale of two markets. *Psychological Science*, *15*(11), 787–793. https://doi.org/10.1111/j.0956-7976.2004 .00757.x.

Jariyasunant, J., Abou-Zeid, M., Carrel, A., Ekambaram, V., Gaker, D., Sengupta, R., and Walker, J.L. (2015). Quantified traveler: Travel feedback meets the Cloud to change behavior. *Journal of Intelligent Transportation Systems*, *19*(2), 109–124. https://doi.org/10.1080/15472450.2013.856714.

Kahneman, D., and Tversky, A. (1972). Subjective probability: A judgment of representativeness. In Staël Von Holstein, C.-A.S. (ed.), *The Concept of Probability in Psychological Experiments* (pp. 25–48). Springer. http://link.springer.com/chapter/ 10.1007/978-94-010-2288-0_3.

Koutsopoulos, H.N., Lau, S., Au, T., Halvorsen, A F., and Zhao, J. (2016). Reducing subway crowding: Analysis of an off-peak discount experiment in Hong Kong. *Transportation Research Record: Journal of the Transportation Research Board*, *2544*(1), 38–46. https://doi.org/10.3141/2544-05.

Larco, N. (2017, August 28). When are AVs coming? (10 car companies say within the next 5 years...). *Urbanism Next*. https://urbanismnext.uoregon.edu/2017/08/28/when -are-avs-coming-10-car-companies-say-within-the-next-5-years/.

Mandayam, C.V., and Prabhakar, B. (2014). Traffic congestion: Models, costs and optimal transport. *The 2014 ACM International Conference on Measurement and Modeling of Computer Systems*, 553–554. https://doi.org/10.1145/2591971 .2592014.

Marchal, F., and Nagel, K. (2005). Modelling location choice of secondary activities with a social network of cooperative agents. *Transportation Research Record: Journal of the Transportation Research Board*, *1935*(1), 141–146. https://doi.org/10 .1177/0361198105193500116.

McCoy, K., Andrew, J., and Lyons, W. (2016). *Ridesharing, Technology and TDM in University Campus Settings: Lessons for State, Regional, and Local Agencies* (DOT-VNTSC-FHWA-16-14). John A. Volpe National Transportation Systems Center. https://rosap.ntl.bts.gov/view/dot/12297.

McMahon, J. (2018, September 6). 5 ways city transit agencies have exploited Uber and Lyft. *Forbes*. https://www.forbes.com/sites/jeffmcmahon/2018/09/06/5-ways -city-transit-agencies-have-found-to-exploit-uber-and-lyft/.

Metro (2018). *MicroTransit Pilot Project*. https://www.metro.net/projects/microtransit/.

Meyer, G., and Shaheen, S. (2017). *Disrupting Mobility: Impacts of Sharing Economy and Innovative Transportation on Cities*. Springer. https://books.google.com/books ?hl=en&lr=&id=3y_XDQAAQBAJ&oi=fnd&pg=PR5&dq=s+shaheen+sharing& ots=f5r1rfhSJi&sig=lM5GzDO8bNSt6I-Wm4oiSnVatpU.

Millard-Ball, A., Hampshire, R.C., and Weinberger, R. (2020). Parking behaviour: The curious lack of cruising for parking in San Francisco. *Land Use Policy*, *91*, 103918. https://doi.org/10.1016/j.landusepol.2019.03.031.

Millonig, A., Wunsch, M., Stibe, A., Seer, S., Dai, C., Schechtner, K., and Chin, R.C. (2016). Gamification and social dynamics behind corporate cycling campaigns. *Transportation Research Procedia*, *19*, 33–39.

Moore, B. (2004, August 28). One less car in Seattle. *EV World*. http://evworld. com/ article.cfm?storyid=740.

Namazu, M., Zhao, J., and Dowlatabadi, H. (2018). Nudging for responsible carsharing: Using behavioral economics to change transportation behavior. *Transportation*, *45*(1), 105–119. https://doi.org/10.1007/s11116-016-9727-1.

Papastergiou, M. (2009). Exploring the potential of computer and video games for health and physical education: A literature review. *Computers and Education*, *53*(3), 603–622.

Perk, V., and Hinebaugh, D. (2016). The bus renaissance—intercity travel, bus rapid transit, technology advances, rural services. *TR News*, May–June 2016. http://www .trb.org/Publications/Blurbs/174689.aspx.

Rassman, C.L. (2014). Regulating rideshare without stifling innovation: Examining the drivers, the insurance gap, and why Pennsylvania should get on board. *Pittsburgh Journal of Technology Law and Policy*, *15*(1), 81–100.

Rayle, L., Dai, D., Chan, N., Cervero, R., and Shaheen, S. (2016). Just a better taxi? A survey-based comparison of taxis, transit, and ridesourcing services in San Francisco. *Transport Policy*, *45*, 168–178. https://doi.org/10.1016/j.tranpol.2015.10 .004.

Redelmeier, D.A., Koehler, D.J., Liberman, V., and Tversky, A. (1995). Probability judgment in medicine: Discounting unspecified possibilities. *Medical Decision Making*, *15*(3), 227–230.

Redelmeier, D.A., and Tversky, A. (1992). On the framing of multiple prospects. *Psychological Science*, *3*(3), 191–193. https://doi.org/10.1111/j.1467-9280.1992 .tb00025.x.

Riggs, W. (2014). Dealing with parking issues on an urban campus: The case of UC Berkeley. *Case Studies on Transport Policy*, *2*(3), 168–176. https://doi.org/10.1016/ j.cstp.2014.07.009.

Riggs, W. (2015). Testing personalized outreach as an effective TDM measure. *Transportation Research Part A: Policy and Practice*, *78*, 178–186. https://doi.org/ 10.1016/j.tra.2015.05.012.

Riggs, W. (2016). Mobile responsive websites and local planning departments in the US: Opportunities for the future. *Environment and Planning B: Planning and Design*, 0265813516656375. https://doi.org/10.1177/0265813516656375.

Riggs, W. (2017a). *The Role of Behavioral Economics and Social Nudges in Sustainable Travel Behavior* (SSRN Scholarly Paper ID 2939404). Social Science Research Network. https://papers.ssrn.com/abstract=2939404.

Riggs, W. (2017b). Painting the fence: Social norms as economic incentives to non-automotive travel behavior. *Travel Behaviour and Society*, *7*, 26–33. https://doi .org/10.1016/j.tbs.2016.11.004.

Riggs, W. (2017c). *Reduced Perception of Safety for Cyclists on Multi-Lane, One-Way and Two-Way Streets: Opportunities for Behavioral Economics and Design* (SSRN Scholarly Paper ID 3011680). Social Science Research Network. https://papers.ssrn .com/abstract=3011680.

Riggs, W. (2019a). *Disruptive Transport: Driverless Cars, Transport Innovation and the Sustainable City of Tomorrow*. Routledge.

Riggs, W. (2019b). Perception of safety and cycling behaviour on varying street typologies: Opportunities for behavioural economics and design. *Transportation Research Procedia*, *41*, 204–218.

Riggs, W. (2020a). The role of behavioral economics and social nudges in sustainable travel behavior. In E. Deakin (ed.), *Transportation, Land Use, and Environmental Planning* (pp. 263–277). Elsevier. https://doi.org/10.1016/B978-0-12-815167-9 .00014-1.

Riggs, W. (2020b). *Telework and Sustainable Travel during the COVID-19 Era* (SSRN Scholarly Paper ID 3638885). Social Science Research Network. https://papers.ssrn .com/abstract=3638885.

Riggs, W., and Beiker, S.A. (2020). Business models for shared and autonomous mobility. In G. Meyer and S. Beiker (eds.), *Road Vehicle Automation 7* (pp. 33–48). Springer International Publishing. https://doi.org/10.1007/978-3-030-52840-9_4.

Riggs, W.W., and Boswell, M.R. (2016). No business as usual in an autonomous vehicle future. *Planetizen*. https://works.bepress.com/williamriggs/53/.

Riggs, W.W., Boswell, M.R., and Zoepf, S. (2017). A new policy agenda for autonomous vehicles: It's time to lead innovation. *Planetizen*. https://works.bepress.com/williamriggs/75/.

Riggs, W., and Gordon, K. (2015). How is mobile technology changing city planning? Developing a taxonomy for the future. *Environment and Planning B: Planning and Design*, 0265813515610337. https://doi.org/10.1177/0265813515610337.

Riggs, W., and Kuo, J. (2015). The impact of targeted outreach for parking mitigation on the UC Berkeley campus. *Case Studies on Transport Policy*, 3(2), 151–158. https://doi.org/10.1016/j.cstp.2015.01.004.

Riggs, W., Marthinsen, E., Mcdougall, J., and Siegman, P. (2011). *University of California, Berkeley. Parking and Transportation Demand Management Master Plan*. UC Berkeley. http://pt.berkeley.edu/sites/pt.berkeley.edu/files/content/UCB%20Parking%20TDM%20Master%20Plan%20-%20FINAL.pdf.

Riggs, W., and Ross, A. (2016). *Hampered by Hardcoreness: How Group Cycling Events Fail to Impact the Everyday Travel Behavior of Novice Cyclists* (SSRN Scholarly Paper ID 2873218). Social Science Research Network. https://papers.ssrn.com/abstract=2873218.

Riggs, W., and Sethi, S.A. (2020). Multimodal travel behaviour, walkability indices, and social mobility: How neighbourhood walkability, income and household characteristics guide walking, biking and transit decisions. *Local Environment*, 25(1), 57–68. https://doi.org/10.1080/13549839.2019.1698529.

Riggs, W., Shukla, S., Ross, A., and Yudowitz, L. (2019a). Transportation demand management and group cycling programs: Lessons for policy and practice. *Focus: The Journal of Planning Practice and Education*, 16. https://papers.ssrn.com/abstract=3507903.

Riggs, W., Shukla, S., Ross, A., and Yudowitz, L. (2019b). *Transportation Demand Management and Group Cycling Programs: Lessons for Policy and Practice* (SSRN Scholarly Paper ID 3507903). Social Science Research Network. https://papers.ssrn.com/abstract=3507903.

Riggs, W., and Yudowitz, L. (2020). *Considering Availability Bias in Parking Decisions* (SSRN Scholarly Paper ID 3765738). Social Science Research Network. https://papers.ssrn.com/abstract=3765738.

Rivadeneyra, A.T., Shirgaokar, M., Deakin, E., and Riggs, W. (2017). Building more parking at major employment centers: Can full-cost recovery parking charges fund TDM programs? *Case Studies on Transport Policy*, 5(1), 159–167. https://doi.org/10.1016/j.cstp.2016.10.002.

Schaller, B. (2017). *Unsustainable? The Growth of App-Based Ride Services and Traffic, Travel and the Future of New York City*. Report by Schaller Consulting, Brooklyn NY.

Senft, G., and Calgary, A. (2005). U-Pass at the University of British Columbia: Lessons for effective demand management in the campus context. *Emerging Best Practices in Urban Transportation Planning (A) Session of the 2005 Annual Conference of the Transportation Association of Canada*.

Shaheen, S. (2018). Shared mobility: The potential of ridehailing and pooling. In D. Sperling, *Three Revolutions* (pp. 55–76). Springer.

Shaheen, S., and Chan, N. (2016). Mobility and the sharing economy: Potential to facilitate the first-and last-mile public transit connections. *Built Environment, 42*(4), 573–588.

Shaheen, S.A., and Cohen, A.P. (2013). Carsharing and personal vehicle services: Worldwide market developments and emerging trends. *International Journal of Sustainable Transportation, 7*(1), 5–34.

Shoup, D.C. (2005). *The High Cost of Free Parking*. Planners Press, American Planning Association. http://www.connectnorwalk.com/wp-content/uploads/The -High-Cost-of-Free-Parking.pdf.

Tang, D., Lin, Z., and Sengupta, R. (2016). A casual analysis of FlexPass: Incentives for reducing parking demand. *TRB 95th Annual Meeting Compendium of Papers*. https://trid.trb.org/view/1394492.

Thaler, R.H., and Sunstein, C.R. (2003). Market efficiency and rationality: The peculiar case of baseball. *Michigan Law Review, 102*(6), 1390–1403.

Thaler, R.H., and Sunstein, C.R. (2008). *Nudge: Improving Decisions about Health, Wealth, and Happiness*. Yale University Press.

Tversky, A. (1972). Elimination by aspects: A theory of choice. *Psychological Review, 79*(4), 281–299.

Tversky, A., and Kahneman, D. (1973). Availability: A heuristic for judging frequency and probability. *Cognitive Psychology, 5*(2), 207–232.

Walmsley, J. (2019, June 26). Watch out, Uber. Berlin is the new Amazon for transportation (with lower fares). *Forbes*. https://www.forbes.com/sites/juliewalmsley/2019 /06/26/watch-out-uber-berlin-is-the-new-amazon-for-transportation-with-lower -fares/.

Ward, J.W., Michalek, J.J., Samaras, C., Azevedo, I.L., Henao, A., Rames, C., and Wenzel, T. (2021). The impact of Uber and Lyft on vehicle ownership, fuel economy, and transit across U.S. cities. *iScience, 24*(1), 101933. https://doi.org/10 .1016/j.isci.2020.101933.

Webber, R. (2018, November 12). Nudge science being used to encourage transit use. *State Smart Transportation Initiative*. https://ssti.us/2018/11/12/nudge-science -being-used-to-encourage-transit-use/.

Weber, J., Azad, M., Riggs, W., and Cherry, C.R. (2018). The convergence of smartphone apps, gamification and competition to increase cycling. *Transportation Research Part F: Traffic Psychology and Behaviour, 56*, 333–343. https://doi.org/10 .1016/j.trf.2018.04.025.

Weinberger, R.R., Millard-Ball, A., and Hampshire, R.C. (2020). Parking search caused congestion: Where's all the fuss? *Transportation Research Part C: Emerging Technologies, 120*, 102781. https://doi.org/10.1016/j.trc.2020.102781.

Weinberger, R., Seaman, M., and Johnson, C. (2008). *Suburbanizing the City: How New York City Parking Requirements Lead to More Driving*. http://trid.trb.org/view .aspx?id=1153742.

Young, M., Allen, J., and Farber, S. (2020). Measuring when Uber behaves as a substitute or supplement to transit: An examination of travel-time differences in Toronto. *Journal of Transport Geography, 82*, 102629. https://doi.org/10.1016/j.jtrangeo .2019.102629.

Zhu, C., Yue, J.S., Mandayam, C.V., Merugu, D., Abadi, H.K., and Prabhakar, B. (2015). Reducing road congestion through incentives: A case study. *TRB 94th Annual Meeting Compendium of Papers*. http://trid.trb.org/view.aspx?id=1336629.

6. Transport innovation theories: a brief overview

Jan Anne Annema

1.   INTRODUCTION

This book is about transport innovations. In Part I of this book, several innovation theories are explained. However, many chapters of this book in Part II use other innovation theories or parts of innovation theories to analyse the success and failure factors of innovations that have not been explained yet. The aim of this chapter is to give a brief overview of seven main modern innovation theories that are discussed in Part II, to compare them, and to reflect on their usefulness. My selection criteria for these seven theories are that in the chapters in Part II, in papers about transport innovations in the past ten years, and in teaching a course about transport innovations (in which master's students had to search for innovation theories in the literature and use them), these theories are often applied.

Basically, a scientific theory is a well-underpinned explanation of 'something'. A theory can include results from empirical research but also hypotheses or even speculations. The key is that a scientific theory is understandable and well-substantiated so that it can be used and tested by fellow scholars but also that it can be contested by them. The vague term 'something' is in this chapter 'transport innovations'. To be precise, this chapter will discuss different kinds of theories that aim to explain the origin of transport innovations, the adoption (or non-adoption) of transport inventions, and the diffusion of transport innovations.

Transport innovation theories are useful for many reasons. According to my view, two reasons stand out. First, the theories give a common ground, a common language, for scientists to research, discuss and develop new knowledge on transport innovations. It will become especially clear that the theories can give inspiration for scientists to include a broad set of potential success and failure factors – not only technical feasibility – when studying the potential of a transport innovation. Second, well-substantiated innovation theories can also give useful insights for practice, for example for policymakers. As will be shown in this chapter, modern innovation theories explaining the origin, diffusion and adoption of transport innovations are complex (or 'rich' to use a friendlier term). This is not because scientists love complexity per se but because scientists have become increasingly aware that to innovate is difficult since it involves many uncertainties, many disappointments and many stakeholders (see Greenacre et al., 2012, for example). At the end of this chapter, I will try to formulate some policy lessons based on modern innovation theories.

To my knowledge, there are not many innovation theories that have been specifically developed for understanding transport innovations. The exception is the political economy model of transport innovations by Feitelson and Salomon (2004) – see theory 3 below. This chapter, therefore, discusses general innovation theories that can be used to understand transport innovations. Interestingly though, almost all crucial papers treated in this chapter developed the theoretical ideas of a generic innovation theory using the transportation system as one of their key sources of inspiration. This is related to the characteristics of the transport system with its complex technology systems and many different stakeholders (users, producers of technology, service providers, governments on all levels, universities, R&D institutes), and, on top of this, it is a system that abounds with policies, informal rules, laws, emotions, habits and so forth.

In the theories explained below, 'innovation' is sometimes clearly defined, other times defined implicitly. In the overview of the theories, I will briefly touch upon definition issues if I think it is required. Broadly speaking, all theories discussed in this chapter define innovations as consisting of an idea, the realization of this idea (the 'invention'), and the exploitation or implementation of the invention (see, for example, Planing, 2017). Successful innovation implies that the invention is widely exploited or implemented. A failed innovation implies the opposite.

In all handbooks and papers presenting or reviewing innovation theories (e.g., Greenacre et al., 2012; Twomey and Gaziulusoy, 2014) there is a discussion about innovation typologies mostly related to the innovation's degree of radicality: incremental versus radical, non-drastic versus drastic, sustaining versus disruptive innovations. In the explanations below I will address explicitly which specific type of innovation the theory is aimed at if I think this is required. In addition, some theories explained below denote themselves as 'transition' theories. It is difficult to precisely explain the difference between modern 'innovation' theories and 'transition' theories. Both acknowledge that 'innovation is a joint activity involving a large number of actors with different interests, perceptions, capabilities and roles' (Twomey and Gaziulusoy, 2014, p. 7). Transition theories more than innovation theories aim to explain the process of the most radical changes in a system, such as the transport system, which are, therefore, most difficult to realize. Transitions are about changes that require the involvement of many actors, that involve not only other production processes or services but also other user and policy practices (Köhler et al., 2019 and Pel, Chapter 2 of this book).

This chapter is not about the history of modern innovation theories nor about the origins of a specific theory. Greenacre et al. (2012) give a brief insightful overview on these topics. Nevertheless, on the highest level – considering all theories discussed in this chapter – it becomes apparent that evolutionary economy is a key founding theory. This theory views the economy as constantly changing, dynamic, perhaps even chaotic. Evolutionary theorists reject the idea of 'rational choice' but instead believe that choices made by firms, individuals, governments and so forth are based on 'bounded rationality'. Bounded rationality implies that decision-making involves psychology and a lack of full information about the consequences of the decision, and that decisions have to be made in uncertainty. With psychology I mean here that actors sometimes ignore or are not able to process optimally information about the consequences of a decision because of certain heuristics (Kahneman, 2011).

The structure of this chapter is that I start with an overview of the seven innovation theories selected (section 2) and, after this, give a brief comparison of the theories, a discussion about their applicability and a conclusion (section 3).

2.   OVERVIEW OF THE SEVEN THEORIES

Theory 1 – The Opportunity Vacuum as a Conceptual Model for the Explanation of Innovation

This theory aims to help individual actors (such as firms) to identify and predict economic opportunities (Planing, 2017). Planing termed his conceptual model 'the opportunity vacuum'. The 'vacuum' can be seen in his words as 'uncharted space on the edges of current knowledge with nothing in it, yet the place where the next ideas emerge quickly' (Planing, 2017, p. 6). The author's opportunity vacuum, which he sees as the true breeding ground for innovation, has three dimensions, which can be thought of as layers stacked on each other:

- The Adjacent Possible (Technology)
- The Adjacent Viable (Economy)
- The Adjacent Acceptable (Society).

According to Planing's theory, successful innovations only occur if there is an intersection amongst the boundaries of possibility within all three dimensions. Planing (2017) states that this overlap can be initiated by changes in every single dimension. So, from the viewpoint of an individual actor, only at the moment when it is technically feasible to realize an idea, financially viable to do so, and when the early majority of the society is ready to adopt the idea, may an innovation become a success.

Theory 2 – Technological Innovation Systems Approach

The so-called innovation systems (IS) approach is already relatively old. The basic assumption of the IS approach is 'that innovation and diffusion of technology is both an individual and a collective act' (Hekkert et al., 2007, p. 415). Different classes of innovation systems are to be found in the literature, such as national, sectoral, regional or technological innovation systems. I focus only on the last – technological innovation systems (TIS) – because in this book we are especially concerned with a specific system (transportation) as a whole without being too much interested in specific national, sectoral or regional innovation systems.

Many papers can be found on TIS. I do not pretend to give a full overview of this large body of literature here. I only took material from papers that I consider key in explaining the main theoretical notions of TIS (Hekkert et al., 2007; Bergek et al., 2008; Wieczorek and Hekkert, 2012; Suurs et al., 2009).

Basically, TIS theory aims to understand and explain innovation processes. Innovation takes place in this theory in systems, or in socio-technical systems, that is, a 'group of components (devices, objects and agents) serving a common purpose. The components of an innovation system are the actors, networks and institutions contributing to the overall function of developing, diffusing and utilizing new products (goods and services) and processes' (Bergek et al., 2008, p. 408).

To understand an innovation process in such a system, it is important in this theory to study both the structure of the socio-technical system in which the particular innovation takes place and what Hekkert et al. (2007) label as 'the functions of this innovation system'. The structural dimensions of an innovation system consist of actors, institutions, interactions and infrastructure. Wieczorek and Hekkert (2012) give a complete overview of structural dimensions of a TIS in their paper on page 77, to which I refer the reader who prefers to have more detail. Next to structural analysis, it is also important to perform a functional analysis according to Hekkert et al. (2007), to understand if an innovation system works as intended. In my own words, these functions represent the activities that take place in an innovation system. The authors propose seven functions to be tested: F1 (entrepreneurial activities), F2 (knowledge development), F3 (knowledge diffusion), F4 (guidance of the search), F5 (market formation), F6 (mobilization of resources) and F7 (creation of legitimacy). Other authors sometimes distinguish slightly different functions, but this is mere detail. By analysing these functions the state of innovation can be shown in a defined moment of time. Both Bergek et al. (2008) and Wieczorek and Hekkert (2012) give hands-on frameworks that can be used by policymakers, consultants and innovation scholars to study the state-of-the-art in innovation systems and to identify problems and subsequent solutions (which can be structural or functional, or both) when innovation does not progress as intended.

In Suurs et al. (2009) an interesting extra feature of the functional analysis is pointed out. Innovation system functions or, in my words, activities may reinforce each other over time, thereby resulting in a so-called virtuous cycle, also termed 'a motor of innovation'. For example, successful knowledge development (F2) (a research project that shows highly successful opportunities) may result in high policy expectations, contributing to guidance of the search (F4), which may, subsequently, lead to a subsidy programme, contributing to resource mobilization (F6), which may induce even more knowledge development (F2), guidance of the search (F4) and so forth. On the other hand, as pointed out also by Suurs et al. (2009), system functions may also reinforce each other 'downwards', resulting in a vicious cycle: 'a motor of decline'.

Theory 3 – The Political Economy of Transport Innovations

In a book chapter by Feitelson and Salomon (2004) a particular angle is chosen, namely transport policy innovations. In the words of the authors, these are innovations advanced by policy entrepreneurs, a term first coined by Kingdon (1984). Policy ideas by these 'entrepreneurs' (e.g., road pricing, new public transport projects) constantly emerge, but there are according to the theory of Kingdon specific moments when these ideas can actually be put on the political agenda, termed 'policy windows'. The theory of Feitelson and Salomon (2004) specifically aims to explain why some of these innovations that are put on the political agenda are adopted and why others are not.

In their theory, or what they call 'a political-economic model of transport innovations', the basic idea is that a policy innovation will only be adopted if it meets all four of the following feasibility criteria: the innovation must be seen as technically, economically, socially and politically feasible. Their 'model' (Figure 6.1) consists of many elements that explain the techno-economic, social and political feasibility. Techno-economic feasibility is in their view composed of two components. An innovation needs to be technically feasible, but they also assume that it is not likely to be seen as feasible unless it can pass a benefit–cost analysis (see in the middle of Figure 6.1). Social feasibility is explained by five elements: the perceived effectiveness of the innovation, the perceived distribution of benefits and costs of the innovation (so, if the innovation is perceived to be economically viable), the perception of problems, the role of non-business interest groups, and the sanctioned discourse.[1] It goes too far to explain the model in detail. Only one extra assumption in their theory is worth mentioning. They see social feasibility as not directly impacting adoption but as an indirect factor that explains adoption only via political feasibility (Figure 6.1). Techno-economic feasibility and political feasibility directly impact the adoption of policy innovation, according to Feitelson and Salomon (2004).

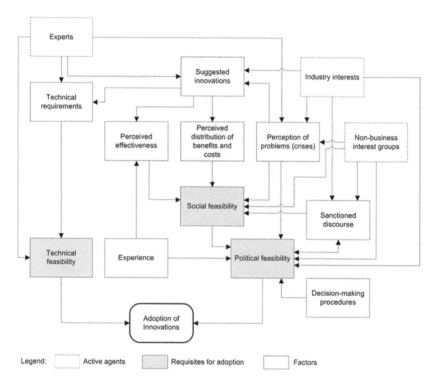

Figure 6.1   *The political economy model of Feitelson and Salomon (2004)*

Theory 4 – The Multi-Level Perspective (MLP)

The MLP is a large topic with several key authors and many interesting papers and books explaining this theoretical perspective (Berkhout, 2002; Unruh, 2000; Elzen et al., 2004; Geels, 2002, 2005a; Kemp, 1994; Schot et al., 1994). The MLP theory is especially aimed at explaining how transitions from one socio-technical system to another socio-technical system can take place.

Socio-technical systems are in this theory (like in the TIS theory) defined as complex systems creating and (re)producing a certain societal function, such as the road transport system which makes sure that people and goods can be moved from one place to another. The automotive industry, fuel producers and distributors, universities, public authorities, users and many more actor groups are part of a socio-technical system (such as road transport in this example) in this theory. In a socio-technical system there is at a certain moment in time a certain technology that is dominant (or a few technologies that are dominant). In this so-called socio-technical 'regime' there are rules which people are used to; there are also user practices that people appreciate, such as an owned car very close to their home. So, in a current regime, there are many cultural/ lifestyle advantages for users, like (feelings of) freedom, and advantages for the industry, which can make good profits without taking too many risks, and so forth. In other words, socio-technical regimes are characterized by lock-in[2] and stability (Geels, 2005b).

Basically, the MLP theory aims to address questions such as: 'How did we get locked in in certain socio-technical regimes?'; 'What kinds of lessons can be learned from this past experience?'; 'How can we escape from undesired lock-ins, such as the fossil fuel-based road transport system we have currently?'

To help answer these questions the theory applies a multi-level perspective (Figure 6.2). Three levels are distinguished in the theory: the meso-level – *the socio-technical regime*; the micro-level – *the niches*; and the macro-level – *the landscape*.

A basic assumption in MLP is that because of the stabilizing character of the meso-level (the current socio-technical regime) radical innovations will not happen there. The idea in MLP is, however, that on the micro-level – in niches – radical change can indeed start to happen. Niches 'act as incubation rooms for radical innovations nurturing their early development' (Geels, 2002, p. 1261). A niche can be a specific market segment (e.g., car racing where new technologies are tried), R&D projects, pilot programmes, and so forth. The crux is that niches are unstable. A niche is small with limited user practices and high uncertainties (e.g., Will subsidy programmes be continued? Is the new technology as good as it first promised to be? Will users and policymakers actually accept the changes in the regime?), and does not yet consist of mature networks between actors.

The macro-level in this theory is often termed the 'socio-technical' landscape. The metaphor 'landscape' refers to characteristics of the wider environment that affect changes or stability in a socio-technical regime. It is important to realize that this macro-level is exogenous to the meso- and micro-level regime. The macro-level is beyond the direct influence of actors within the meso- and micro-levels and cannot be changed by them at will. For example, the road transport regime is influenced by landscape developments such as long-term social, macro-economic, cultural and demographic developments which cannot easily nor quickly be diverted in other 'desired' directions by policymakers, lobby groups or firms.

Crucial in the MLP theory is that transitions can only be explained by studying the interplay of processes that happen at all the three different levels distinguished. For example, perhaps there is pressure from the landscape (macro) on the socio-technical regime (meso) to open up. This could be growing worries about climate change in society, for example. If at the same time there are some niches (micro-level) where transport innovations with low carbon emissions are tried or used, perhaps one or some of them might penetrate the current regime and change it drastically.

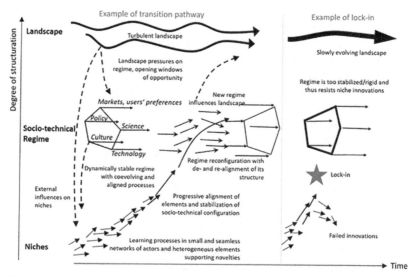

Note: This particular figure is taken from Ollivier et al. (2018) because it gives a nice schematic distinction between a transition pathway (left) and a lock-in (right)

Figure 6.2   The multi-level view of transitions of Geels (2005b)

Theory 5 – Rogers's Theory on Diffusion of Innovations

The innovation theory by Rogers (2003) is perhaps the most well-known (and cited) of them all. The theory of Rogers aims to explain how innovations spread across society. An important idea in his theory is the so-called innovation-decision process: 'an information-seeking and information-processing activity, where an individual is motivated to reduce uncertainty about the advantages and disadvantages of an innovation' (Rogers, 2003, p. 172). He distinguishes five steps in this process: (1) knowledge, (2) persuasion, (3) decision, (4) implementation and (5) confirmation. He proposes five attributes of innovations that explain how individuals perceive the uncertainty reduction potential of a specific innovation. This was a bit vague to me – to be frank – but in my interpretation of his theory I assume that people are always somewhat afraid of innovation (or to put it more broadly 'to do something new') because it brings uncertainty compared to the status quo situation. The five attributes of an innovation that reduce these feelings of uncertainty – and, thus, promote the adoption of innovation – are (1) relative advantage ('the degree to which an innovation is perceived as being better than the idea it supersedes', p. 229), (2) compatibility ('the degree to which an innovation is perceived as consistent with the existing values, past experiences and needs of potential adopters', p. 15), (3) complexity ('the degree to which an innovation is perceived as relatively difficult to understand and use', p. 15), (4) trialability ('the degree to which an innovation may be experimented with on a limited basis', p. 16) and (5) observability ('the degree to which the results of an innovation are visible to others', p. 16). Rogers stresses that all these five factors speed up an innovation adoption process, not only the most obvious one: 'relative advantages'.

The most well-known part of Rogers's theory is that he distinguishes different innovation adopter categories based on their 'innovativeness'. In his idea, successful innovations generate a very specific adoption category curve: a normal distribution curve consisting of successive innovators groups (Figure 6.3, black). According to Rogers, this idea also implies that the adoption rate of successful innovations can be conceptualized as being an S-curve (grey). At the start only a relatively small part of the population or of firms adopt the innovation so the rate is relatively slow; later, the big chunks of the early and the late majority start adopting so the adoption rate is fast; finally, only the relatively small group of laggards are still adopting, slowing down the rate at the end until market saturation is reached.

Theory 6 – Hype Cycle

The idea of hype cycles is included in this chapter because in more practical innovations research it is often alluded to as a notion that can explain a novelty's diffusion path from its 'birth' to full adoption. The notion of innovation hype cycles originates from the consultancy group Gartner (see, for example, their site, https://www.gartner.com/en/research/methodologies/gartner-hype-cycle, accessed October 2021). Their idea is that characteristics of novelties (e.g., technologies, new management philosophies, new services) tend to get overstated at the very early stages of the development of a new idea when,

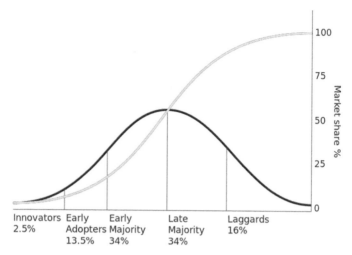

Note: Successive groups of consumers adopt an innovation (black curve). The innovation's market share (grey curve) eventually reaches the saturation level.
Source: https://en.wikipedia.org/wiki/Diffusion_of_innovations, accessed February 2022.

Figure 6.3   *The diffusion of innovations according to Rogers (2003)*

relatively, much attention in the media, on social media, in conferences and even in scientific journals is paid to the new idea. This attention results in 'inflated expectations' (a so-called hype phase) in Gartner's view. Because these expectations are not yet realistic, a phase of disillusionment, as Gartner terms it, sets in, which over time will be overcome when the novelty reveals its true value and, subsequently, is widely adopted ('enlightenment'). This cycle is used by Gartner, amongst others, to specify for their clients 'the right time' for investments in novelties.

Scholars in the sociology of expectations field have a far more subtle view on the role of expectations in socio-technical transitions (e.g., see Kriechbaum et al., 2021) compared to Gartner. They see indeed an important role for expectations in the formation and function of socio-technical niches (see theory 4 above). However, they do not see the hype cycle just explained as a kind of 'physical law' that will always occur and always in the same shape, and, thus, will have some kind of predictive power. Expectations at the beginning of developing something new might indeed play an important role in forming a niche according to these scholars, but they do not always have to be inflated; nor do they have to be deceptive. Other reasons, too, may mean that the attention towards a novelty disappear over time. Also, these scholars note that 'cycles of hype and disillusionment may be repetitive, and the phase

Enlightenment may not always be reached' (Kriechbaum et al., 2021, p. 3). So, the suggested dynamics theory by Gartner – 'hype–disillusionment–enlightenment' – may indeed explain some innovations' development, but the dynamics will often be far more complicated and unpredictable.

Theory 7 – Role of Industries in Disruptive Innovations

There is a huge field of theories and frameworks that aim to explain the role of industries in innovations. It is without a doubt that private companies have to be innovative in order to reduce their production costs and to keep or gain market share by continuously offering better products. There are many business theories explaining which factors determine this 'standard' innovation in a company. Relevant for this book are theories explaining the role of companies in the field of radical or disruptive transport innovations. Hardman et al. (2013) define innovations as being disruptive if two out of the following three characteristics are fulfilled: the innovation is disruptive to (1) market leaders, (2) end-users or (3) infrastructure. Electric vehicles, hydrogen vehicles and autonomous vehicles are the obvious examples, but also an idea such as 'mobility-as-a-service' (MaaS) can be regarded disruptive as it affects incumbent market leaders and users of the transport system. There are many papers in this field and I do not pretend that I can cover them all. I focus on three theoretical observations from this literature.

The first observation is that most papers mention the phenomenon of incumbent companies resisting a new technology or service, as in the MLP. The authors of these papers use different names for this phenomenon: 'the incumbent curse', 'familiarity trap', 'path dependency' (van den Hoed, 2007; Sick et al., 2016; Dijk et al., 2016; Hardman et al., 2013). Another phenomenon is the so-called 'sailing ship effect', which describes that incumbent companies have the tendency to react to new technology by improving the old technology. Broadly speaking two main reasons can be found in the literature for this reluctant behaviour of companies: path dependency and second-mover advantages. Path dependency means that instead of doing something completely new, industries tend to focus on reproduction and on smaller steps of improvement because the 'new thing' does not align with their routines, their prior knowledge or their proven technologies, amongst others. Second-mover advantages mean that incumbents may strategically wait for new entrants on their market being the first movers (electric car producer Tesla being the obvious example) and only after a while start to 'imitate' (or even buy) the new entrants, which implies that they can save on their R&D spendings and can learn from the first-mover market successes and failures. However, companies that choose the second-movers strategy can, of course, be too late.

The second observation is that many factors can be found in this literature that aim to explain whether a certain innovation will become dominant in the market. One stream of literature conceptualizes this striving for dominance as 'a battle' between two (or even more) competing technologies (e.g., Suarez, 2004; van de Kaa et al., 2011, 2017). Although the specific factors mentioned can be somewhat different, the outcomes of a battle for dominance can basically be explained according to these theories by factors related to the firms that are part of the battle, such as their financial strength, their brand reputation and credibility, their learning orientation, their technological superiority, and so forth. Additionally, so-called environmental factors (such as policies) determine the outcomes of a battle for dominance.

The third observation is that another stream of literature conceptualizes successful adoption of disruptive innovations by firms not as a battle between companies striving for dominance per se but as a consequence of a broader set of factors that explains how a company escapes its path dependency (Dijk et al., 2016; Hardman et al., 2013; van den Hoed, 2007; Cowan and Hulten, 1996). Generally speaking, the factors relate to expectations about consumer acceptance, expectations about making profits, external shocks or crises, company-related characteristics (see the 'battle literature', previous paragraph), and many more.

3.   COMPARISON, DISCUSSION AND CONCLUSION

The theories selected in this chapter show the importance of a broad approach when studying transport innovations (or innovations in general). Although they differ (see Table 6.1), the overarching feature is that innovations' success or failure is dependent on many factors: technical factors, user acceptance factors, factors related to economic viability, firm-related factors, factors related to societal and political feasibility (so understanding innovations' societal impacts is also key), and so forth. Another observation is the role of innovation systems (TIS perspective) and of context and time in explaining innovations' success or failure and adoption speed. The landscape developments in the MLP theory or the sanctioned discourse in the political economy approach point at the role of contextual factors in explaining the success (or failure) of an innovation. The MLP theory, expectation theory (the hype cycle idea) and Rogers's theory are theories where time plays an important role. These theories conceptualize that adoption can be disappointingly slow after the invention of the idea or that an adoption path can be a bumpy cycle of attention, disillusion, attention again and so forth until the innovation is finally broadly adopted or never heard of again.

Table 6.1   *Comparing the seven theories selected*

Theory	Aim	Core	Application fields
The opportunity vacuum	To identify and predict economic opportunity	Only at the moment when it is technically feasible to realize an idea, financially viable to do so and when the early majority of the society is ready to adopt the idea, may an innovation become a success.	Analysing opportunity of early innovations from an individual perspective (e.g., a firm wants to develop flying cars)
The technological innovation systems approach	To understand and explain innovation processes within a system	Understanding can only be reached if both the structure of the socio-technical system and the functions within the system are studied.	Analysing the strengths and weaknesses within an ongoing innovation system (e.g., electric vehicles uptake, biofuels market penetration, and so forth)
Political economy of transport innovations	To explain transport policy innovations adoption	A policy innovation will only be adopted if it is technically, economically, societally and politically feasible.	Analysing adoption or non-adoption of transport policy innovations that are put on the political agenda (e.g., road pricing)
Multi-level perspective	To explain transitions from one socio-technical system to another	Transitions can only be explained by studying the interplay of processes that happen at the niche, regime and landscape level.	Analysing transitions from one ('locked-in') socio-technical regime (e.g., horse-powered transport) to another (e.g., fossil fuel-based vehicles)
Rogers's theory on diffusion of innovations	To explain how innovations spread across society	Five factors speed up an innovation adoption process. In innovation diffusion across society different successive groups of adopters can be distinguished.	Analysing innovations' adoption diffusion (e.g., for electric cars) and conceptualizing the potential speed of this process (S-curve)

Theory	Aim	Core	Application fields
Hype cycle	To explain a novelty's diffusion path from its 'birth' to full adoption	Novelties follow a path from hype to disillusionment to enlightenment. This may indeed explain some innovations' development but the dynamics will often be far more complicated and unpredictable.	Analysing innovation adoption pathways (e.g., for autonomous vehicles)
Theories on the role of industries in innovations	To explain the role of industries in radical innovations	Incumbent industries tend to resist radical innovations. Many factors are distinguished in the literature that can explain why new entrants or incumbent firms still engage in radical innovations.	Analysing why firms adopt (or do not adopt) radical innovations (e.g., the role of the big car manufacturers in producing electric cars)

When comparing the theories, no theory with the exception of Gartner's hype cycle has the pretence, however, that it can help decision-makers or firms to predict at an early stage which transport innovation will become a success, or will be never heard of again. As all theories have a connection with evolutionary economics, they all acknowledge the dynamic – perhaps even chaotic – and unpredictable character of innovation progress. The theories accept innovation failure in economic progress. It is not that the theories believe that striving for failure is important and, thus, should be aimed for, but most theories see innovation failure as something inevitable in order to progress. Failures of transport innovations have happened, will happen and even have to happen in order to learn. Failure is also inevitable because it is so hard to predict which candidate innovations will succeed and to suggest a recipe for success. Perhaps an important lesson of almost all innovation theories in this chapter is that picking one 'winning idea' based on assumingly rational arguments, and betting your money on it, is risky.

So if not as predictors, then how can the theories be used? I see four main fields of application:

- Theories can provide the basis for policy development and innovation management. For example, the MLP theory (theory 4) is strong in explaining how we escaped lock-in in transport socio-technical regimes in the past. General lessons can be learned from these historic analyses about factors that are important to consider for bringing about future regime changes such as those towards a fossil fuel-free road transport system. The MLP assumes that pressure from the landscape is required, socio-technical regimes should open up to some extent, and niches should exist that might use such openings for changing the system. On top of this, the developments on these three levels should coincide somewhere in time. The MLP is somewhat connected to the notion of 'strategic niche management' (SNM), which is an approach that suggests that transitions can be supported by niche creation (Schot and Geels, 2008). In the paper of Schot and Geels (2008) the authors argue that SNM cannot be seen as a technological fix nor as a technology push. Niches and experiments are to be perceived as crucial for bringing about transitions, they agree, but not as the sole force in doing so. It seems more opportune according to these authors that 'governments focus on endogenous steering of niches or on steering from within' (p. 538).
- Theories can help to understand the social and technical processes of innovation in transport. This is especially the strength of the TIS approach (theory 2). Are the right actors and networks in place? Is there a strong infrastructure where actors and networks can grow, meet and exchange ideas and learn from each other? Do the functions in the innovation system

reinforce or work against each other? The answers to these kinds of questions can help policymakers 'to steer within'.

- Theories can help to understand which factors could positively contribute to the success of a specific innovation. The theories show that a broad perspective is required to analyse the potential success of an innovation. Only focussing on some positive expectations may result in disappointments (theory 6). Planing (theory 1), Feitelson and Salomon (theory 3), Rogers (theory 5) and the role of industry in disruptive innovation (theory 7) all show the need to include a broad spectrum of factors in the analysis to understand failure (no adoption) or success (adoption). The theories show that the potential impacts of the innovation in an early stage should be estimated not only for the economy but also for the wider society in relation to environmental, privacy and safety issues, amongst others. The crux is that on its own technical feasibility is not enough for success according to all these theories. Success is also determined by factors such as consumer acceptance, social feasibility, industries' willingness (or daring) to step in, economic feasibility, and many more. Both the theories of Planing (2017) and those of Feitelson and Salomon (2004) stress also that all requirements related to technical, economic, financial and social feasibility have to be met before an innovation may become a success.
- Theories can help to understand time and place in innovation adoption. In these analyses, how and which context factors influence the success or failure of candidate innovations can be researched. In other words, these kinds of analyses can help answer questions such as 'under which conditions/context could a candidate innovation be successful?' and 'are current conditions/context factors in accordance with a specific innovation's impacts and expectations?'

I conclude that many useful theories exist that can help to understand the success and failure of transport innovations, as also shown in many of the other chapters in this book. The disappointment might be that theories do not have the power to predict. Modern innovation theories – broadly speaking – reject the idea of the possibility of an easy innovation fix. Innovations in transport involve many actors with large and opposite interests. Additionally, subsystems within the transport system are often locked in. So, to think that a silver-bullet innovation exists that with large sums of subsidies or other policies will conquer a subsystem easily might be called naïve according to the insights of modern innovation theories. Still, the theories can help policymakers by changing or even transitioning the transport system in their desired direction. The four suggested applications of the theories just mentioned can show the strengths and weaknesses within innovation systems and of a specific innovation. These insights on factors for the success and failure of innovations can be used by policymakers to steer the system carefully in the direction they aim for.

NOTES

1. The discourse sanctioned by decision-makers as being politically feasible.
2. The literature on 'path dependence' and 'lock-in' states that when one technology at a certain moment in time gains a lead, it benefits from increasing returns. Increasing returns can be caused by economies of scale, learning by using, network externalities, informational increasing returns and technological interrelatedness (Arthur, 1988). Lock-in means something like 'we cannot or can hardly change ("escape") the current situation'.

REFERENCES

Arthur, W.B. (1988). Competing technologies: an overview. In G. Dosi, C. Freeman, R. Nelson, and G. Silverberg (eds), *Technical Change and Economic Theory* (pp. 590–607). London: Pinter.

Bergek, A., Jacobsson, S., Carlsson, B., Lindmark, S., and Rickne, A. (2008). Analyzing the functional dynamics of technological innovations systems: a scheme of analysis. *Research Policy, 37*, 407–429.

Berkhout, F. (2002). Technological regimes, path dependency and the environment. *Global Environmental Change, 12*, 1–4.

Cowan, R., and Hulten, S. (1996). Escaping lock-in: the case of the electric vehicle. *Technological Forecasting and Social Change, 53*(1), 61–79.

Dijk, M., Wells, P., and Kemp, R. (2016). Will the momentum of the electric car last? Testing a hypothesis on disruptive innovation. *Technological Forecasting and Social Change, 105*, 77–88.

Elzen, B., Geels, F., and Green, K. (2004). *System Innovation and the Transition to Sustainability: Theory, Evidence and Policy.* Cheltenham, UK and Northampton, MA, USA: Edward Elgar Publishing.

Feitelson, E., and Salomon, I. (2004). The political economy of transport innovations. In M. Beuthe, V. Himanen, A. Reggiani, and L. Zamparini (eds), *Transport Developments and Innovations in an Evolving World* (pp. 11–26). Berlin and Heidelberg: Springer Verlag.

Geels, F.W. (2002). Technological transitions as evolutionary reconfiguration processes: a multi-level perspective and a case study. *Research Policy, 31*, 1257–1274.

Geels, F. (2005a). *Technological Transitions and System Innovations: A Co-evolutionary and Sociotechnical Analysis.* Cheltenham, UK and Northampton, MA, USA: Edward Elgar Publishing.

Geels, F. (2005b). The dynamics of transitions in socio-technical systems: a multi-level analysis of the transition pathway from horse-drawn carriages to automobiles (1860–1930). *Technology Analysis and Strategic Management, 17*(4), 445–476.

Greenacre, P., Gross, R., and Speirs, J. (2012). Innovation theory: a review of the literature. Imperial College Centre for Energy Policy and Technology Working Paper.

Hardman, S., Steinberger-Wilckins, R., and van der Horst, D. (2013). Disruptive innovations: the case for hydrogen fuel cells and battery electric vehicles. *International Journal of Hydrogen Energy, 38*(35), 15438–15451.

Hekkert, M., Suurs, R., Negro, S., Kuhlmann, S., and Smits, R. (2007). Functions of innovation systems: a new approach for analysing technological change. *Technological Forecasting and Social Change*, *74*, 413–432.

Kahneman, D. (2011). *Thinking, Fast and Slow*. New York: Farrar, Straus and Giroux.

Kemp, R. (1994). Technology and the transition to environmental sustainability: the problem of technological regime shifts. *Futures*, *26*, 1023–1046.

Kingdon, J. (1984). *Agendas, Alternatives and Public Policies*. New York: Harper Collins.

Köhler, J., Geels, F., Kern, F., Markard, J., Wieczorek, A., Onsongo, E., Alkemade, F., Avelino, F., Bergek, A., Boons, F., Fünfschilling, L., Hess, D., Holtz, G., Hyysalo, S., Jenkins, K., Kivimaa, P., Martiskainen, M., McMeekin, A., Mühlemeier, M.S., Nykvist, B., Pel, B., Raven, R., Rohracher, H., Sanden, B., Schot, J., Sovacool, B., Turnheim, B., Welch, D., and Wells, P. (2019). An agenda for sustainability transitions research: state of the art and future directions. *Environmental Innovation and Societal Transitions*, *31*, 1–32.

Kriechbaum, M., Posch, A., and Hauswiesner, A. (2021). Hype cycles during socio-technical transitions: the dynamics of collective expectations about renewable energy in Germany. *Research Policy*, *50*, 1–18.

Ollivier, G., Magda, D., Maze, A., Plumecocq, G., and Laimine, C. (2018). Agroecological transitions: what can sustainability transition frameworks teach us? An ontological and empirical analysis. *Ecology and Society*, *23*(2), 5.

Planing, P. (2017). On the origin of innovations – the opportunity vacuum as a conceptual model for the explanation of innovation. *Journal of Innovation and Entrepreneurship*, *6*(5), 1–18.

Rogers, E. (2003). *Diffusion of Innovations*. Fifth edition. New York: Free Press.

Schot, J., and Geels, F. (2008). Strategic niche management and sustainable innovation journeys: theory, findings, research agenda, and policy. *Technology Analysis and Strategic Management*, *20*(5), 537–554.

Schot, J., Hoogma, R., and Elzen, B. (1994). Strategies for shifting technological systems: the case of the automobile system. *Futures*, *26*, 1060–1076.

Sick, N., Nienaber, A.-M.L., vom Stein, N., Schewe, G., and Leker, J. (2016). The legend about sailing ship effects – is it true or false? The example of cleaner propulsion technologies diffusion in the automotive industry. *Journal of Cleaner Production*, *137*, 405–413.

Suarez, F. (2004). Battles for technological dominance: an integrative framework. *Research Policy*, *33*(2), 271–286.

Suurs, R., Hekkert, M., and Smits, R. (2009). Understanding the build-up of a technological innovation system around hydrogen and fuel cell technologies. *International Journal of Hydrogen Energy*, *34*(24), 9639–9654.

Twomey, P., and Gaziulusoy, I. (2014). Review of system innovation and transition theories. Working paper for the Visions & Pathways project.

Unruh, G. (2000). Understanding carbon lock-in. *Energy Policy*, *28*, 817–830.

Van de Kaa, G., Scholten, D., Rezaei, J., and Milchram, C. (2017). The battle between battery and fuel cell powered electric vehicles: A BWM approach. *Energies*, *10*(11), 1707.

Van de Kaa, G., van den Ende, J., de Vries, H.J., and van Heck, E. (2011). Factors for winning interface format battles: a review and synthesis of the literature. *Technological Forecasting and Social Change*, *78*(8), 1397–1411.

Van den Hoed, R. (2007). Sources of radical technological innovations: the emergence of fuel cell technology in the automotive industry. *Journal of Cleaner Production, 15*(11–12), 1014–1021.

Wieczorek, A., and Hekkert, M. (2012). Systemic instruments for systemic innovation problems: a framework for policy makers and innovation scholars. *Science and Public Policy, 39*, 74–87.

PART II

Transport innovations

7. Technological innovation systems and transport innovations: understanding vehicle electrification in Norway[1]

Ove Langeland, Cyriac George and Erik Figenbaum

1. INTRODUCTION

Transforming the transport sector to a more sustainable system is a key global challenge. While there has been a steady increase in the demand to move people and goods in recent decades, the transport sector is still heavily reliant on carbon-based fuels. Despite improvements in vehicle efficiency, transport is the only major European sector in which greenhouse gas (GHG) emissions have been increasing as compared with 1990 levels (EEA, 2020). This chapter focuses on road transport for passenger vehicles and on innovations related to vehicle and fuel technologies that can contribute to a shift to a more sustainable transport system.

A more sustainable transport system may develop along several pathways and include numerous technological changes. In Norway, electrification of vehicles represents the most important innovation for more sustainable passenger road transport. There are four main electric drivetrains to consider: battery electric vehicles (BEVs), plug-in hybrid electric vehicles (PHEVs), hybrid electric vehicles (HEVs) and hydrogen fuel-cell electric vehicles (FCEVs).

This chapter studies the emergence of these technologies using a technological innovation system (TIS) framework to enhance our understanding of how such technologies are developed, introduced and upscaled. It focuses on the strong growth of BEVs and the corresponding lack of growth for FCEVs and, to a lesser extent, considers the moderate growth of HEVs and PHEVs to provide a comprehensive understanding of the transition away from fossil fuel driven vehicles. Put simply: why have BEVs been such a success whereas the development of FCEVs is still in an experimental phase, that is, a failure? And how can the TIS framework throw light on the emergence and growth of such innovations, and on obstacles to transport innovation processes.

The chapter is organised as follows. In the next section, we give a short overview of the development of vehicle electrification in Norway, mainly focusing on BEVs and FCEVs. In section 3, we present the innovation systems approach, mainly related to the TIS framework but supplemented by the multi-level perspective (MLP). The fourth section describes the development of vehicle electrification in Norway by applying the TIS functions. Section 5 focuses on dynamics and interactions between battery and hydrogen TISs. The final section concludes with some theoretical reflections and implications for further research.

2.   VEHICLE ELECTRIFICATION IN NORWAY

Electromobility development in Norway has been on a different scale than in all other European countries. By the end of 2020 BEVs constituted 12 per cent of the total passenger vehicle fleet with more than 340,000 BEVs registered while PHEVs constituted another 5 per cent, having passed 140,000 units (SSB, 2021). The BEV share of new vehicle registrations reached 42 per cent in 2019 and increased to 54 per cent in 2020 (OFV, 2021). The PHEV market share was 14 per cent (OFV, 2021). The FCEV market, on the other hand, is almost non-existent with no more than a few dozen vehicles sold per year.

The BEV development is the result of a long-term stable policy framework that started in 1990 when the first imported BEV was registered, and continued as the first incentives were introduced through the 1990s to allow market experimentation and development. More incentives to support BEVs in Norway were introduced during the periods 1999–2002 and 2007–2010. These developments had little effect on sales until 2010. Since 2011, the market and policies have been in a roll-out mode, supporting increasingly ambitious climate policy targets. The national BEV fleet has increased rapidly year by year as seen in Figure 7.1. This is partly due to new cheaper long-range BEVs coming on the market (e.g. Tesla Model 3) and that some popular models temporarily disappeared from the market when the new World Harmonised Light-Duty Vehicle Test Procedure (WLTP) became obligatory. FCEV sales effectively came to a standstill in 2019 after an explosion at a hydrogen refuelling station (HRS) near Oslo. New initiatives for vehicles and infrastructure target a restart of the FCEV market in the coming years.

The international automobile industry develops and supplies an increasing number of attractive BEV models to meet legal obligations to reduce CO_2 emissions from new vehicles in Europe beyond what is possible with internal combustion engine vehicles (ICEVs) alone. Norwegian vehicle importers and dealers take advantage of large incentives and policies that were mostly in place prior to 2011, which make BEVs competitive with ICEVs in terms of both total cost of ownership and purchase price. The building of a network

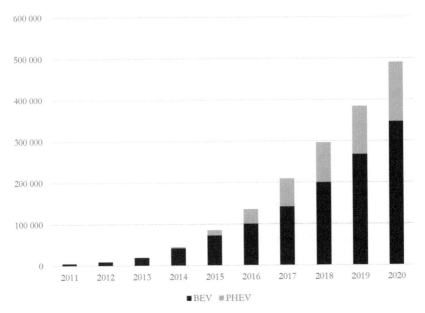

Source: OFV as cited by the Norwegian EV Association (2021).

Figure 7.1   *Total number of registered BEVs and PHEVs in Norway,*
             2011–2020

of fast chargers in and between cities across Norway has further supported this development. Despite having the same incentives as BEVs, FCEVs have not reached the same level of market development. Apart from a few stations, hydrogen refuelling infrastructure has not been built and automobile manufacturers have understandably been cautious with developing FCEVs for consumer markets (see section 4 – F1). As of 2020 there were approximately 160 FCEVs registered in Norway (Hybil, 2020) with 2018 being a peak year with 51 FCEVs sold (OFV, 2021).

An overview of the vehicle electrification developments in Norway is presented in Appendix Table 7A.1.

3.   THE INNOVATION SYSTEMS APPROACH AND THE TIS FRAMEWORK

The extensive electrification of the Norwegian passenger vehicle fleet in recent years represents an important innovation with regard to decarbonisation of the transport system. This section takes this development as its point of departure for discussing the main features of an innovation systems approach within a sustainability transition perspective. It first presents the main characteristics of a technological innovation system (TIS) and then discusses the functions and dynamics of a TIS, and the interactions and relations between technologies and contexts. The TIS framework is well suited for this purpose since it was developed to understand and explain the nature and rate of profound or radical technological changes such as the electrification of transport systems.

Innovation System and Sustainability Transition Research

Moving towards a 'decarbonised' transport system requires deep structural changes or a *transition* from established *socio-technical regimes* (Grin et al., 2010). However, existing systems tend to be very difficult to 'dislodge' because they are stabilised by various lock-in processes that lead to path-dependent developments and 'entrapment' (Fagerberg, 2000). This makes it difficult for innovative sustainable solutions to develop and bring about radical structural changes. Innovation system concepts within transition research focus on how and why existing systems endure and re-produce; what mechanisms destabilise them and create windows of opportunity for successful *experiments* and new development pathways (Geels et al., 2008).

Although there is no universal definition of innovation system, there are some common features related to the concept. Most authors point out that the flow of knowledge, information and technology among actors such as firms and organisations are pivotal to innovation processes (Freeman, 1988, 1995; Lundvall, 1992; Nelson, 1993; Edquist, 2005). The system concept also stresses that innovation does not take place in a vacuum, but among actors and institutions that are related to each other in a coherent whole. The innovation system should also have a (societal) function, that is, it should be geared towards carrying out activities to achieve something useful. The main function of an innovation system is to pursue innovation processes (Carlsson et al., 2002; Lundvall, 1992; Edquist, 2005), and such processes require different types of capabilities. An innovation system should also have defined boundaries so as to be distinguished from its context (Carlsson et al., 2002).

The TIS framework has increasingly been applied to study the emergence and growth of clean-tech sectors (Bergek et al., 2008a, 2015) and as such TISs fall into the line of the continually growing sustainability transitions research (Markard et al., 2012). Studies of measures and innovations for decarbonising the transport sector represent an important part of such clean-tech transition research (Langeland et al., 2018).

The TIS Framework

The TIS approach has proven useful for identifying key transition players and institutions, activities and processes carried out by these, weaknesses in the innovation systems, and key drivers and inhibiting factors (Jacobsson and Bergek, 2011). It is primarily applied to radical innovations but can also be applied to incremental innovations that cumulatively lead to systemic change. This TIS approach can be used across sectors and technologies and forms a basis for comparative studies of the barriers and drivers that are central to the stages of technology development, how they are removed or stimulated by the players involved, and how innovation systems emerge (Hekkert et al., 2007; Bergek et al., 2008a). The approach has an active policy dimension and engages in developing targeted tools and policy mixes that can help reduce or remove barriers (Negro and Hekkert, 2010).

The TIS framework consists of structural *elements* and specific *functions* that should be maintained for systems to evolve. Structural elements consist of *actors* (individuals, organisations and networks), *institutions* (habits, routines, norms and strategies), *interactions* (cooperative relationships) and *infrastructures* (physical, financial and knowledge) (Carlsson and Stankiewicz, 1991; Wieczorek and Hekkert, 2012). The main TIS functions are entrepreneurial activities, knowledge development and dissemination, influence on the direction of search (also called guidance of search), market formation, resources mobilisation, legitimacy, and the development of positive externalities (Bergek et al., 2010). In sum, the TIS framework represents an analytical framework for examining a fairly rigorous set of interlinking 'functions' that are thought to be key to the development and diffusion of a new technology (Bergek et al., 2008a; Hekkert et al., 2007). These interlinkages have been described via the metaphor 'cumulative causation' (Jacobsson and Bergek, 2011). An overview of functions, definitions and indicators that can be used for assessing the TIS functions is presented in Table 7.1. A discussion of interlinking functions and TIS dynamics follows in the analysis in section 5.

The Dynamics of a TIS

Having explained the structural elements and functions of a TIS, a more complete understanding of TIS dynamics can be summarised by five main relationships (see Figure 7.2): (1) the dynamics between structural elements; (2) the internal dynamics of the functions; (3) the influence of the structural elements on the functions; (4) the feedback from the functions to the structure and (5) the influence of exogenous factors on the functions. The mutual dynamic interaction between structural elements (1) and functions (2) as represented by the loop constituted by relationships (3) and (4) is captured through a coupled

Table 7.1   *Overview of functions in innovation systems, definitions and indicators*

Function	Definition	Indicators to track the functions
Entrepreneurial activities (F1)	Experimental activities that seek to appropriate knowledge and translate it into business opportunities, all within a context of market-based risk	Number of new entrants, number of diversification activities of incumbent actors, number of experiments with the new technology
Knowledge development and diffusion (F2)	Activities that lead to the creation of knowledge through processes of learning, e.g. learning by searching, learning by doing	Extent of R&D projects, investment or patents in a specific field
	Activities that lead to exchange of information but also learning by interacting and learning by using in networks	Number of workshops and conferences, network size and intensity
Influence on the direction of search (F3)	Activities that positively affect the visibility of wants of actors (users) and that may have an influence on further investments in the technology	Vision, regulations, targets set by governments or industries, number of press papers that raise expectations
Market formation (F4)	Activities that contribute to the creation of a demand or the provision of protected space for the new technology	Number of niche markets, specific tax regimes, environmental standards
Resource mobilisation (F5)	Activities that are related to the allocation of basic inputs such as financial, material or human capital	Quantitative measures of financial, material and human capital allocation and qualitative assessments of whether or not resource access is problematic
Legitimacy (F6)	Activities that counteract resistance to change or contribute to taking a new technology for granted	Rise and growth of interest groups and their lobby actions
Development of positive externalities (F7)	Outcomes of investments or of activities that cannot be fully appropriated by the investor; free utilities that increase with number of entrants and the co-location of firms	Extent of interlinkages between system functions, structural elements, other technologies or contextual elements

Source: Adapted from Bergek et al. (2008a) and Hekkert et al. (2007).

functional-structural analysis. Although system elements are considered in this chapter, we mainly follow a functional approach because, as argued by Bergek et al. (2008a), functions are more directly related to system performance than system structure. System performance is usually interpreted as the development, diffusion and utilisation of new products and processes. A functional analysis of battery and hydrogen vehicles follows in section 4.

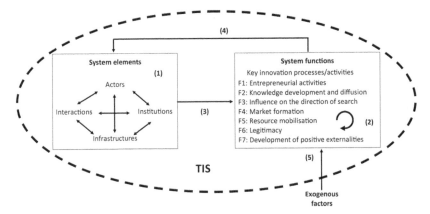

Source: Adapted from Hellsmark (2011).

Figure 7.2   A schematic representation of the dynamics of TISs

Problems with system dynamics can be rooted in problems with either the system elements or the system functions (Carlsson et al., 2002; Wieczorek and Hekkert, 2012). In the case of the former, and simpler, problem, system elements can be either missing or lack the capacity or interest to introduce the focal innovation(s). The development of a new technology can be slow, or the new technology can even fail to appear, because important actors are absent or lack competence, or because specific institutions are not in place or not able to support the new technology. Such situations can also arise due to interaction problems, such as cognitive distance between actors, lack of trust and lack of interactive learning in the innovation system. Finally, infrastructural problems may hamper the development of a new technology because necessary resources (e.g. physical, knowledge, financial) are not sufficiently provided. To summarise: systemic problems can be defined as factors that negatively influence the direction and speed of innovation processes and hinder the development and functioning of innovation systems.

Identifying generative mechanisms that drive and/or change the development of an innovation system is at the core of a TIS analysis (Bergek, 2019; Köhler et al., 2020). This knowledge enables us to specify what characterises the dynamic interplay of actors and functions in successive phases of systems emergence. These types of dynamic, non-linear patterns of development, building on cumulative causation or virtuous cycles, have been conceptualised as motors of innovation in TIS research (Suurs and Hekkert, 2009). Insight into how and why the systems work or not is based on the integration of systemic functions with each other as well as with structural elements, including incumbent elements of the existing socio-technical regime.

When focusing on factors stimulating or hindering innovation for transition, *entrepreneurial activities*, therefore, are of particular interest as they encompass the role of experiments and demonstration projects in the emergence of innovation systems related to renewable technologies such as transport.

Interactions between Technologies

The phrase 'development of positive externalities' (F7), or 'spillovers', has been used to describe different types of positive interactions between technological systems. Spillovers are linked to the 'relatedness' of technological systems, which can take multiple forms, for example (1) the extent to which systemic functions integrate with one another, (2) the extent to which they 'share structural elements: actors and networks, technology or institutions' (Bergek et al., 2008b, p. 585), and (3) how the new technology relates to other emerging technologies. Spillovers are facilitated by, among other things, the entry of new types of firms into a given technological field with the effect that some of the functions listed above are strengthened (Bergek et al., 2008a). The entry of new firms can, for example, influence other actors' search functions, support market creation, improve technological legitimacy, and support the creation of pooled labour markets (Bergek et al., 2008b). Another type of spillover is linked to the way in which different technological systems interact. The classification of solar photovoltaics and biofuels as renewable technologies, for instance, can boost their legitimacy in that they are both seen as constituents of a low-carbon energy future, despite the fact that they share little in common from a technical perspective (Bergek et al., 2008b). This type of spillover can be useful for the advocacy of favourable policies by coalitions of like-minded actors.

Bergek et al. (2015) further explore different interactions between structural factors and technologies, and between a TIS and its context. They distinguish between 'external links' and 'structural couplings'. External links (see arrow 5 in Figure 7.2) resembles the landscape in the MLP framework whereas structural couplings refers to shared elements (actors, networks, institutions, technologies) between a TIS and specific context structures. Structural couplings is the most prominent and may explain why and how synergies *and* conflicts can arise between technological systems. The authors note, for instance, that photovoltaic cells enjoy synergistic interactions with battery technologies, but conflict with wind turbines. Value chain configurations are one source of interactions between technological systems, particularly when different technologies make up complex products. This can be problematic when upstream actors control critical resources or inputs. When these types of interactions, or 'structural couplings' occur, downstream actors may seek to collaborate along the value chain via joint ventures or strategic alliances in a way that supports some of the TIS functions outlined above. Furthermore, technological systems can compete for inputs and resources (e.g. land for energy crops/food) and for institutional resources (e.g. financial resources distributed via government policies).

Sandén and Hillman (2011) elaborate on the different types of interactions that can arise between technological systems in a transition process. Technologies are depicted as socio-technical systems made up of heterogeneous elements that are organised in value chains, and interactions between technologies arise because of overlapping value chains. The authors outline six types of interaction between technologies: competition, symbioses, neutralism, parasitism, commensalism and amensalism (Sandén and Hillman, 2011, Table 1, p. 407). Competition refers to the way technologies compete in markets and for resources. Symbiosis occurs between technologies that occupy different positions in the same value chain or between products that complement one another. Neutralism refers to technologies that do not influence one another, either because they utilise different resources and provide different services, or because they utilise commonly available resources. Parasitism occurs when a new technology benefits from the existence of an older technology, at the expense of that older technology. Commensalism occurs when non-exclusive resources associated with one technology can benefit another technology (e.g. non-patented knowledge). Amensalism occurs when a new technology is locked out via the existence of an older technology. According to Markard and Truffer (2008) the basic modes of interaction between technologies are competition and complementation. This will also be the main focus in our analysis of the modes of interaction between the different pathways of vehicle electrification in Norway – BEVs verses FCEVs – in sections 4 and 5.

TIS in Context

Research on technological change related to sustainability transition has been growing rapidly over the past decade (Köhler et al., 2019). This research is particularly driven by global climate and environmental problems, and it prescribes radical changes in socio-technical systems, such as transport, for responding to such grand challenges (Geels, 2012). There are several analytical frameworks within the field of transition research, the most prominent of which are TIS, multi-level perspective (MLP), strategic niche management (SNM) and transition management (TM). SNM and TM primarily focus on governance processes in transitions (Walrave and Raven, 2016) whereas the

MLP framework and the TIS approach both focus on emerging technologies but in different ways; and they may well complement each other in studies of far-reaching technological change (Markard and Truffer, 2008). Both the TIS and MLP strands have emerged and developed largely independent of each other but share a common ground rooted in evolutionary economics, acknowledging the importance of networks, learning processes, institutions, path dependency, lock-in, interdependence, non-linearity and coupled dynamics. Both frameworks also apply an interdisciplinary perspective, they account for the particularities of spatial and historical contexts, and they provide insights for transition and innovation policy making.

The MLP (Rip and Kemp, 1998; Geels, 2002, 2005; Smith et al., 2005) argues that transitions come about through interacting processes within and between three analytical levels – niches, socio-technical regimes and an exogenous socio-technical landscape – and that structural changes start as radical innovations or transition experiments in niches. A main challenge in transition processes in sectors like transport is to induce upscaling of new technologies and processes in order to bring about deep structural changes. Regarding vehicle electrification in Norway, the BEV has managed to do this and is already becoming a part of the transport regime, while the FCEV on the other hand is still at an emerging phase and constitutes a niche. Applied to innovation and transformation processes related to vehicle electrification in Norway, the MLP framework can be used to illustrate the different development for BEVs and FCEVs as niches at different stages of integration with the regime context. As Figure 7.3 indicates, the BEV TIS has reached the growth phase whereas the FCEV TIS is still in its formative phase.

The TIS approach may benefit from elements of the broader MLP framework in analysis of transition processes, in understanding how technological change and diffusion of emerging niche technologies (radical innovations) take place, and in identifying which factors can speed up sustainability transitions. In this way, the MLP framework can be used to contextualise TIS. The strength of TIS is that it allows an understanding of the functioning of an innovation system related to a specific technology, which barriers may hamper the innovation process and which drivers may enhance it. There is a strong focus on technology-specific factors and internal functions of the innovation system, but less attention has been paid to contextual interactions between a TIS and the environment. A broader and more contextual understanding can be obtained by combining the TIS approach and the MLP framework. Such a combined approach may enhance our understanding of how a specific technology develops in the setting of a larger sectoral or societal change (Markard and Truffer, 2008).

A similar approach could be to put more emphasis on TIS context structures and interactions as suggested by Bergek et al. (2015). They operate

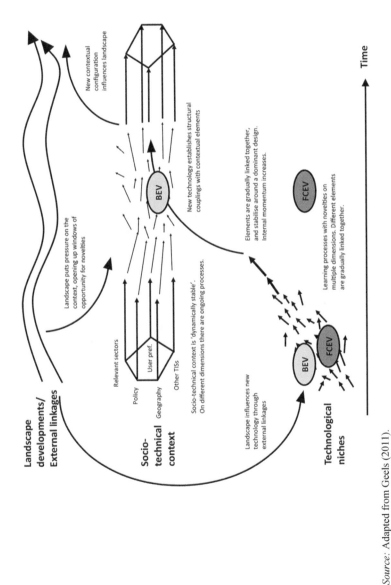

Source: Adapted from Geels (2011).

Figure 7.3   *Battery electric and fuel cell electric vehicle TISs within the MLP*

with three generic types of contextual structures and modes of interactions. First are structural relations with other TISs, for example that between BEVs and FCEVs. Second are structures related to pre-existing infrastructures and institutions, such as the incumbent electric power sector or national emission reduction frameworks. Third are context structures related to the provision of specific system-level assets to the TIS, such as technology-specific policies (e.g. a national hydrogen strategy), competence or venture capital. Both a combined TIS–MLP approach and a TIS contextual approach seek to remedy the potential weakness of TIS analysis as an inward-looking approach.

Boundaries of the TIS

The discussion of interaction between technologies and the TIS context highlights the issue of boundaries of the TIS. A system operates in a context, and it should be possible to distinguish an innovation system from its environment. However, there is no simple recipe for drawing boundaries around an innovation system as one can delimit a TIS within narrow or wide boundaries. The most common way is to identify boundaries spatially/geographically, sectorally or functionally. A TIS can operate within national or regional borders or sometimes even on a global basis, it can be sectorally delimited to a specific technology field or product area, or it can be functionally delimited in terms of what societal functions and activities take place within the innovation system (Carlsson and Stankiewicz, 1991; Edquist, 2001, 2005; Bergek et al., 2008a; Klein and Sauer, 2016; Edsand, 2019).

4. TIS FUNCTIONS AND THE DEVELOPMENT OF VEHICLE ELECTRIFICATION IN NORWAY

This section describes the TIS functions related to the development of vehicle electrification in Norway, in order to assess the system performance of the emerging technologies. The analysis focuses on FCEVs on the one hand and BEVs (and PHEVs to a lesser extent) on the other, in the passenger vehicle segment in Norway. The networked relationship of actors and institutions working with these technologies constitute the two focal TISs, whereas international activities are considered exogenous.

Actors in both TISs consist of public authorities, that is, politicians, ministries, civil services (e.g. Norwegian Public Roads Administration, Norwegian Environment Agency, Norwegian Directorate for Civil Protection, Norwegian Water Resources and Energy Directorate), government-owned enterprises (e.g. Enova, Transnova), local governments, environment organisations (e.g. Bellona, Zero) and, importantly, users. Important institutions include targets (GHG emission for vehicles, average CO_2 emission for vehicles, zero emission vehicle [ZEV] shares), policies and incentives, guidelines and standards, laws, public opinion and sentiment, and the interactions they create.

The BEV TIS also consists of companies that have been involved in vehicle development (PIVCO/Think, Kewet/ElbilNorge, Miljøbil Grenland), parts suppliers (Saft, Siemens), most of the traditional vehicle importers, dealers and workshops, charging infrastructure providers (Statkraft, Fortum, Grønn kontakt, Ionity, Circle K) and other stakeholders (energy companies, the EV Association). The hydrogen TIS also consists of infrastructure providers (Uno-X, HyNor, Hyop, St1, NEL), components producers (Hexagon), large industries (Statoil, Hydro) and some vehicle manufacturers (Mazda, Mercedes, Hyundai, Toyota).

F1: Entrepreneurial Activities

Entrepreneurial activity can be broken up into two main categories: vehicles and infrastructure. Norway has not been a major vehicle producer but has made efforts to develop BEVs, such as the PIVCO/Think two-seater prototype tested out in Norway and California from 1990 to 1996. A production version was developed with international automotive expertise from 1997 to 1998 and was produced from 1999 to 2002 when Ford owned Think. Ford sold Think in 2003 due to changes to the California ZEV mandate. Think re-emerged with a new model from 2008 to 2010, partly produced in Norway and partly in Finland, but went bankrupt during the concurrent financial crisis. A total of 3,500 Think BEVs had been produced since 1999. PureMobility, a company that had taken over the Danish Kewet factory in 1999 and moved it to Norway, suffered the same destiny after having produced a total of 1,100 Kewet/Buddy BEVs. Other initiatives to establish BEV production in the Grenland area (Miljøbil Grenland) with TATA (India) as an investor also failed.

Other activities involved new and second-hand vehicle imports, mainly French BEVs produced between 1997 and 2002; the latest generation of BEVs with lithium-ion batteries from 2011; and the build-out of normal chargers in Oslo since 2007 – from 2009 a financial crisis countermeasure led to the installation of some 2,500 normal chargers across Norway. Fast chargers have been built since 2011 with economic support from the national agencies Transnova and Enova as well as some local and regional governments. Tesla chargers, and fast chargers in and around cities, have been built without public support. Fast charging in cities is considered a fully commercial market.

At the end of 2020, Norway had 2,330 fast chargers, 820 Tesla Superchargers and 13,800 normal chargers (Nobil, 2021). A huge advantage for BEVs has been the ability to charge at home from available power sockets mounted in garages and outside buildings. This allows users to adopt BEVs with little effort, but is now only recommended for occasional charging.

The first hydrogen vehicles in Norway were a pair of adapted Mercedes Sprinter internal combustion engine (ICE) vans in 2002. Miljøbil Grenland introduced a hydrogen ICE car (an adapted Toyota Prius HEV) in Norway in 2009, and Think modified a few BEVs to operate on hydrogen, whereas the Mercedes importer for Norway tested ten B-Class FCEVs in the Oslo region starting in 2010. Mazda delivered only 4 out of a planned fleet of 30 RX8 hydrogen ICE cars to Norway in 2008 for use in the so-called HyNor 'hydrogen highway' project, which was designed to provide several HRSs along the 580 km road corridor connecting Oslo with Stavanger. The aim of the project was to gain experience with refuelling infrastructure, but it failed to generate significant market development for either HRSs or FCEVs. The number of stations has declined since the HyNor project concluded – as of 2020, there is only one commercially available HRS for passenger cars in the country, located just outside of Oslo. More recent efforts in Norway have focused on FCEVs such as Hyundai's Nexo and ix35, and the Toyota Mirai, which constitute the majority of Norway's FCEV fleet of about 160 vehicles.

F2: Knowledge Development and Diffusion

Activities related to market introduction and fora that enable actors within the BEV and hydrogen systems to interact and work together are indicators of diffusion of knowledge. There were early EV conferences in the 1990s to increase awareness. From 2006, conferences on low emissions (with BEV and hydrogen topics) have been arranged annually by non-governmental organisations (NGOs) and agencies. EV-specific conferences have been arranged from 2015, and hydrogen conferences from 2016. Electric car rally events were frequent from 1993 to 2014, some also with hydrogen vehicle classes. EV test drives were arranged as early as 1993, an EV centre was set up in Stavanger in 1996, and another in Oslo for light commercial vehicles with a battery swap service, and there was an EV rental scheme in Oslo from 1996 to 1997. EV leasing was tested out in several places in the late 1990s.

The Norwegian EV Association (Norsk elbilforening) was founded in 1995 as an industrial lobby and development forum with 17 members, and gradually transitioned into a consumer organisation with 78,000 members and 30 employees by 2019. This growth was made possible after EV dealers started paying for first-year membership of the association for all EV buyers from 2010. The association supports members and works for incentives and improved charging infrastructure. The Norwegian Hydrogen Association (NHA: Norsk hydrogenforum) was established in 1997. It seeks to promote the technology broadly, but much of its activity focuses on transport applications. The Norwegian Hydrogen Vehicle Association (Hybil: Norsk hydrogenbilforening) was formally established in 2018. In general, the level of knowledge development activity for hydrogen passenger vehicles is much lower than that for battery technology in Norway. Efforts to promote fuel cell technology for maritime and industrial applications, however, have been much greater in Norway; prominent initiatives and organisations that seek to establish hydrogen value chains include H2 Cluster, Maritime CleanTech and Ocean Hyway Cluster.

F3: Influence on the Direction of Search

The purpose of the vehicle tax policy in Norway up to 1990 was to generate government income (Ministry of Finance, 2003). From 1990, the vehicle tax system and incentives were increasingly used to allow BEV experimentation (Ministry of Finance, 1989) and industrial development inspired by the ZEV mandate in California. In the 1997 Kyoto Protocol, Norway's target was to keep emission growth to 1 per cent over 1990 levels for 2008–2012 (Ministry of Environment, 2001). The climate policy bill in 2001 (Ministry of Environment, 2001) and the following governments (Sem, 2001; Soria Moria, 2005) focused increasingly on domestic emission reductions. The BEV policy and incentives were extended to FCEVs in the Parliament for 2004–2006. In the 2007 Climate Policy Proposition (Ministry of Environment, 2007), Norway's target was to be climate neutral by 2050 and to reduce emissions by 30 per cent in 2030 compared to 1990. In 2012, the Parliament added an ambitious goal to reduce average new passenger vehicle emissions, that is, new vehicle average CO_2 emission was to be reduced to 85 g/km by 2020. This was not possible without electrification (Figenbaum et al., 2013).

It was also decided that all ZEV incentives should remain in place until 50,000 cars had been sold, or through 2017. In 2017, this was extended to 2020 for some incentives and to 2023 for others (Ministry of Finance, 2017), and the incentives for FCEVs should now last until 2025 or 50,000 vehicles. The Parliament also approved the National Transport Plan (Ministry of Transport, 2017), containing the target of only selling ZEVs in the new vehicle market from 2025. In 2020, Norway's Paris Agreement obligation was revised to a 50–55 per cent reduction by 2030 (Paris Agreement, 2020).

FCEVs have virtually the same incentives as BEVs, but the infrastructure is a bigger barrier. BEVs can be charged from any outdoor socket. Hydrogen vehicles need public infrastructure and thus a coordinated infrastructure strategy. The first hydrogen strategy was at the municipal/regional level, for Akershus County and Oslo for 2014–2025. The 2020 National Hydrogen Strategy was criticised for not having clear targets for hydrogen use in passenger vehicles. Follow-up initiatives by the Government signal towards a more formal commitment to hydrogen fuel, but there is still no clear support for mass market passenger car use.

The number of press articles for BEVs has been much higher than for FCEVs, as seen in Figure 7.4 (Retriever search, 2020), due to the greater entrepreneurial and market activities. For hydrogen the main trajectory now seems to be heavy-duty vehicle adoption, which could lead to passenger car spillover effects if and when the infrastructure is in place.

F4: Market Formation

BEV market experiments were initialised through an exemption from the vehicle registration tax from 1990 and further supported by exemptions from the annual tax (1996), road tolls (1997) and public parking fees (1999). A VAT exemption for BEVs came into effect in 2001, during the national BEV industrialisation effort when the Norwegian BEV producer Think was owned by Ford. Bus lanes were opened to ZEVs in 2003 in the Oslo area and nationwide from 2005. From 2005, the incentives were extended to FCEVs and then to hydrogen ICEVs in 2006 to allow the on-road testing of vehicles and HRSs between Oslo and Stavanger for the HyNor project. Ferry charges were reduced from 2008, and the VAT exemption extended to leasing of vehicles and batteries from 2015. An exemption from re-registration tax came in 2017. During the 2016 negotiations in Parliament for the 2017 budget, it was decided that ZEV owners may need to pay up to half of the full price for toll roads, ferries and parking (Stortinget, 2016). The other incentives are still in place, extended and protected through political agreements in the Parliament and climate policy targets.

The early BEV market in the 1990s consisted of a few fleets and the occasional enthusiast. The industrialisation activities led to knowledge building but did little for sales through to 2010. The market took off when the traditional vehicle producers, importers and dealers started selling BEVs from 2011. By 2020, most brands offered BEVs, with the best-selling vehicles being sold in the thousands per year. The situation is very different for hydrogen vehicles. They did not become available to market experimentation as test vehicles until 2005, and from 2014 only came to market in small numbers through 'small series' production (a few thousand annually globally). Somewhat larger production has begun in more recent years for one Toyota and one Hyundai model (tens of thousands). The Norwegian hydrogen vehicle consumer market is, furthermore, in limbo due to a lack of refuelling options following the 2019 HRS explosion. Additionally, the Norwegian Directorate for Civil Protection, which regulates the production, storage and distribution of hydrogen fuel with respect to safety, uses guidelines that are stricter than that of gasoline/diesel (DNVGL, 2019), making it comparatively more difficult to establish HRSs and transport hydrogen fuel.

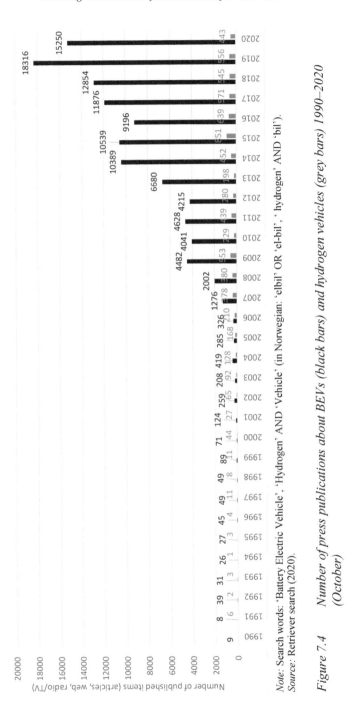

Note: Search words: 'Battery Electric Vehicle', 'Hydrogen' AND 'Vehicle' (in Norwegian: 'elbil' OR 'el-bil', 'hydrogen' AND 'bil').
Source: Retriever search (2020).

Figure 7.4   *Number of press publications about BEVs (black bars) and hydrogen vehicles (grey bars) 1990–2020 (October)*

F5: Resource Mobilisation

In 1996, Think had to go outside Norway to find vehicle design and construction expertise for their BEV (Lotus Engineering UK), and to France, Germany and Italy to find experienced automobile industry suppliers (Siemens, Saft, Fiat, respectively), but found domestic aluminium expertise at Hydro Aluminium. The body concept was a Norwegian technology based on rotational mould-ing. Think's strategy was to learn from experienced automotive suppliers and designers to build up competences in Norway for future development of vehicles and intellectual property. The story of Think was one of continually trying to obtain enough cash to get to market, apart from when Think was part of Ford.

The Buddy EV production in Norway came about when the Norwegian importer/dealer Kollega Bil bought the Danish EV company Kewet and moved the production equipment and parts to Norway in 1999. The Buddy EV was modified to be a four-wheel motorcycle in 2005, with an improved version coming in 2009, aided by a Portuguese R&D partner.

The attempt to create EV production from gliders in the Grenland area was a daunting task for a company that leased out BEVs together with electric ducted fans (EDFs). Yet from a seemingly impossible starting point, they attracted international interest from TATA and then the battery supplier Electrovaya, as successive owners.

In the hydrogen TIS, the target has been to deliver HRS solutions, parts and services in the hydrogen value chain. The HyNor project, with large industrial actors such as Statoil and Hydro participating, built a hydrogen test road between Oslo and Stavanger with Norwegian-made HRS components. When Statoil pulled out of HyNor, it resulted in a huge loss of large-scale industry resources for, and confidence in, HRSs in Norway. The exit of Uno-X from the HRS market after the explosion at their station in 2019 was even worse. The government funding agency for climate-friendly solutions, Enova, is willing to support HRS, as evidenced in their support for Uno-X (among others) over the years, but few want to build HRSs for passenger cars at the moment. The Norwegian hydrogen company NEL and the Toyota importer may be the exception, but plans are still vague. Heavy-duty and maritime operators seem more willing to invest in infrastructure.

The BEV TIS grows organically within the national policy and incentive framework. Hydrogen needs a coordinated plan. The National Hydrogen Strategy from June 2020 dedicated NOK 120 million to hydrogen develop-ment. The budget for 2021 allocated an additional NOK 100 million. Most of this is expected to go to pilot projects, infrastructure and value chains, outside the passenger car segment.

F6: Legitimacy

An NGO, Bellona, was the first proponent of EVs, importing a BEV in 1989. It was also responsible for importing the first two hydrogen vehicles in Norway back in 2002. In both instances, the goal was to bring attention to the technology.

The Norwegian EV Association, established in 1995, grew out of a need to create an ecosystem around electromobility, and was a loose network of collaborators arranging a few conferences and working for improved framework conditions and incentives, such as the VAT exemption from 2001. The exemption was introduced to aid Ford in establishing a viable home market for Think. New incentives were again introduced in 2008 as Think again approached production, this time with national and international investors as owners. The EV Association became the dominant NGO in the segment from 2010 when dealers started to give buyers a one-year membership for free.

Hydrogen not only lacks the scale of legitimacy-building activities as compared with batteries, but suffers from negative legitimacy stemming from the HRS explosion in 2019, when all HRS operations were halted pending official investigation and review. During this time, the operator Uno-X announced that it would suspend its HRS activities irrespective of the investigation. The approximately 150 FCEV owners in the country were left without refuelling options until a new station (operated by Hynion) was established outside of Oslo in late 2019. The explosion and the subsequent downscaling of refuelling infrastructure sent a clear message that the FCEV market for passenger cars in Norway is far from stable.

F7: Development of Positive Externalities

Norwegian industrialisation and lobbying activities had led to substantial incentives being available to and exploitable by car importers as they started importing BEVs from 2011. The market would not have been there without this historical heritage. The rapidly increasing fleet of BEVs has made it possible to install expanding networks of fast chargers across Norway. After 2010, Think employees became experts in maritime battery and electric propulsion applications, EV charging, EV systems design and EV research. Although Buddy has given up on vehicle production, it remains a second-hand EV importer, and has kept the existing older Buddies on the road.

The regional company Asko Midt Norge, part of Norway's largest grocery wholesaler, has a dozen forklifts and four trucks operating using the same HRS, which is also used to refuel more than a dozen FCEVs used by employees. There is currently only one publicly available HRS for passenger cars in

Norway, located just outside of Oslo. Upscaling of hydrogen fuel would be more likely if inter-segment users could use the same infrastructure.

We also see spillover effects between battery and hydrogen technology in a ferry piloting project in South-West Norway. The Hjelmeland–Nesvik–Skipavik route is actually one that is perfectly suited for battery operation, but this car ferry will be the first in the world to use liquid hydrogen fuel, while having a battery backup. There are also potential positive externalities with industrial uses of hydrogen.

5.   DYNAMICS AND INTERACTIONS BETWEEN BATTERY AND HYDROGEN TISs

In the following section, we further develop our analysis of the internal dynamics within the two TISs, after which we look at the structural couplings between the respective TISs and contextual elements (Bergek et al., 2015). The section concludes with a consideration of external linkages and landscape forces.

Functional Dynamics and Integration

Given the overall Norwegian climate policy and the related incentive schemes to promote alternative fuels, the guidance of search and market formation functions have largely been the same for both TISs. The interrelationship between these and other functions, however, has been very different. The links between entrepreneurial experimentation and guidance of search as well as market formation are clear for battery technology in Norway. Similar but weaker links exist for hydrogen. The links to F3 and F4 show how public-sector activities influence entrepreneurial activities at different levels.

At the national level, the long-term climate targets that are part of 'guidance of search' provide the broad framework within which actors can consider new fuels and drivetrains. It has been clear for at least two decades now that the Norwegian Government is committed to promoting the adoption of ZEVs to achieve its climate targets. At a more micro level, the set of incentives for ZEVs has played out differently for the two TISs. BEV actors have been able to increase the availability of vehicles and infrastructure since 2010 and consumers have been eager to adopt. Despite hydrogen being included in the national incentive scheme, entrepreneurial experimentation remains tepid – still at the piloting stage whereas BEVs are essentially becoming part of the private transport regime in Norway (Figenbaum, 2017). For hydrogen, the climate targets and incentives were not sufficient to jump-start the market.

Functional overlaps are evident within the two focal TISs, perhaps more clearly in the case of the BEV TIS and legitimacy. All of the functions have, in some form or another, contributed to bolstering the BEV TIS's legitimacy. Although there has been limited research and development (i.e. knowledge development) domestically, knowledge diffusion has been a key driver of the success of BEVs and PHEVs. In particular, the evolution of the Norwegian EV Association from an industry organisation to one focused more on consumers increased levels of awareness and acceptance among the general public. This is especially relevant with passenger cars as they are largely consumer goods, as opposed to vehicles purchased by governments or fleet operators. While the hydrogen TIS does have actors working on knowledge diffusion, it does not compare with the scale and breadth of that for the BEV TIS. Furthermore, much of the Norwegian activity for promoting hydrogen fuel for transport is geared towards the maritime and heavy-duty segments.

Similarly, the greater levels of entrepreneurial activity and legitimacy for BEVs and PHEVs led to a positive feedback loop whereby more of these vehicles were purchased, and more charging infrastructure was developed. FCEVs never managed to initiate such positive feedbacks. As such, the automobile industry lacks interest in mass producing FCEVs while hydrogen refuelling infrastructure remains largely underdeveloped, and vice versa.

Furthermore, the hydrogen TIS has suffered from singular instances of delegitimisation. The departure of Statoil from the HyNor project represented a failure in resource mobilisation, and reduced the confidence the public and industry had in the technology. Similarly, when the National Hydrogen Strategy allocates little money for passenger vehicle adoption, it sends a message that the public commitment is not there. The Government has since announced more aggressive hydrogen initiatives with additional funding – this may have a positive impact, but mainly in the maritime and heavy-duty segments. An explosion at an HRS in June 2019 froze the market overnight with all stations, public and private, being closed and sales of FCEVs dropping effectively to zero. Uno-X, the operator of the HRS in question, has since abandoned all hydrogen investments, despite having planned a network of at least twenty HRSs. Although one public and two private HRSs have resumed operations since the explosion, the incident has played a pivotal role in diminishing the legitimacy of hydrogen in the passenger vehicle segment. Although there have also been incidents of fires involving battery-powered vehicles and vessels, these have had no significant impact on the legitimacy of BEVs or the BEV market in general.

Structural Coupling

The hydrogen efforts of early BEV companies in Norway must be noted as examples of positive *TIS–TIS* interaction. Both Think and Miljøbil Grenland developed hydrogen demonstrators based on their BEV platforms. There are complementarities between these TISs at a technological as well as operational level. As Bjørnar Kruse of the environmental organisation Zero stated, 'Fuel Cells vehicles benefit greatly from the general electrification of the car industry' (Dalløkken, 2010). This complementarity is driven in large part by the many common components for BEVs and FCEVs, including the lithium-ion battery.

In terms of operation, if and when fossil fuels are phased out, there may be trips that are not suited for BEVs, such as long-distance travel. In such cases, hydrogen may be the best alternative. The complementarity here would be macro, which is to say, at the level of the national fleet. Additionally, in a ZEV world, FCEVs can be used in combination with BEVs in multiple vehicle households whereby the former is reserved for longer trips and the latter is used more for shorter trips on a day-to-day basis. The same logic would apply for transport operators, who could choose to balance their fleet based on distance, payload and refuelling/charging opportunities. The rapid improvements and cost reductions for battery technologies have however rapidly narrowed the range advantage that FCEVs have over BEVs.

Potential TIS–TIS interaction could also be seen if we were to consider other vehicle segments to be their own TISs. If maritime and heavy-duty applications of hydrogen were to upscale in the future, this could provide the foundation upon which the passenger car market could also grow. The physical nature of hydrogen and the need for special refuelling infrastructure makes it more suitable for heavy-duty and long-distance travel, such as for use with long-haul trucks, buses and ferries. This could put in place a hydrogen value chain and critical infrastructure, which may later benefit the passenger vehicle segment. An example of this can be seen at Asko's HRS outside of Trondheim. Although the fuel is primarily intended for Asko's trucks and forklifts, surplus fuel is made available to local users of passenger FCEVs. Subsequent analyses of hydrogen fuel may want to consider consolidated hydrogen and BEV TISs with subordinate sub-TISs for each vehicle segment.

In terms of *geography*, BEVs have been disproportionally successful in the larger urban areas of Norway. It should be noted that Norway's geography may hold potential for FCEVs in the future. Norway has the lowest population density among the countries on the European continent and rugged mountains throughout. BEVs may be less suitable for such long distances and energy-intensive routes, but so long as a fossil fuel alternative exists, the incentive to adopt FCEVs remain marginal. FCEVs also meet competition from the newest generation of PHEVs that offers battery electric drive modes sufficient to cover most users' daily driving needs (Plötz et al., 2017) while using existing infrastructure.

The BEV TIS has a strong structural coupling with the electric power *sector* in Norway. The plentiful supply of renewable electricity (96 per cent hydroelectric based and 2 per cent wind power based) provided an incumbent infrastructure that was suitable for the introduction of BEVs. Over 75 per cent of the population live in detached or semi-detached houses and are therefore able to charge a BEV or PHEV at home. Basic electromobility, for instance in multi-vehicle households, can thus be supported throughout Norway even without public charging infrastructure. Simply put, BEVs work anywhere the grid is available, and they run cheaply. The annual energy cost of operating a BEV in Norway is about one-fifth as compared with ICEVs. Hydrogen, on the other hand, must be made available from public HRSs for users to adopt FCEVs; this, combined with a much higher energy cost than for BEVs, represents a main barrier for the upscaling of FCEVs. Sectoral couplings may be possible for hydrogen in the future in such energy- and carbon-intensive industries as steel and cement production, oil and gas refining, and agriculture.

The *political* contexts for the battery and hydrogen TISs have been similar in Norway. As mentioned earlier, the climate targets and incentives schemes were the same for both, albeit with some differences in timing. An important difference though is the safety regulations that govern the transport and storage of compressed gas fuels. HRSs must have a much larger buffer zone around them as compared with other refuelling stations that provide unpressurised liquid fuel (i.e. gasoline and diesel). Such safety regulations limit the number of potential sites for establishing HRSs, especially in populated areas. Charging infrastructure is, by contrast, possible to install in most locations where electricity is or can be made available.

External Linkages

The guidance of search and market formation functions are themselves linked with international events and activities. Climate targets are developed within international frameworks, for example the Kyoto Protocol, Paris Agreement, European Union (EU). Furthermore, the incentives that fall under the market formation function were inspired by similar ZEV efforts in California in the early 1990s. It is no coincidence that California was one of the main piloting sites for Think's PIV3 prototype in the late 1990s.

Think could also exploit innovation that had occurred in France (Saft Ni-Cd battery, charging system), Germany (Siemens drive system) and Italy (gearbox and drive system integration). Later on, the EU CO_2 regulations for new vehicles and the innovation around lithium-ion batteries have provided Norway with a wide selection of mass-produced BEVs with reliable, long-lasting batteries. Furthermore, the BEV TIS was able to benefit from lithium-ion battery innovations that took place in the consumer electronics industry.

With the exception of the early BEV manufacturing companies in Norway, both the battery and hydrogen TISs are subject to the availability of vehicles that are developed and manufactured internationally. There are currently dozens of BEVs and PHEVs available in Norway from virtually all major manufacturers. For FCEVs, on the other hand, there are only two options – the Hyundai Nexo and the Toyota Mirai. The diversity of supply from international manufacturers has made BEVs and PHEVs a more attractive option than FCEVs.

These international links can be thought of as external linkages or links to landscape events (Bergek et al., 2015). In the case of BEVs, these external linkages were able to open up windows of opportunity within the Norwegian context with respect to broader climate policy targets as well as specific incentives for ZEVs. BEVs are currently high on the agenda throughout Europe, leading to a growing market that likely will be sustained in the long run.

Another key external linkage that may have long-term impacts on the Norwegian hydrogen TIS is the increased attention to hydrogen in Europe. Norway's own hydrogen strategy came out in 2020, within months of similar strategies in Germany, Spain, the Netherlands and Portugal. Even the EU launched its own hydrogen strategy in the same year. Increases in international activity related to hydrogen may influence domestic activity in Norway, especially as Norway is well positioned to be an exporter of hydrogen based on its renewable energy and natural gas resources.

6.   CONCLUDING REMARKS

Electrification of vehicles has been the most important transport innovation for decarbonisation of road passenger transport in Norway. In this respect, the growth and market share of electric vehicles show that BEVs have been a success whereas FCEVs can be considered (for now) a relative failure as they are still in an experimental phase. In this chapter, we have tried to explain this development using the TIS framework, discussed the framework's strengths and weaknesses, and incorporated complementary approaches and contextual factors.

The TIS framework has allowed for a comparative analysis of two prominent technological alternatives. A fundamental strength of the TIS framework is in identifying key players and institutions in transition processes, focusing on barriers and drivers. As the present study clearly indicates, the BEV TIS exhibits greater levels of activity as well as integration for most of the systemic functions in the passenger vehicle segment. In general, actors within the BEV TIS were able to develop the positive externalities that have helped Norway become the most advanced country in the world in terms of BEV uptake.

Entrepreneurial activities among BEV producers occurred earlier and were more intensive than those of their hydrogen counterparts in Norway. Although all the Norwegian EV makers eventually went bankrupt, their pioneering efforts helped develop the institutional setup for electromobility in Norway. Developing a refuelling infrastructure was the most ambitious set of activities in the hydrogen TIS. Much of this momentum was lost following the HRS explosion in 2019. A basic infrastructure of outdoor and garage-mounted domestic sockets provided an infrastructure that made it possible for consumers to start adopting BEVs, thereby avoiding the need to develop an entirely new infrastructure for charging. A supportive public network of normal and fast chargers which has been built since 2011 has further supported BEVs.

Knowledge diffusion activities have been more intense in Norway than knowledge development for both TISs. Prominent environmental organisations have promoted both battery and hydrogen solutions for years. A key difference is the transformation of the Norwegian EV Association from an industry organisation to a consumer organisation, which marks a level of outreach that has no equivalent in the hydrogen TIS. Knowledge diffusion activities for hydrogen are more focused on maritime and heavy-duty applications. Guidance of search and market development are areas in which the activities were largely the same for both TISs, although BEV and PHEV users were able to capitalise to a greater extent.

Given that there is an ample supply of BEVs and related technology from abroad, as well as an incumbent infrastructure in Norway with which BEVs can operate, resource mobilisation activities have been more crucial for the hydrogen TIS, especially with respect to the development of refuelling infrastructure. The National Hydrogen Strategy launched in 2020 is a broad intersectoral initiative that does not place great emphasis on the passenger vehicle segment. The hydrogen TIS experienced a singular setback in 2019 with the HRS explosion, which led to a rapid decline in legitimacy from which the passenger FCEV segment has not recovered. The growth of the BEV market in Norway, on the other hand, has been in itself a legitimating activity, as is the continued government support for incentives.

Contextual analysis using the TIS framework highlights the external linkages as well as structural couplings that either enable the focal TISs to enter the mainstream or block them from doing so. Broadly speaking, we view each TIS as interacting with the contextual elements similar to the way in which niche technologies interact with the socio-technical regime within the MLP framework. With the upscaling of the BEV TIS continuing, it has in many ways already become part of the context or regime in Norway. The combined TIS–MLP framework also considers landscape forces that can interact directly with the TIS, as was the case with California ZEV legislation influencing the BEV TIS in Norway.

A potential weakness of the TIS framework is its inward orientation. In this study, therefore, we have supplemented the TIS framework with the MLP approach in order to remedy this weakness. Context analysis and consideration alongside the MLP allow for broader investigation. The contextual analysis also suggests that although the prospects for FCEVs in the passenger vehicle segment are not promising, this does not seal the fate of the technology as a whole. Potential structural couplings with other sectors, for example steel and cement, and the introduction of hydrogen fuel in the maritime and heavy-duty segments shows more promise. Future analyses of hydrogen fuel for Norwegian transport must consider the heavy-duty and maritime segments, either as distinct TISs, or as part of an integrated hydrogen TIS that includes all segments. A broader approach may even include hydrogen for non-transport applications, which may have spillover effects to the transport system. The relationship between the two TISs must also be reassessed as the share of fossil fuel vehicles continues to diminish in Norway. More research is also needed regarding regulations and public acceptance of hydrogen fuel, especially with respect to safety and security.

NOTE

1.　This paper received support from the Research Council of Norway as part of the following projects: Electromobility Lab Norway (ELAN), under the ENERGIX programme (Grant No. 267848); Safe Hydrogen Fuel Handling and Use for Efficient Implementation (SH2IFT), under the ENERGIX programme (Grant No. 280964); and Digitalisation and Mobility: Smart and Sustainable Transport in Urban Agglomerations (DIGMOB), under the TRANSPORT programme (Grant No. 283331).

REFERENCES

Bergek, A. (2019). Technological innovation systems: A review of recent findings and suggestions for future research. In: Boons, F. and McMeekin, A. (eds), *Handbook of Sustainable Innovation*. Cheltenham, UK and Northampton, MA, USA: Edward Elgar Publishing, pp. 200–218.

Bergek, A., Hekkert, M. and Jacobsson, S. (2010). Functions in innovation systems: A framework for analysing energy system dynamics and identifying goals for system-building activities by entrepreneurs and policy makers. RIDE/IMIT Working Paper No. 84426-008.

Bergek, A., Hekkert, M., Jacobsson, S., Markard, J., Saanden, B. and Truffer, B. (2015). Technological innovation systems in context: Conceptualizing contextual structures and interaction dynamics. *Environmental Innovation and Societal Transitions*, 16, 51–64.

Bergek, A., Jacobsson, S., Carlsson, B., Lindmark, S. and Rickne, A. (2008a). Analyzing the functional dynamics of technological innovation systems: A scheme of analysis. *Research Policy*, 37 (3), 407–429.

Bergek, A., Jacobsson, S. and Sandén, B. (2008b). 'Legitimation' and 'development of positive externalities': Two key processes in the formation phase of technological innovation systems. *Technology Analysis and Strategic Management*, 20 (2008), 575–592.

Carlsson, B., Jacobsson, S., Holmén, M. and Rickne, A. (2002). Innovation systems: Analytical and methodological issues. *Research Policy*, 31 (2), 233–245.

Carlsson, B. and Stankiewicz, R. (1991). On the nature, function and composition of technological systems. *Evolutionary Economics*, 1, 93–118.

Dalløkken, P.E. (2010). Første fullblods hydrogenbil er her. *Teknisk Ukeblad*. 28.09.2010. https://www.tu.no/artikler/industri-forste-fullblods-hydrogenbil-er-her/239918.

DNVGL (2019). *Vurdering av risiko ved anlegg for farlig stoff: Vedlegg 6 Sikkerhetsavstand for fylleanlegg for hydrogen som drivstoff til lette kjøretøy.* Report number: 2018-1200, Rev. 1. Document number: 244734. Oslo. https://www.dsb.no/globalassets/dokumenter/horinger-og-konsekvensutredninger/forslag-til-temaveiledning-om-sikkerhetsavstander/vedlegg-6----sikkerhetsavstand-for-h2-fyllestasjon-for-lette-kjoretoy.pdf.

Edquist, C. (2001). The systems of innovation approach and innovation policy: An account of the state of the art. Lead paper presented at the DRUID Conference, Aalborg, June 12–15, 2001, under theme F: 'National Systems of Innovation, Institutions and Public Policies'.

Edquist, C. (2005). Systems of innovation: Perspectives and challenges. In Fagerberg, J., Mowery, D. and Nelson, R. (eds), *Oxford Handbook of Innovation.* Oxford: Oxford University Press, pp. 181–208.

Edsand, H.-E. (2019). Technological innovation system and the wider context: A framework for developing countries. *Technology in Society*, 58 (2019), 101150. https://doi.org/10.1016/j.techsoc.2019.101150.

EEA (European Environment Agency) (2020). Transport: Increasing oil consumption and greenhouse gas emissions hamper EU progress towards environment and climate objectives. 03.02.2020. https://www.eea.europa.eu/publications/transport-increasing-oil-consumption-and/increasing-oil-consumption-and-ghg.

Fagerberg, J. (2000). Introduction. In Fagerberg, J., Mowery, D.C. and Nelson, R.R. (eds), *The Oxford Handbook of Innovation.* Oxford: Oxford University Press, pp. 10–11.

Figenbaum, E. (2017). Perspectives on Norway's supercharged electric vehicle policy. *Environmental Innovation and Societal Transitions*, 25, 14–34. https://doi.org/10.1016/j.eist.2016.11.002.

Figenbaum, E., Eskeland, G.S., Leonardsen, J.A. and Hagman, R. (2013). *85 g CO_2 per km i 2020 – Er det mulig?* TØI rapport 1264/2013. Transportøkonomisk institutt, Oslo. https://www.toi.no/publications/85-g-co2-km-in-2020-is-that-achievable-article31927-29.html.

Freeman, C. (1988). Japan, a new national system of innovation. In Dosi, G., Freeman, C., Nelson, R.R., Silverberg, G. and Soete, L. (eds), *Technical Change and Economic Theory.* London: Pinter, pp. 330–48.

Freeman, C. (1995). The national system of innovation in historical perspective. *Cambridge Journal of Economics*, 19, 5–24.

Geels, F. (2002). Technological transitions as evolutionary reconfiguration processes: A multi-level perspective and a case-study. *Research Policy*, 31 (2002), 1257–1274.

Geels, F. (2005). *Technological Transitions and System Innovations: A Co-evolutionary and Socio-technical Analysis.* Cheltenham, UK and Northampton, MA, USA: Edward Elgar Publishing.

Geels, F.W. (2011). The multi-level perspective on sustainability transitions: Responses to seven criticisms. *Environmental Innovation and Societal Transitions*, 1, 24–40. https://doi.org/10.1016/j.eist.2011.02.002.

Geels, F.W. (2012). A socio-technical analysis of low-carbon transitions: Introducing the multi-level perspective into transport studies. *Journal of Transport Geography*, 24, 471–482. https://doi.org/10.1016/j.jtrangeo.2012.01.021.

Geels, F., Hekkert, M. and Jacobsson, S. (eds) (2008). The dynamics of sustainable innovation journeys. *Technology Analysis and Strategic Management*, 20 (5), 521–536.

Grin, J., Rotmans, J. and Schot, J.W. (2010). *Transitions to Sustainable Development: New Directions in the Study of Long Term Transformative Change*. New York: Routledge.

Hekkert, M.P., Suurs, R.A.A., Negro, S.O., Kuhlmann, S. and Smits, R.E.H.M. (2007). Functions of innovation systems: A new approach for analysing technological change. *Technological Forecasting and Social Change*, 74 (4), 413–432.

Hellsmark, H.R.A. (2011). Unfolding the formative phase of gasified biomass in the European Union. The role of system builders in realizing the potential of second-generation transportation fuels from biomass. PhD thesis. Chalmers University of Technology, Sweden.

Hybil (Norsk Hydrogenbilforening – Norwegian Hydrogen Vehicle Association) (2020). Statistics. https://www.hybil.no/statistikk/.

Jacobsson, S. and Bergek, A. (2011). Innovation system analyses and sustainability transitions: Contributions and suggestions for research. *Environmental Innovation and Societal Transitions*, 1, 41–57.

Klein, M. and Sauer, A. (2016). Celebrating 30 years of innovation system research: What you need to know about innovation systems. Hohenheim Discussion Papers in Business, Economics and Social Sciences, No. 17-2016.

Köhler, J. et al. (2019). An agenda for sustainability transition research: State of the art and future directions. *Environmental Innovation and Societal Transitions*, 31, 1–32.

Köhler, J., Raven, R. and Walrave, B. (2020). Advancing the analysis of technological innovation systems dynamics: Introduction to the special issue. *Technological Forecasting and Social Change*, 158 (2020) 120040. https://doi.org/10.1016/j.techfore.2020.120040.

Langeland, O., Andersson, M., Julsrud, T.E., Sarasini, S., Schnurr, M. and Tongur, S. (2018). *Decarbonising the Nordic Transport System: A TIS Analysis of Transport Innovations*. TØI report 1678/2018. Institute of Transport Economics, Oslo. https://www.toi.no/getfile.php?mmfileid=50109.

Lundvall, B. (1992). *National Systems of innovation: Towards a Theory of Innovation and Interactive Learning*. London: Pinter

Markard, J., Raven, R. and Truffer, B. (2012). Sustainability transitions: An emerging field of research and its prospects. *Research Policy*, 41 (6), 955–967.

Markard, J. and Truffer, B. (2008). Technological innovation systems and the multi-level perspective: Towards an integrated framework. *Research Policy*, 37 (4), 596–615.

Ministry of Environment (2001). *Norsk klimapolitikk*. St. meld. nr. 54. (2000–2001). Det Kongelige Miljøverndepartement. https://www.regjeringen.no/contentassets/91b54f03dc224f3397c95b04be350f49/no/pdfa/stm200020010054000dddpdfa.pdf.

Ministry of Environment (2007). *Norsk klimapolitikk*. St. meld. nr. 34 (2006–2007). Det Kongelige Miljøverndepartement. https://www.regjeringen.no/en/dokumenter/report-no.-34-to-the-storting-2006-2007/id473411/.

Ministry of Finance (1989). *St. prp nr. 1 Statsbudsjettet 1990 (1989–90), dokument 1: Statsbudsjettet medregnet folketrygden, dokument 2: Skatter og avgifter til statskas-*

sen. https://www.stortinget.no/no/Saker-og-publikasjoner/Stortingsforhandlinger/Lesevisning/?p=1989-90&paid=1&wid=a&psid=DIVL801&pgid=a_0047.

Ministry of Finance (2003). *Bilavgifter.* Rapport fra en arbeidsgruppe. Avgitt til Finansdepartementet 30. april 2003. https://www.regjeringen.no/contentassets/aef d9d12738d43078cbc647448bbeca1/no/pdfs/stm201620170029000dddpdfs.pdf.

Ministry of Finance (2017). *Notification of Tax Measures for Electric Vehicles.* Letter to EFTA Surveillance Authority 2017. 06.11.2017. www.regjeringen.no/contentassets/7b86180aaa8c4980ac60700c5ded97cd/notifikasjon.pdf.

Ministry of Transport (2017). *Nasjonal transportplan 2018–2029.* Meld. St. 33 (2016–2017). Samferdselsdepartementet, Oslo. https://www.regjeringen.no/en/dokumenter/meld.-st.-33-20162017/id2546287/.

Negro, S.O. and Hekkert, M.P. (2010). Seven typical system failures that hamper the diffusion of sustainable energy technologies. Paper presented at the International Schumpeter Society Conference, Aalborg, 2010.

Nelson, R.R. (ed.) (1993). *National Systems of Innovation: A Comparative Study.* Oxford: Oxford University Press.

Nobil (2021). Data extracted from the Norwegian database of chargers, www.nobil.no.

Norwegian EV Association (Norsk elbilforening) (2021). Elbilbestand. https://elbil.no/om-elbil/elbilstatistikk/elbilbestand/.

OFV (Opplysningsrådet for veitrafikken – Norwegian Road Federation) (2021). Bilåret 2020. https://ofv.no/aktuelt/2021/bil%C3%A5ret-2020.

Paris Agreement (2020). Norway's second NDC 2020. Update of Norway's Nationally Determined Contribution to the Paris Agreement. Sent 07.02.2020. https://www4.unfccc.int/sites/ndcstaging/PublishedDocuments/Norway%20First/Norway_up datedNDC_2020%20(Updated%20submission).pdf.

Plötz, P., Funke, S.A., Jochem, P. and Wietschel, M. (2017). CO_2 mitigation potential of plug-in hybrid electric vehicles larger than expected. *Scientific Reports* (Nature), 7, 16493. https://www.nature.com/articles/s41598-017-16684-9.

Retriever search (2020). Press article archive and search service. https://www.retrievergroup.com/.

Rip, A. and Kemp, R. (1998). Technological change. In Rayner, S. and Malone, E.L. (eds), *Human Choice and Climate Change*, Volume 2. Columbus, OH: Battelle Press, pp. 327–399.

Sandén, B.A. and Hillman, K.M. (2011). A framework for analysis of multi-mode interaction among technologies with examples from the history of alternative transport fuels in Sweden. *Research Policy*, 40 (3), 403–414.

Sem (2001). Government political declaration. Information found in: Samarbeidsregjeringens statusrapport. Oppfølging av Sem-erklæringen etter 1 år 19. oktober 2002. Downloaded from the national library. www.nb.no.

Smith, A., Stirling, A. and Berkhout, F. (2005). The governance of sustainable socio-technical transitions. *Research Policy*, 34 (10), 1491–1510.

Soria Moria (2005). Government political declaration for 2005–2008 (in Norwegian). https://www.regjeringen.no/globalassets/upload/smk/vedlegg/2005/regjeringsplatform_soriamoria.pdf.

SSB (Statistics Norway) (2021). Registered vehicles. https://www.ssb.no/en/transport-og-reiseliv/landtransport/statistikk/bilparken.

Stortinget (2016). Account of the debate and decision over the national budget for 2017, during the 5 December 2016. https://www.stortinget.no/no/Saker-og-publikasjoner/Publikasjoner/Referater/Stortinget/2016-2017/refs-201617-12-05/?all=true.

Suurs, R.A.A. and Hekkert, M.P. (2009). Cumulative causation in the formation of a technological innovation system: The case of biofuels in the Netherlands. *Technological Forecasting and Social Change*, 76 (8), 1003–1020. https://doi.org/10.1016/j.techfore.2009.03.002.

Walrave, B. and Raven, R. (2016). Modelling the dynamics of technological innovation systems. *Research Policy*, 45 (9), 1833–1844.

Wieczorek, A.J. and Hekkert, M.P. (2012). Systemic instruments for systemic innovation problems: A framework for policy makers and innovation scholars. *Science and Public Policy*, 39 (1), 74–87.

APPENDIX

Table 7A.1   *Overview of policies, incentives and market developments 1990–2020*

Year	Policy/Incentive	Vehicle series production (first time)	Other events	Market share (%)	Fleet share (%)
1990	Registration tax exemption (temporary)			~0	~0
1991		Kewet	Lithium-ion battery used in Sony Camcorder	~0	~0
1992				~0	~0
1993			California ZEV mandate	~0	~0
1994			PIVCO (Think) prototype tested at Lillehammer Olympics	~0	~0
1995			The Norwegian EV Association founded PIVCO demonstration fleet	~0	~0
1996	Registration tax exemption (permanent) Exemption from annual tax			~0	~0
1997	Free toll roads		Norsk hydrogenforum established	~0	~0
1998			Ford buys Think Nordic	~0	~0
1999	Free parking law change enters into force BEV-specific number plates introduced	Think City	BEV-specific number plates Ford opens Think factory	0.1	~0
2000	Reduced company car benefit tax	Peugeot 106, Citroën Saxo	Kewet moved to Norway	0.3	~0
2001	Exemption from VAT			0.2	~0
2002			California ZEV mandate revised, resulting in weaker incentives for BEV production	0.3	0.1

Year	Policy/Incentive	Vehicle series production (first time)	Other events	Market share (%)	Fleet share (%)
2003	Bus lane access for ZEVs in Oslo area	REVA	Ford sells Think Nordic HyNor HRS corridor launched	0.2	0.1
2004				0.2	0.1
2005	FCEVs get exemption from registration and annual taxes, equal treatment to BEVs Bus lane access national for BEVs/FCEVs	Buddy EV	Buddy production in Oslo	0.1	0.1
2006	Hydrogen ICE vehicles get temporary exemption from registration taxes			0.3	0.1
2007				0.2	0.1
2008	Climate Policy Settlement Reduced ferry rates incentive introduced	Think City (new model)		0.5	0.1
2009			EVS 26 Stavanger Parliament election EV resource group report financial crisis	0.3	0.1
2010	Charging infrastructure (normal) support programme, chargers installed		Funds for industry support Klimakur 2020 report	0.5	0.2
2011	First support programme for fast chargers	The birth of the modern BEV: Mitsubishi i-Miev, Nissan Leaf, Peugeot iOn, Citroën C-Zero	First fast charger installed	1.4	0.2
2012	Climate Policy Settlement firms up policy, incentives to remain in place until 50,000 BEVs on road or through 2017		5 FCEV buses in Oslo (Ruter)	2.9	0.3

Year	Policy/Incentive	Vehicle series production (first time)	Other events	Market share (%)	Fleet share (%)
2013		Tesla Model S, VW e-Up; BMW i3	Parliament election	5.5	0.7
2014		VW E-Golf, Hyundai ix35 FCEV		12.5	1.5
2015	National fast charger support programme for network along main roads Enova support for HRS begins	Toyota Mirai FCEV		17.1	2.6
2016				15.7	3.7
2017	National Transport Plan vehicle target: only sell ZEVs from 2025 onwards	Tesla Model X	Parliament election	20.9	5.1
2018	Exemption ownership change tax, BEVs shall pay max. 50% of rates for ICEVs for parking fees, toll roads and ferries	Hyundai Nexo (replaces ix35)		31.2	7.1
2019	Fast charger support for municipalities without fast chargers	Tesla Model 3, Audi e-Tron	SVV-Norled sign 10-year contract for FCEV ferry in Rogaland HRS explosion in Sandvika (Oslo area) NEL-Yara Green-H2 plant	42.4	9.3
2020	Support for fast chargers in Northern Norway	Mercedes EQC, VW ID.3, VW ID.4, and numerous others, including Chinese BEVs	> 10% BEVs in the national car fleet 4 FCEV trucks in Trøndelag (Asko) National Hydrogen Strategy launched NOK 25 billion for Langskip CCS	54.3	12.1

8. Beyond market success: unpacking the societal implications of the e-bike[1]

Qi Sun

1. INTRODUCTION

Cycling has long been widely recognized as a benign mode of transport thanks to low pollution and greenhouse gas emissions, health benefits, space efficiency, flexibility and affordability (Handy et al., 2014; Heinen et al., 2010). Notwithstanding these apparent assets, the promotion of cycling remains a challenging mission. The development of electric bikes (e-bikes) overcomes some barriers associated with conventional cycling, since riding an e-bike is a less physically demanding activity, which allows for travel at higher speeds and over longer distances or in hilly geographies (Behrendt et al., 2021; Bourne et al., 2020; Cairns et al., 2017). *E-bikes* can generally be defined as two-wheeler vehicles supported by an electric motor (Weiss et al., 2015). Different types of e-bikes, with varying design and performance characteristics, can be found in different jurisdictions. Their exact nature depends on local technical requirements and levels of law enforcement. The e-bike spectrum ranges from the scooter style of e-bikes where pedalling is not required, notably common in China, to the pedal-assisted type of e-bike (or pedelec), which only functions while the rider pedals (although they can switch on battery-powered assistance to reduce the physical effort up to a maximum speed of 25 km/hour) (Cairns et al., 2017; Fishman and Cherry, 2016; Jones et al., 2016). The latter are most common in European countries.

The e-bike is arguably a successful transport innovation as it represents the most rapidly adopted alternative-fuel vehicle in the history of motorization (Fishman and Cherry, 2016). The e-bike boom started in the early 21st century (Weiss et al., 2015). In 2015 an estimated 40 million e-bikes were sold worldwide (Fishman and Cherry, 2016). China makes up approximately 90 per cent of the global market (Zuev, 2018), with annual sales exceeding 35 million units (Lin et al., 2017). In 2019 the Chinese e-bike fleet reached nearly 300 million (GOV.cn, 2019). The world's second-largest e-bike market is the European Union (EU), which registered an annual sale of 1.67 million units in 2016, a 17-fold growth compared to a decade earlier. Germany and the Netherlands lead the way, accounting for more than half of the sales in the EU (CONEBI, 2017).

Despite the e-bike's considerable market penetration rate, visions of future low-carbon mobility frequently overlook the e-bike as a transport innovation. In research, a car-centric focus has dominated the e-mobility field. At the policy level, the electric car has received widespread governmental support across the world, which is in stark contrast to the few favourable policy interventions, or even disadvantageous regulations, for the e-bike. For example, in some Chinese cities e-bikes were banned due to road safety and traffic management considerations (Wells and Lin, 2015; Zuev, 2018). While electric cars do perform better than conventional cars when it comes to pollution and CO_2 emissions, they do not effectively tackle problems regarding congestion, excessive road and parking space, traffic safety, health risks related to inactive lifestyles, and so on. The e-bike can however address many of these issues and thus can offer a more sustainable alternative compared to electric cars.

The purpose of this chapter is to explain why the e-bike is becoming popular and to gain a better understanding of the societal impacts of this transport innovation. By drawing together a wide range of relevant material, this chapter investigates (1) the factors that explain the market success of the e-bike and (2) the socio-environmental implications of increased e-bike popularity. Then using a mobility biography approach, this chapter empirically sheds light on how mode choices around e-bikes occur and the accompanying societal impacts.

2. THE SUCCESS OR FAILURE OF E-BIKE FROM THE TRANSITIONS PERSPECTIVE

The transformations towards sustainable socio-technical systems, also referred to as sustainability transitions, is a rapidly expanding research field. Sustainability transitions literature has covered diverse topics like mobility, energy, buildings, agri-food, cities and waste management (Köhler et al., 2019). Transitions theory offers an insightful lens to unpack the factors contributing to the success or failure of e-bikes as a transport innovation. A transitions perspective stresses the importance of systemic approaches (Pel, Chapter 2 in this volume). In the context of mobility, it places transport innovations in the larger context of a socio-technical system, which consists of not merely the technological artefact per se, the e-bike in this case, but several broader interrelated elements including road infrastructure and traffic systems, regulations and policies, cultural meanings, maintenance and distribution networks, user practices and markets (Geels, 2005). Features and dynamics of a web of these elements can act in favour of or against a certain transport innovation.

When there is a major shift in the configuration of the most important elements that make up the mobility system, it would imply a socio-technical transition (Geels, 2012).

Furthermore, the multi-level perspective maintains that transitions are non-linear processes which result from the interplay of multiple developments at three analytical levels: niches, as protected spaces where radical innovations emerge; regimes of well-established practices and rules; and an exogenous landscape, which is the wider context beyond the control of individual actors (Geels, 2012). The niche of e-bikes and the existing regimes do not exist in isolation but constantly interact with each other within and between different levels (Lin et al., 2018). Those existing regimes include both the dominant automobility regime and, contrastingly, the subaltern regimes, which capture only a small share of the transport market and the overall mobility, such as the conventional bicycle in some regions (Geels, 2012).

One of the factors that account for the market success or failure of the e-bike is its interactions with dominant regimes in a mobility system, especially the automobile regime. From a global point of view, the e-bike is more successful in certain parts of the world than in others. Notably, China stands out as the world's largest producer and consumer of e-bikes (Gu et al., 2020), with an e-bike ownership of more than 20 per cent (GOV.cn, 2019). Perhaps most importantly, what sets China apart from other major e-bike countries is the combination of an absence of an established automobility regime and a decline of the conventional bicycle regime when the e-bike appeared in the transport system as a niche around 2000. The e-bike emerged and grew rapidly during a period of urban expansion coupled with a growing demand for a faster individual mode of transport. At that time, automobility was a niche, and a private car remained too expensive for most households. In addition, against the background of China being an ex-kingdom of bicycles, ownership of human-powered bicycles peaked in the 1990s and steadily decreased after 2000 (Gu et al., 2020). The e-bike thus experienced an explosive growth and gradually reached a regime level. However, as automobility is becoming more dominant in the mobility system due to several supporting conditions, including accommodating road infrastructure, favourable policy orientations and increased income level, Lin et al. (2018) argue that the e-bike may eventually reflect a dying regime in the Chinese context.

The relative success of the e-bike seems to largely depend on the role that cycling plays in the dominant mobility regime. Figure 8.1 indicates that there tends to be an increased uptake of the e-bike where conventional cycling is already popular. Part of the reason might relate to the infrastructure and traffic elements in a socio-technical system. The e-bike flourishes when it can take advantage of existing cycling infrastructures, such as cycle paths and bicycle parking facilities. According to road regulations in China and European coun-

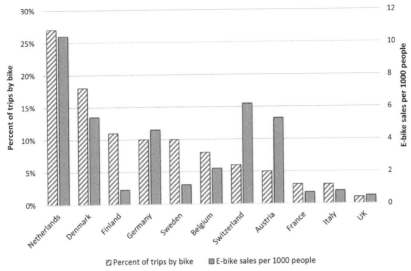

Note: Bicycle mode share data are from around the early 2000s – Netherlands (2005), Denmark (2001), Finland (1998), Germany (2002), Sweden (2000), Belgium (1999), Switzerland (2000), Austria (1995), France (1994), Italy (2000), UK (2005). It is assumed that e-cycling was marginal back in those years and that the bicycle shares consist primarily of trips by conventional bicycle, so that there's hardly any self-correlation effect between the share of trips by conventional bicycle and e-bike sales.

Source: Drawn by the author using bicycle share of trips data from Pucher and Buehler (2008) and e-bike sales per 1000 people data from Fishman and Cherry (2016).

Figure 8.1   *Bicycle share of trips and e-bike sales per 1000 people in European countries*

tries, e-bikes are categorized as non-motorized vehicles and are permitted on infrastructures designed for conventional bicycles (Jones et al., 2016; Zuev, 2018). Many Chinese, Dutch, Danish and German cities enjoy an extensive bicycle infrastructure system that provides favourable conditions for using e-bikes (Pucher and Buehler, 2017; Weiss et al., 2015). Moreover, countries like the Netherlands, Denmark and Belgium are building cycle highways to promote cycling further. These dedicated high-quality bicycle routes often go beyond municipal borders, which corresponds well to the advantages an e-bike offers, including travelling at higher speeds over longer distances. Conversely, in countries where cycling is still a minority practice, a lack of appropriate infrastructures and cohabitation with motorized traffic may cause people to perceive cycling as unsafe. Thus, this is a barrier to upscale cycling, whether electric assistance is provided or not (Rérat, 2021). Nevertheless, the market success of the e-bike innovation does not necessarily depend on a strong presence of cycling. It is worth noting the remarkably high e-bike ownership in countries like Switzerland and Austria compared to the relatively low cycling levels in those countries (Figure 8.1). One possible explanation could be that the e-bike is a particularly useful innovation in places with hilly terrain since it can help to 'flatten' the topography, thereby improving overall enjoyability of the cycling experience in hilly areas (Behrendt et al., 2021; Castro et al., 2019; Marincek and Rérat, 2020).

Additionally, improvement in the price/performance ratio is a critical driver in a wider adoption of the e-bike (Geels, 2005). E-bikes prices span a broad range from EUR 100 in China up to EUR 5600 in Europe. One of the main factors contributing to the price differences is battery technology, which arguably makes up the largest share of manufacturing costs (Weiss et al., 2015). It is estimated that 95 per cent of e-bikes in China depend on traditional lead-acid batteries (Fishman and Cherry, 2016); whereas most European e-bikes use lithium-ion batteries, a more advanced yet expensive technology (Weinert et al., 2007; Weiss et al., 2015). Thus, in China e-bikes serve as an affordable mode of transport. Notably, e-bike adoption surged when prices dropped with technological advancements (Weiss et al., 2015). In contrast, however, in European countries the purchase price is identified as a significant barrier to adopting e-bikes (Cairns et al., 2017). Perhaps this partially explains why e-bike ownership is higher in Switzerland and richer western European countries. Furthermore, the cost factor can be influenced by means of incentives and subsidies. In Lausanne, Switzerland, a municipal subsidy for the purchase of an e-bike has existed since the year 2000. It covers 15 per cent of the price of an e-bike up to a maximum of CHF 500 and an additional subsidy for purchasing an e-bike battery (Marincek and Rérat, 2020). In Sweden, for example, sales of e-bikes registered a peak in 2018, making up 20 per cent of the total bicycle market share, when the government introduced a subsidy of 25 per cent, or up to SEK 10 000, back on the purchase price. When the rebate programme was withdrawn the following year, the proportion fell to 16 per cent (Andersson et al., 2021).

The niche of e-bikes benefits from both public policies that supports it directly and legislation that supports it indirectly. For instance, several public administrations in Austria facilitated the market diffusion of e-bikes by providing information and market access (Wolf and Seebauer, 2014). Contrastingly, by restricting motorcycles in more than 150 cities, Chinese local governments unintentionally created windows of opportunity in the market, which serves as a niche space for the e-bike to gather momentum and scale-up, a positive 'side-effect' (Gu et al., 2020; Yang, 2010).

Regarding cultural aspects, the symbolic meaning of the e-bike co-exists and interacts with the other elements of the established mobility regimes. On the one hand, e-bike users are generally well-received in cycling-friendly countries where cycling is seen as a normal daily practice (Simsekoglu and

Klöckner, 2019). On the other hand, Behrendt (2018) points out that a double stigma is attached to e-cyclists, from both non-assisted cyclists and car users, as in the case of the UK, where such a social stigma partially explains the failure of the e-bike. In North American and some European countries, e-cyclists have to legitimize themselves among cyclists because e-bikes are often associated with old age, physical disability, laziness or 'cheating' (Behrendt, 2018; Jones et al., 2016; Popovich et al., 2014). Meanwhile, in places where cycling is only a subaltern regime whereas automobility dominates socially and spatially, it is disrespected by many motorists (Behrendt, 2018; Rérat, 2021). This illustrates how dominant regimes tend to reinforce stability and prevent change.

3. SOCIETAL IMPLICATIONS

Although the e-bike can be seen as a successful transport innovation in terms of its rapid uptake despite some stigmas, it is unclear whether such a success also holds true when it comes to its societal implications, including both environmental and social aspects. A thorough understanding of an innovation's impacts on various social aspects is crucial in order to move beyond the technological fix mentality. 'Innovation may be at the focus of many analyses, but it is not a goal in itself. In the ultimate instance, innovation is only secondary to the key concerns of system transformation and system *sustainability*' (Pel, Chapter 2 in this volume, p. 28). Instead of assuming that innovation means progress, a detailed sustainability analysis is needed to assess whether a certain innovation benefits society.

3.1 Environmental Impacts

The e-bike's environmental impacts span its entire life cycle, which consists of the production stage, usage phase and end-of-life treatment. A meta-analysis of life cycle assessment reveals that in terms of both energy use (Figure 8.2a) and greenhouse gas emissions (Figure 8.2b), e-bikes are generally more sustainable than other modes of transport, except for conventional bicycles (Weiss et al., 2015).

In particular, the environmental impacts of e-bikes during the usage phase critically depend on the power mix of the local electricity sector. Energy carbon intensiveness varies substantially from power plants based on green and renewable energy sources like hydro and wind power to those based on fossil fuels, such as coal and gas. Therefore, the life cycle energy usage of e-bikes can benefit from decarbonization of the energy sector (Weiss et al., 2015). Moreover, human exposure to pollution reduces significantly as e-bikes relocate the pollutants of motorized transport away from numerous and difficult-to-manage vehicles to fewer concentrated sources (power plants)

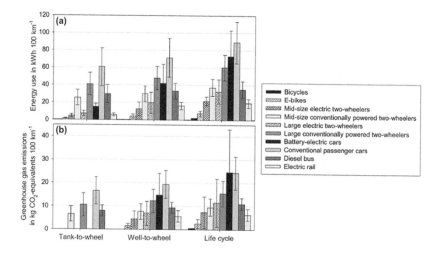

Figure 8.2   *Indicative distance-specific tank-to-wheel, well-to-wheel and life cycle energy use (a) and greenhouse gas emissions (b) of selected vehicles for passenger transport*

in less densely populated areas (Cherry, 2007; Ji et al., 2012). Additionally, the energy efficiency of e-bikes is relatively high thanks to their lightweight and electric drive technology, resulting in an energy consumption that is about one-tenth of a small electric car travelling the same distance (Fishman and Cherry, 2016).

Perhaps the most problematic phase of the life cycle is its end-of-life treatment, especially concerning the batteries, which could lead to environmental contamination and adverse health impacts if they are not appropriately handled. In China, the dominant type of battery is lead-acid, partly because of cost considerations. However, less than 30 per cent of lead-acid batteries are properly recycled (Zuev, 2018), resulting in 95 per cent of total lead emissions being released in the end-of-life phase due to the lack of strict regulations in the recycling industry (Liu et al., 2015). In Europe, most e-bikes are equipped with lithium-ion batteries (Weiss et al., 2015). Thus, lead pollution is less of a concern, but extraction of lithium and battery waste management still present environmental hazards.

3.2 Social Impacts

From a public health point of view, the pedelec type of e-bike, where human pedalling is required, could contribute to physical fitness. Considering oxygen uptake, metabolic equivalents, energy expenditure per minute, heart rate and power output, moderate evidence was found that e-cycling provides physical activity of at least moderate intensity. The e-bike offers a physically active alternative to the largely sedentary behaviour associated with motorized travel (Bourne et al., 2018). There is a significant increase in weekly energy expenditure for individuals switching from a private motor vehicle or public transport trips to e-bikes (Castro et al., 2019). Compared to conventional bicycles, e-bikes elicit a lower level of physical intensity over the same distance (Bourne et al., 2018). Nevertheless, physical activity levels are similar among e-bikers and cyclists. The difference in physical intensity is compensated by the fact that people usually cycle longer distances and with higher frequency on an e-bike compared to a conventional bicycle (Bourne et al., 2020). Therefore, even when an e-bike replaces journeys made by a conventional bicycle, individuals still accrue sufficient physical activity to meet the guidelines for significant health benefits (Castro et al., 2019). Additionally, e-bikes seem to be more appealing among those with little interest in and a lower level of physical activity, rather than those who have an active lifestyle (Sundfør and Fyhri, 2017). Among the least active individuals, gains in health outcomes are the greatest from active travel by e-cycling (Bourne et al., 2018). Unsurprisingly, the aforementioned health benefits do not apply to scooter-style e-bikes, which can be ridden by an electric motor alone.

Notwithstanding the potential fitness contributions, the growing popularity of e-bikes raises safety concerns. When comparing injury risks (both crash rate and crash severity) between e-bikes and mechanical bicycles in a European context, differences diminish after controlling for riding exposure (in terms of the number of kilometres ridden) and demographic variables (Cherry and MacArthur, 2019). Nonetheless, there are a considerable number of e-bike related road accidents, partly because older adults are overrepresented among e-cyclists in Europe. Compared with the mental workloads of cyclists in their middle adulthood, that of the elderly cyclists were found to be relatively high, especially in complex traffic situations. Coupled with the higher speed of e-cycling, these factors may contribute to an increase in the accident risk of elderly e-cyclists (Vlakveld et al., 2015).

Another possible factor leading to e-bike related road conflicts is that other road users often do not anticipate the speed of the e-bike. E-bikes resemble conventional bicycles but are often travelling faster while making little sound. Car drivers both have less time to notice them and underestimate their speed. Similarly, pedestrians may not expect the higher speed of an e-bike. Making e-bikes more distinctive so that they are easily distinguished from conventional bicycles may improve road safety, although this may reinforce the social stigma associated with the e-bike (Dozza et al., 2016). Another factor that might play a role in mitigating conflicts between e-cyclists and other road users is the so-called safety-in-numbers effect. As the number of e-cyclists grows, there would be a less than proportional increase in the number of accidents involving them (Elvik and Bjørnskau, 2017), presumably thanks to increasing awareness and the adapted behaviour of other road users.

3.3   Substitution Effect

A growing consensus has emerged among researchers that the key to understanding the environmental and social impacts of e-bikes is the extent to which they replace travel by other modes of transport, especially cars. Thus, the mode substitution effects of e-bikes have attracted extensive research interest. The review below is not exhaustive but rather presents some up-to-date research findings on this topic.

Based on a meta-analysis of e-bike mode substitution studies around the world, Bigazzi and Wong (2020) show that median mode substitution reported in the literature is highest for public transport (33 per cent), followed by the conventional bicycle (27 per cent), car (24 per cent) and walking (10 per cent), although large variations exist in specific displacement patterns across regions. Nevertheless, a general trend can be discerned that the mode shift effect is largely influenced by the primary mode of transport prior to the introduction of the e-bike (Castro et al., 2019; Kroesen, 2017; Sun et al., 2020).

As the first country to adopt e-bikes on a large scale, China has attracted investigations on mode shift effects since early on. Based on an intercept survey in Nanjing, Lin et al. (2017) reveal that e-bikes are not a replacement for cars on a substantial scale, but they are instead substituting the benign modes of cycling by conventional bicycle (38 per cent), using the bus (36 per cent) and walking (28 per cent). This is consistent with findings from travel diary surveys among e-bike users in Kunming, the surveys reporting that more than 50 per cent of displaced trips were formerly made by bus (Cherry et al., 2016).

In European countries, the growing research interest in e-bikes goes hand in hand with the increasing popularity of e-bike usage over the past decade (Bigazzi and Wong, 2020; Bourne et al., 2020). A survey among e-bikers across seven European cities reports that e-bikes exclusively substituted 25 per cent of private motorized vehicle trips (car or motorbike), 23 per cent of conventional bicycle trips and 15 per cent of public transport trips (Castro et al., 2019). In the Netherlands, a prospective observational study indicates that conventional bicycle use reduces significantly, while car use reduces less strongly following e-bike ownership (Sun et al., 2020). The results are corroborated by findings from a cross-sectional study which compares the travel behaviour of e-bike owners with that of non-e-bike owners (Kroesen, 2017). A survey among Danish e-bikers shows that e-bike usage substituted conventional bicycle trips (64 per cent), car trips (49 per cent) and bus trips (48 per cent) (multiple choice in the travel survey, therefore mode shares do not add up to 100 per cent) (Haustein and Møller, 2016). In Sweden, a study shows that e-bike trips predominantly replace car journeys, while respondents from urban areas also tend to replace conventional bicycle journeys and journeys taken on public transit with e-bike trips (Winslott Hiselius and Svensson, 2017).

Car-dependent countries such as the USA and Australia also witnessed market penetration by e-bikes. A study among over 1700 e-bike owners across North America suggests that of all the e-bike trips, 45.6 per cent replaced car trips, 27.3 per cent replaced trips by active transport and public transit, and 25.3 per cent were newly generated trips (MacArthur et al., 2018). Among Australian e-bike owners, mode shift from a private motor vehicle was highest (Johnson and Rose, 2015).

It should be noted that there are several prospective experimental e-bike studies targeting the car user group (e.g. Cairns et al., 2017; de Kruijf et al., 2018; Andersson et al., 2021), which represent meaningful policy intervention instruments to cut carbon emissions in transport, but their findings on substitution effects should not be generalized to the whole e-bike user group. These experimental studies commonly provide people with access to e-bikes to try out for a certain period. The trial research is usually targeted specifically at frequent car users, with the aim to reduce car travel among the participants. Such studies often report higher replacement rates of car journeys. However, the results need to be interpreted with caution, because by research design, the car was the primary travel mode before introducing e-bikes. Thus, among all the e-bike users in real-world settings where people have different primary travel modes prior to e-bike access, e-bike adoptions cannot trigger car replacement at such high rates.

4.   FROM 'WHAT' TO 'HOW'

The explorations mentioned above present important insights into what transport modes e-bikes substitute and to what extent. Existing literature, by and large, demonstrates a mixed picture of e-cycling substituting both conventional cycling and car usage to varying degrees. This reveals changes in travel patterns in relation to e-bike adoption, but the process of change itself is left opaque. This leaves scope to develop an understanding of how changes in daily travel occur. Specifically, why some people change from a car to an e-bike, and under which conditions one would decide to ride an e-bike instead of a conventional bicycle. Such an endeavour is indispensable to gain a comprehensive understanding of the success/failure and societal implications of e-bikes. Thanks to its strength of capturing travel pattern changes longitudinally, a mobility biography approach can offer a plausible theoretical underpinning to investigate the mechanisms behind the dynamics in daily mobility.

4.1   Theoretical Background

Mobility biographies research (MBR) originated as a way to address the gap resulting from dominant cross-sectional research designs, aiming to capture how and why mobility behaviour evolves across the life course (Rau and Scheiner, Chapter 4 in this volume). The MBR approach is interdisciplinary in nature, drawing upon knowledge from the fields of psychology, sociology and geography. It enables comprehensive assessments of travel behaviour evolvement in an integrated manner along the dimensions of (1) life events including both private and professional career; (2) adaptation of long-term mobility decisions such as residential relocation and car purchase; (3) exogenous interventions like infrastructure changes, incentives and mobility campaigns; and (4) long-term life processes including age effects and socialization (Müggenburg et al., 2015).

Travel behaviour is constantly driven by forces of both change and stability, including habits and heuristics (Rau and Scheiner, Chapter 4 in this volume). In particular, MBR puts the emphasis on important life events, which play a significant role in weakening routines and activating a re-evaluation of mobility behaviour (Müggenburg et al., 2015). Van der Waerden et al. (2003) further distinguish life events between (1) key events which are intentional, such as residential relocation and the birth of a child, and (2) critical incidents which usually, and contrary to key events, occur unexpectedly, such as involvement in an accident. What tends to be overlooked, however, are the gradual changes not linked to discrete events, for instance age effects and socialization. These long-term processes may trigger the breaking of routines thanks to learning processes and coping strategies used to adapt to changing circumstances over an individual's biography (Müggenburg et al., 2015; Scheiner, 2020).

Additionally, mobility change and stability can be distinguished on three levels: individual, interpersonal and societal (Rau and Scheiner, Chapter 4 in this volume). The individual level focuses on personal behaviour and attitude. The interpersonal level highlights the role of interpersonal communication and interactions. Regarding the societal level, the emphasis is placed upon communities, organizations and institutions. It is worth noting that the three levels are intrinsically connected. As Rau and Scheiner (Chapter 4 in this volume, p. 75) point out,

> [C]ommunities and organisations are made up of individuals and individuals' actions cannot be properly understood without reference to the social, economic, administrative and political settings within which they occur. Taken together, they shape networks of actors who negotiate stability and change of mobility practices on the individual and collective level. It is this networked, interconnected character that fosters inertia but also presents chances for fast and wide-ranging innovation in the transport sector whenever the opportunity arises.

The conceptual approach for the analysis in section 5 is embedded in the MBR literature elaborated above. While acknowledging the importance of habits and routines, this chapter pays more explicit attention to change, as the subject of this research is centred around people adopting e-bikes.

4.2   Interviews

To better understand the biographical reasons and individual motives that lead people to change travel modes, a qualitative method involving semi-structured interviews was employed. The empirical analysis focuses on the Netherlands, since its e-bike ownership is one of the highest among European countries. Moreover, the Netherlands makes a compelling case to study as it has a dominant automobile regime while cycling is also rather prominent in the mobility system. The fieldwork involves a sample of 21 Dutch e-bike users with diverse socio-demographic backgrounds. Participants were recruited and interviewed during the period from November 2020 to February 2021. The whole recruitment and interview process was conducted online due to the precautions of the COVID-19 pandemic.

The research aims were best served by a combination of two recruitment methods: convenience sampling and purposive sampling. First, a flyer with participation criteria was disseminated via online social media platforms including Facebook, Twitter and LinkedIn. Second, purposive sampling was employed to balance convenience-sampling participants regarding age cohort and gender distribution, particularly at a later stage of the recruitment process. The resulting sample was meant to illustrate a broad range of e-bike user practices by a diverse study group with balanced gender distribution as well as variation in age, level of education, occupational background and household income level (see Table 8.1 for an overview of participants). Each interview took approximately 40 minutes and was digitally audio-recorded. Interview data was transcribed and then coded using the qualitative analysis data package NVivo 12. All participants gave verbal consent to take part in the study, which was granted ethical approval by the Eindhoven University of Technology's Ethical Review Board.

All interviews were based on the same set of guiding questions, which were semi-structured, consisting of both close-ended and open-ended questions. One of the key objectives of the interviews was to identify the factors motivating people to acquire an e-bike as well as changes in travel mode. Thus, retrospective inquiries were a central component of the interviews looking into dynamics in mobility practices around the e-bike. A retrospective method may cause scepticism due to possible recall bias. Nonetheless, a qualitative approach encourages participants to give more detailed information and could therefore potentially better reconstruct past travel practice developments than a quantitative research design (Marincek and Rérat, 2020). In addition, the credibility of such reconstructions is enhanced since this study is centred around the pivotal moment of e-bike acquisition, which although had taken place several years back, was more easily remembered and is instrumental to the research aim.

5.   ANALYSIS

The analysis of the interview data is structured according to the three levels of change – individual, interpersonal and societal – discussed in section 4.1. For each level, the main themes were identified and illustrated by comments from the interview participants. The analysis also shows how the different levels are closely intertwined with one another.

5.1   Individual Level

Life events
A shift in mobility practice can be triggered by critical life events such as a change in commute distance because of a new job/school (no. 12, 18, 19). For example, in the case below, the interviewee found a new job in another city. His commute distance increased from 6 km to 20 km. After trying out his son's e-bike, he bought another e-bike and used it for commuting.

> I live in The Hague. I've been working in Leiden for over three years. When I went to apply there, we already had a second-hand e-bike in the house for my son, who has to cycle a little further to school. Then I went to the job interview on his [e-] bike and when I got the job, I very quickly bought him another e-bike, within a few days also via Marktplaats [an online platform trading second-hand goods], so also second-hand, to cycle to work … . Before, I worked in Delft, and I went by conventional bicycle, which was only 6 kilometres. (Luca, 50–59)

Table 8.1   *Socio-demographics of participants (N = 21)*

No.	Name (anonymized)	Gender	Age	Education	Work status	Household car ownership	Household income level
1	Anneke	F	50–59	Low	Employed	Y	High
2	Betty	F	60–69	High	Retired	Y	Low
3	Caroline	F	40–49	High	Volunteer	Y	High
4	Daniel	M	40–49	High	Employed	Y	High
5	Emily	F	50–59	Low	Employed	Y	Low
6	Fiona	F	50–59	Low	Employed	Y	High
7	Greta	F	40–49	High	Employed	Y	High
8	Henny	F	60–69	High	Retired	N	Low
9	Ingrid	F	50–59	High	Employed	Y	Medium
10	Julia	F	40–49	Low	Employed	Y	High
11	Karen	F	18–29	High	Student	N	Low
12	Luca	M	50–59	High	Employed	Y	Medium
13	Marjolijn	F	40–49	High	Employed	Y	Medium
14	Natalie	F	≥70	Low	Retired	Y	Low
15	Peter	M	50–59	High	Retired	Y	Low
16	Ruben	M	40–49	Low	Employed	Y	Medium
17	Simon	M	50–59	Low	Employed	Y	Medium
18	Tracey	F	18–29	High	Employed	Y	Medium
19	Vincent	M	40–49	High	Employed	Y	Medium

No.	Name (anonymized)	Gender	Age	Education	Work status	Household car ownership	Household income level
20	Victoria	F	30–39	High	Employed	N	High
21	Willem	M	18–29	High	Employed	Y	Medium

Note: Educational level: low = secondary school education or lower, high = college/university education or higher; household income level: low = below the national benchmark income, medium = between the national benchmark income and two times the national benchmark income, high = two times the national benchmark income and above.

In contrast to interventions of change in professional career or residential relocation, some unintentional abrupt changes, including (road) accidents and specific medical conditions (no. 11, 20), may serve as windows of opportunities activating a re-evaluation of mobility practices (Rau and Scheiner, Chapter 4 in this volume). This is also reflected in this primary research:

> At first, I was struggling with a mountain bike. We don't have a car. So I went to work with my mountain bike. And we did everything on the bike. And then I broke my hand actually, in a [road] accident. And I couldn't sit on the mountain bike, because there was too much pressure on my hands. I borrowed the e-bike from my mother-in-law. And when I was using it, I was like, 'Oh, why didn't I think of this sooner?' Because you can travel with the same speed but without the effort. (Victoria, 30–39)

> [For] medical reasons I couldn't ride on a normal bike anymore. … Too much strain on the muscles. So I couldn't make it to school. So that's why I switched [from a conventional bicycle to an e-bike]. (Karen, 18–29)

Gradual change
Besides the abrupt changes mentioned above, gradual changes also appear to play a role when studying mobility biographies around e-bike adoptions. For several people, the e-bike provided them with a way to continue cycling since riding a conventional bicycle became too physically demanding due to reasons such as aging, accompanied with deterioration of physical strength (no. 6, 9, 10, 14, 15, 17).

> I first had a Gazelle [a Dutch bicycle brand] mother bike, but it was really heavy. And my children were already too big to sit on it. I wanted another bike. And my condition is not so good that I thought 'if I buy a conventional bicycle, am I really going to cycle on it?' So I wanted a bike with support and that's why I chose an electric bike. (Julia, 40–49)

5.2   Interpersonal Level

Interpersonal interactions are often implicated in e-bike uptake. Individuals' life courses are not isolated from those of others. Instead, they are embedded in family, friends and neighbourhood networks, and often share a particular relation with their significant others (Rau and Scheiner, Chapter 4 in this volume).

As a common activity in the Netherlands, cycling is part of social life. Cycling together with family and friends at a similar pace was a motivation for some to adopt an e-bike (no. 1, 3, 8, 15, 17).

> Well, because I had friends that I could ride with. They had an e-bike. And that wasn't fun when you're riding a conventional bicycle with someone who has an

> e-bike. ... I live in South Limburg and then you have hills. And it's no fun when you arrive huffing and puffing and she's already at the top. (Henny, 60–69)

In several instances, differences in cycling speeds were often due to a decline in the physical strength of either the participants themselves and/or one's cycling partner. Particularly, it is not uncommon for a couple to adopt e-bikes together in order to cycle side by side.

> It [e-bike adoption] is a bit [because] of worse fitness. And one of the things was that I have two kids who are 12 and 14 and they ride a bike like hell. And I just couldn't ride a bike with my family anymore. So I just had a bad condition and I thought 'yeah, come on, I'm 44 [years old]. I'm not going to sit at home like an old person'. I just want to be able to cycle along so I thought this is how it would be. (Caroline, 40–49)

Another participant explained how his wife could not ride a conventional bicycle anymore because of her reduced stamina. She bought an e-bike since she wanted to continue cycling. However, riding a conventional bicycle together with a person on an e-bike caused problems. Eventually, he adopted an e-bike himself too.

> So she had bought an electric bike and I thought: I would ride on a conventional bicycle with [my wife riding on] an electric bike. But that was so disappointing. Uh, I was like, guys, this just isn't it. Plus my knees hurt when I cycled a long way. And then when I stopped cycling, I really had problems with my legs and my knees. ... So I bought an electric bike too. Then we could cycle electrically together. Yes, that's just perfect. That's why we bought the same [e-]bikes. So that we can be sure that we ride the same speed, because it is very annoying when one pedals easily and goes fast, whereas the other has difficulties in riding. (Simon, 50–59)

5.3   Societal Level

Wider social, cultural, economic, political and administrative structures also serve as crucial sources of socialization, providing individuals with the behavioural and communicative tools to participate in society, including the transport system (Rau and Scheiner, Chapter 4 in this volume).

Since adopting an e-bike usually involves relatively high acquisition costs, supportive financial schemes leverage the affordability of the e-bike. In the Netherlands, employers provide fiscal benefits through the so-called 'bicycle plan', which offers employees tax benefits when they purchase a bike mainly for commuting purposes. Such a fiscal scheme also covers e-bikes.

> I worked in a corporation. And they also had the bicycle plan. So I thought now I'm gonna do this. And yeah, it's better. Still, you have to pay a few hundred euros, but yeah, I really like it. (Emily, 50–59)

Obviously, the wider social context does not always work in favour of the e-bike but could sometimes present a barrier. For the Dutch case, this is particularly relevant when it comes to cultural aspects of the e-bike. As discussed in section 2, the symbolic meaning attached to the e-bike was rather negative, albeit it has improved in recent years. Several participants mentioned the stereotypes associated with the e-bike of being lazy and a vehicle for older people (no. 4, 5, 6, 7, 9, 11).

> I don't want to have a Stella [a popular Dutch brand of] e-bike. Because they have the image that it's really like, for older people, or pregnant mothers or whatever. (Emily, 50–59)

> Well, now it's for everyone actually. Indeed, a few years ago it was for older people who couldn't cycle properly anymore. (Ingrid, 50–59)

5.4   Interplay of Different Levels

The analysis above looks at the uptake of an e-bike on each of the three levels respectively. It can be observed from the mobility biographies of many participants that the three levels often interact and intertwine with each other in the course of e-bike adoptions.

In the case of Karen (no. 11) mentioned above, on an individual level she had to switch from a conventional bicycle to an e-bike due to medical reasons. However, she was reluctant to do so in the beginning because of the cultural stigma.

> I didn't want it [the e-bike] at first. That was something my doctor said [I was advised to use an e-bike]. And I was like, I don't want that [the e-bike], because that's for old people and people will think I'm lazy. But that was back then six years ago. And right now I see everyone using it [the e-bike]. So I don't have a problem with it [e-bike] now. (Karen, 18–29)

A few participants (no. 6, 7) directly inherited an e-bike from one of their parents, which represents influence on an interpersonal level. They admitted that if the e-bike was not given by a family member, they would not have acquired an e-bike themselves, again, due to the social stigma attached to the e-bike. Then it became apparent that the e-bike was an asset for them.

> I have a very severe form of asthma myself. And I actually found, say seven years ago, ten years ago, if you took an electric bike you were a bit lazy. And it was for old people. And that's what it really looked like. Really, I'm not going to use an e-bike anyway. But in the meantime, I could hardly ride a bike anymore. So now I have the e-bike of my dad and I think 'Oh I should have done that years ago, because I have fun cycling again and I can use it without being out of breath.' (Fiona, 50–59)

Emily used to commute to work by car, a mode which she was not satisfied with partly due to expensive car parking at her workplace. For economic and health reasons, she preferred to cycle rather than drive. However, Emily was dependent on her car as she needed to transport her dog and take her child to school on her way to work. The e-bike option only became viable when her dog died, and her child was old enough to go to school by himself. Additionally, she happened to come across a relative cheap e-bike online. This switch was made possible partially thanks to the bike plan.

6.   CONCLUSIONS

Internationally, the e-bike is on the rise as an innovation in transport. Taking into account its rapid market penetration and its potential to serve as a component in the transition towards low-carbon mobility, the e-bike deserves more attention from both researchers and policymakers. The e-bike is seemingly a successful transport innovation at first glance given its popularity, but a deeper understanding of its societal impacts is essential to go beyond mere sale numbers. In this chapter the success or failure of the e-bike was analysed via a transitions theory lens at a higher-order systemic level. This chapter then looked into the societal impacts of the e-bike by reviewing existing literature and subsequently zoomed in at a micro-level to study e-bike mobility with a biographical approach. By employing plausible theoretical frameworks, this chapter provides a comprehensive overview of the e-bike innovation and generates some new insights.

Little work to date applies transitions theory in the field of e-bikes, except for a few attempts to study e-bikes via a transitions theory lens in a Chinese context (see, for example, Lin et al., 2018). This chapter showcases that harnessing the explanatory power of transitions theory opens fruitful avenues to shed light on the success or failure of the e-bike. Besides the various elements in the mobility system which directly shape the trajectory of e-bike diffusion, the success/failure of this transport innovation is significantly influenced by the interactions between the e-bike niche and the dominant regimes like automobility and cycling. It can be posited that the e-bike is becoming part of the cycling regime, which would expand the domain outside the dominant automobile regime. Further research could explore how other factors play a role in the adoption of e-bikes, such as environmental awareness, health considerations and risk concerns. Based on a scoping review of e-cycling, Bourne et al. (2020) conclude that younger adults are more motivated to use the e-bike because of sustainability reasons and cost considerations of saving money. Older generations, on the other hand, attach more importance to e-bikes' potential to maintain or increase physical activity and fitness. However, the influences of these factors are not country-specific, and so far research is lacking regarding the roles that these factors play in different contexts.

In terms of the societal impacts of e-bikes, this chapter mainly looked into the environmental and social aspects. Generally speaking, e-bikes are more energy-efficient and less polluting than fossil fuel powered motor vehicles (Weiss et al., 2015). Nonetheless, comprehensive environmental assessments should not neglect issues such as the power mix of the local electricity sector and the conduct of battery recycling, which is largely a grey area. In the EU, collection and recycling of batteries are regulated by the Battery Directive, which bans disposal of batteries in landfills or by incineration, and states that all collected batteries should be recycled (European Commission, 2006). However, little is known about the enforcement of battery recycling, nor about actual recycling practices. When it comes to the social aspect, while the e-bike contributes to fitness, it brings potential road safety concerns and risks. Given its various environmental and social implications, the e-bike is an exemplar showing that innovation should not simply be considered as progress but rather as a double-edged sword (Pel, Chapter 2 in this volume).

The empirical analysis part of this chapter employs a biographical approach to study the dynamics of e-bike adopters. MBR complements the cross-sectional type of research design which traditionally dominates mobility research but overlooks the longitudinal dimension of the decision-making process (Greene and Rau, 2018; Müggenburg et al., 2015). Furthermore, 'qualitative data much better captures underlying and new, not hypothesized factors and helps in understanding the complexity of multiple factors, whereas quantitative approaches often simplify such interrelations' (Müggenburg et al., 2015, p. 160). Thus, this study employs a longitudinal qualitative research design with retrospective biographical interviewing when exploring the shift from other transport modes to the e-bike. It contributes to the existing body of e-bike literature by investigating how mode choices around e-bikes occur and their consequent societal impacts. Such insights are crucial in order to comprehend the underlying mechanisms of e-bike mode choice and to make meaningful policy interventions.

Moreover, the empirical work illustrates that unpacking change and stability on three levels helps to structure the understanding of mobility biographies. However, this does not imply that different levels of social organization are isolated; instead, they are intrinsically linked to each other (Rau and Scheiner, Chapter 4 in this volume). This point also links to conceptual developments in MBR thanks to increasing recognition that over and above individual events, life courses are also socially embedded (Scheiner, 2020). A practice-oriented approach moves beyond a focus on individual travel behaviour but instead posits mobility practices in the context of broader social and material conditions (Rau and Scheiner, Chapter 4 in this volume). Further studies in this field can benefit from practice-centred inquiries into e-bike mobility biographies. Additionally, one direction of future research priorities is to include non-e-bike users to investigate why people, particularly current car users, do not change their travel behaviour, so as to identify barriers preventing an upscaling of the e-bike's societal benefits.

NOTE

1. This work was supported by the 'Bicycle Challenges: Past, Present, and Future' project, funded by Eindhoven University of Technology, PON Holding, and The Dutch Public Works Administration.

REFERENCES

Andersson, A., Adell, E., and Winslott Hiselius, L. (2021). What is the substitution effect of e-bikes? A randomised controlled trial. *Transportation Research Part D: Transport and Environment* 90, 102648. https://doi.org/10.1016/j.trd.2020.102648.

Behrendt, F. (2018). Why cycling matters for electric mobility: towards diverse, active and sustainable e-mobilities. *Mobilities* 13, 64–80. https://doi.org/10.1080/17450101.2017.1335463.

Behrendt, F., Cairns, S., Raffo, D., and Philips, I. (2021). Impact of e-bikes on cycling in hilly areas: participants' experience of electrically-assisted cycling in a UK study. *Sustainability* 13, 8946. https://doi.org/10.3390/su13168946.

Bigazzi, A., and Wong, K. (2020). Electric bicycle mode substitution for driving, public transit, conventional cycling, and walking. *Transportation Research Part D: Transport and Environment* 85, 102412. https://doi.org/10.1016/j.trd.2020.102412.

Bourne, J.E., Cooper, A.R., Kelly, P., Kinnear, F.J., England, C., Leary, S., and Page, A. (2020). The impact of e-cycling on travel behaviour: a scoping review. *Journal of Transport and Health* 19, 100910. https://doi.org/10.1016/j.jth.2020.100910.

Bourne, J.E., Sauchelli, S., Perry, R., Page, A., Leary, S., England, C., and Cooper, A.R. (2018). Health benefits of electrically-assisted cycling: a systematic review. *International Journal of Behavioral Nutrition and Physical Activity* 15, 116. https://doi.org/10.1186/s12966-018-0751-8.

Cairns, S., Behrendt, F., Raffo, D., Beaumont, C., and Kiefer, C. (2017). Electrically-assisted bikes: potential impacts on travel behaviour. *Transportation Research Part A: Policy and Practice* 103, 327–342. https://doi.org/10.1016/j.tra.2017.03.007.

Castro, A., Gaupp-Berghausen, M., Dons, E., Standaert, A., Laeremans, M., Clark, A., Anaya-Boig, E., Cole-Hunter, T., Avila-Palencia, I., Rojas-Rueda, D., Nieuwenhuijsen, M., Gerike, R., Panis, L.I., de Nazelle, A., Brand, C., Raser, E., Kahlmeier, S., and Götschi, T. (2019). Physical activity of electric bicycle users compared to conventional bicycle users and non-cyclists: insights based on health and transport data from an online survey in seven European cities. *Transportation Research Interdisciplinary Perspectives* 1, 100017. https://doi.org/10.1016/j.trip.2019.100017.

Cherry, C.R. (2007). Electric two-wheelers in China: analysis of environmental, safety, and mobility impacts. PhD dissertation. University of California, Berkeley, USA.

Cherry, C.R., and MacArthur, J.H. (2019). E-bike safety. A review of empirical European and North American studies. White Paper. Light Electric Vehicle Education and Research Initiative.

Cherry, C.R., Yang, H., Jones, L.R., and He, M. (2016). Dynamics of electric bike ownership and use in Kunming, China. *Transport Policy* 45, 127–135. https://doi .org/10.1016/j.tranpol.2015.09.007.

CONEBI (2017). *European Bicycle Market*. Confederation of the European Bicycle Industry, Brussels.

de Kruijf, J., Ettema, D., Kamphuis, C.B.M., and Dijst, M. (2018). Evaluation of an incentive program to stimulate the shift from car commuting to e-cycling in the Netherlands. *Journal of Transport and Health* 10, 74–83. https://doi.org/10.1016/j .jth.2018.06.003.

Dozza, M., Bianchi Piccinini, G.F., and Werneke, J. (2016). Using naturalistic data to assess e-cyclist behavior. *Transportation Research Part F: Traffic Psychology and Behaviour* 41, 217–226. https://doi.org/10.1016/j.trf.2015.04.003.

Elvik, R., and Bjørnskau, T. (2017). Safety-in-numbers: a systematic review and meta-analysis of evidence. *Safety Science* 92, 274–282. https://doi.org/10.1016/j.ssci .2015.07.017.

European Commission (2006). Directive 2006/66/EC of the European Parliament and of the Council of 6 September 2006 on batteries and accumulators and waste batteries and accumulators and repealing Directive 91/157/EEC.

Fishman, E., and Cherry, C. (2016). E-bikes in the mainstream: reviewing a decade of research. *Transport Reviews* 36, 72–91. https://doi.org/10.1080/01441647.2015 .1069907.

Geels, F.W. (2005). The dynamics of transitions in socio-technical systems: a multi-level analysis of the transition pathway from horse-drawn carriages to automobiles (1860–1930). *Technology Analysis and Strategic Management* 17, 445–476. https:// doi.org/10.1080/09537320500357319.

Geels, F.W. (2012). A socio-technical analysis of low-carbon transitions: introducing the multi-level perspective into transport studies. *Journal of Transport Geography* 24, 471–482. https://doi.org/10.1016/j.jtrangeo.2012.01.021.

GOV.cn (2019). Domestic bicycle ownership is nearly 400 million, number one in the world (我国自行车社会保有量近4亿辆稳居世界第一). http://www.gov.cn/ xinwen/2019-11/22/content_5454675.htm.

Greene, M., and Rau, H. (2018). Moving across the life course: a biographic approach to researching dynamics of everyday mobility practices. *Journal of Consumer Culture* 18, 60–82. https://doi.org/10.1177/1469540516634417.

Gu, T., Kim, I., and Currie, G. (2020). The two-wheeled renaissance in China—an empirical review of bicycle, e-bike, and motorbike development. *International Journal of Sustainable Transportation* 1–20. https://doi.org/10.1080/15568318 .2020.1737277.

Handy, S., van Wee, B., and Kroesen, M. (2014). Promoting cycling for transport: research needs and challenges. *Transport Reviews* 34, 4–24.

Haustein, S., and Møller, M. (2016). Age and attitude: changes in cycling patterns of different e-bike user segments. *International Journal of Sustainable Transportation* 10, 836–846. https://doi.org/10.1080/15568318.2016.1162881.

Heinen, E., van Wee, B., and Maat, K. (2010). Commuting by bicycle: an overview of the literature. *Transport Reviews* 30, 59–96. https://doi.org/10.1080/ 01441640903187001.

Ji, S., Cherry, C.R., Bechle, M.J., Wu, Y., and Marshall, J.D. (2012). Electric vehicles in China: emissions and health impacts. *Environmental Science and Technology* 46, 2018–2024. https://doi.org/10.1021/es202347q.

Johnson, M., and Rose, G. (2015). Extending life on the bike: electric bike use by older Australians. *Journal of Transport and Health* 2, 276–283. https://doi.org/10.1016/j .jth.2015.03.001.

Jones, T., Harms, L., and Heinen, E. (2016). Motives, perceptions and experiences of electric bicycle owners and implications for health, wellbeing and mobility. *Journal of Transport Geography* 53, 41–49. https://doi.org/10.1016/j.jtrangeo.2016.04.006.

Köhler, J., Geels, F.W., Kern, F., Markard, J., Onsongo, E., Wieczorek, A., Alkemade, F., Avelino, F., Bergek, A., Boons, F., Fünfschilling, L., Hess, D., Holtz, G., Hyysalo, S., Jenkins, K., Kivimaa, P., Martiskainen, M., McMeekin, A., Mühlemeier, M.S., Nykvist, B., Pel, B., Raven, R., Rohracher, H., Sandén, B., Schot, J., Sovacool, B., Turnheim, B., Welch, D., and Wells, P. (2019). An agenda for sustainability transitions research: state of the art and future directions. *Environmental Innovation and Societal Transitions* 31, 1–32. https://doi.org/10.1016/j.eist.2019.01.004.

Kroesen, M. (2017). To what extent do e-bikes substitute travel by other modes? Evidence from the Netherlands. *Transportation Research Part D: Transport and Environment* 53, 377–387. https://doi.org/10.1016/j.trd.2017.04.036.

Lin, X., Wells, P., and Sovacool, B.K. (2017). Benign mobility? Electric bicycles, sustainable transport consumption behaviour and socio-technical transitions in Nanjing, China. *Transportation Research Part A: Policy and Practice* 103, 223–234. https:// doi.org/10.1016/j.tra.2017.06.014.

Lin, X., Wells, P., and Sovacool, B.K. (2018). The death of a transport regime? The future of electric bicycles and transportation pathways for sustainable mobility in China. *Technological Forecasting and Social Change* 132, 255–267. https://doi.org/ 10.1016/j.techfore.2018.02.008.

Liu, W., Sang, J., Chen, L., Tian, J., Zhang, H., and Olvera Palma, G. (2015). Life cycle assessment of lead-acid batteries used in electric bicycles in China. *Journal of Cleaner Production* 108, 1149–1156. https://doi.org/10.1016/j.jclepro.2015.07.026.

MacArthur, J., Cherry, C., Harpool, M., and Schepke, D. (2018). *Electric Boost: Insights from a National E-bike Owner Survey*. Transportation Research and Education Center. https://doi.org/10.15760/trec.197.

Marincek, D., and Rérat, P. (2020). From conventional to electrically-assisted cycling. A biographical approach to the adoption of the e-bike. *International Journal of Sustainable Transportation* 1–10. https://doi.org/10.1080/15568318.2020.1799119.

Müggenburg, H., Busch-Geertsema, A., and Lanzendorf, M. (2015). Mobility biographies: a review of achievements and challenges of the mobility biographies approach and a framework for further research. *Journal of Transport Geography* 46, 151–163. https://doi.org/10.1016/j.jtrangeo.2015.06.004.

Popovich, N., Gordon, E., Shao, Z., Xing, Y., Wang, Y., and Handy, S. (2014). Experiences of electric bicycle users in the Sacramento, California area. *Travel Behaviour and Society* 1, 37–44. https://doi.org/10.1016/j.tbs.2013.10.006.

Pucher, J., and Buehler, R. (2008). Making cycling irresistible: lessons from the Netherlands, Denmark and Germany. *Transport Reviews* 28, 495–528. https://doi .org/10.1080/01441640701806612.

Pucher, J., and Buehler, R. (2017). Cycling towards a more sustainable transport future. *Transport Reviews* 37, 689–694. https://doi.org/10.1080/01441647.2017.1340234.

Rérat, P. (2021). The rise of the e-bike: towards an extension of the practice of cycling? *Mobilities* 16, 423–439. https://doi.org/10.1080/17450101.2021.1897236.

Scheiner, J. (2020). Changes in travel mode use over the life course with partner inter-actions in couple households. *Transportation Research Part A: Policy and Practice* 132, 791–807. https://doi.org/10.1016/j.tra.2019.12.031.

Simsekoglu, Ö., and Klöckner, C.A. (2019). The role of psychological and socio-demographical factors for electric bike use in Norway. *International Journal of Sustainable Transportation* 13, 315–323.

Sun, Q., Feng, T., Kemperman, A., and Spahn, A. (2020). Modal shift implications of e-bike use in the Netherlands: moving towards sustainability? *Transportation Research Part D: Transport and Environment* 78, 102202. https://doi.org/10.1016/j.trd.2019.102202.

Sundfør, H.B., and Fyhri, A. (2017). A push for public health: the effect of e-bikes on physical activity levels. *BMC Public Health* 17, 809. https://doi.org/10.1186/s12889-017-4817-3.

van der Waerden, P., Timmermans, H., and Borgers, A. (2003). The influence of key events and critical incidents on transport mode choice switching behaviour: a descriptive analysis. Paper presented at the 10th International Conference on Travel Behaviour Research, 10–15 August, Lucerne.

Vlakveld, W.P., Twisk, D., Christoph, M., Boele, M., Sikkema, R., Remy, R., and Schwab, A.L. (2015). Speed choice and mental workload of elderly cyclists on e-bikes in simple and complex traffic situations: a field experiment. *Accident Analysis and Prevention* 74, 97–106. https://doi.org/10.1016/j.aap.2014.10.018.

Weinert, J.X., Burke, A.F., and Wei, X. (2007). Lead-acid and lithium-ion batteries for the Chinese electric bike market and implications on future technology advance-ment. *Journal of Power Sources* 172, 938–945. https://doi.org/10.1016/j.jpowsour.2007.05.044.

Weiss, M., Dekker, P., Moro, A., Scholz, H., and Patel, M.K. (2015). On the electrifi-cation of road transportation – a review of the environmental, economic, and social performance of electric two-wheelers. *Transportation Research Part D: Transport and Environment* 41, 348–366. https://doi.org/10.1016/j.trd.2015.09.007.

Wells, P., and Lin, X. (2015). Spontaneous emergence versus technology manage-ment in sustainable mobility transitions: electric bicycles in China. *Transportation Research Part A: Policy and Practice* 78, 371–383. https://doi.org/10.1016/j.tra.2015.05.022.

Winslott Hiselius, L., and Svensson, Å. (2017). E-bike use in Sweden – CO2 effects due to modal change and municipal promotion strategies. *Journal of Cleaner Production* 141, 818–824. https://doi.org/10.1016/j.jclepro.2016.09.141

Wolf, A., and Seebauer, S. (2014). Technology adoption of electric bicycles: a survey among early adopters. *Transportation Research Part A: Policy and Practice* 69, 196–211. https://doi.org/10.1016/j.tra.2014.08.007.

Yang, C.-J. (2010). Launching strategy for electric vehicles: lessons from China and Taiwan. *Technological Forecasting and Social Change* 77, 831–834. https://doi.org/10.1016/j.techfore.2010.01.010.

Zuev, D. (2018). *Urban Mobility in Modern China*. Cham, Switzerland: Springer International Publishing. https://doi.org/10.1007/978-3-319-76590-7.

9. Explaining the growth in light electric vehicles in city logistics

Ron van Duin, Walther Ploos van Amstel and Hans Quak

BOX 9.1   CARGO BIKE MARKET SNAPSHOT

Research finds that the global market value of cargo bikes will hit 2.4 billion euros by 2031. Analysts with Future Market Insights assessing the growth of cargo bikes have placed the parcel courier industry as a key buyer of electric cargo bikes, forecasting that 43 per cent of sales could go to this industry. This growth is driven by city logistics trends, particularly as studies emerge showing the high efficiency and cost saving of the cargo bike versus the delivery van. It will not solely be direct incentives that drive uptake, however. The policy that restricts motoring and emissions is expected to be a key driver for businesses that seek profitability, with three-wheeled electric cargo bikes making up nearly half the market. The advance of e-bike technology has seen a strong rise in market share for assisted cargo bikes, now accounting for a 73 per cent market share. Potentially limiting the growth is the legislation governing the output and range of electric cargo bikes (FMI, 2021).

To deal with the issues of faster delivery, clean delivery (low/zero emission) and less space in dense cities, the light electric freight vehicle (LEFV) can be – and is used more and more as – an innovative solution. The way logistics in urban areas is organized is being challenged, as the global growth of cities leads to more jobs, more businesses and more residents. As a result, companies, workers, residents and visitors demand more goods and produce more waste. More space for logistics activities in and around cities is at odds with the growing need for accommodation for people living and working in cities. Therefore, logistics real estate has been pushed out of the cities in the past decades, that is, logistics sprawl, and less space is available for logistics activities within and around the cities (Dablanc, 2011; Boer et al., 2017; WEF, 2020). The digitization of the 'customer journey' in business-to-consumer (B2C) and business-to-business (B2B) channels leads to more, smaller and more time critical shipments in all segments of city logistics, leading to a further growth in the number of delivery vans; more than 80 per cent of commercial traffic in urban areas now comprises delivery vans. City logistics represents 5 to 10 per cent of all vehicle movements in cities today (Boer et al., 2017; Deloison et al., 2020; DfT, 2021), and this share is even higher for the city centres. Urban freight transport vehicles contribute to deteriorating public-space quality, air quality and road safety, and cities worldwide are focusing on car-free public spaces in inner cities, campuses and residential areas, thus 'curbing traffic' (Bruntlett and Bruntlett, 2021). Within car-free areas, deliveries can be done with (autonomous) light electric vehicles or by foot, but curbing traffic will impact city logistics, since less public space will be available for supplying the city.

The LEFV can be an innovative solution for dealing with these challenges. LEFVs take up less road space, are zero emission, and are less intrusive than vans in city logistics. Companies like UPS, TNT, PostNL, Hermes, GLZ and DHL, along with many small and medium enterprises (SMEs), started first by experimenting with cargo bike deliveries in European cities on a small scale as an alternative to delivery vans after 2010. The experiences with LEFVs have been generally positive, with, for example, lower costs, better manoeuvrability, less space occupation on the roads and faster delivery. Cargo bikes can make urban freight more efficient by reducing delivery distances and times (Verlinghieri et al., 2021).

Moreover, these companies understand that by using smart processes, such as containerization, standardization and automation, extra handling costs in local hubs can be minimized. Dutch food retailers Albert Heijn, Plus and Picnic (online) successfully introduced LEFVs; Dutch electronics online retailer Coolblue, in the three years since introducing cargo bikes in 2018, has used them to deliver over a million orders; and Deutsche Post, Ocado and JD are testing ultra-compact LEFVs (with truck-robot solutions) that autonomously follow delivery staff.

A LEFV is a bike, moped or compact vehicle with an electric assistance or drive mechanism, equipped for the delivery of goods, or for the transportation of people, with limited speed. In general, LEFVs are (very) quiet, flexible in usage, emission-free, and need less space than conventional delivery vehicles (Balm et al., 2018). Three types of LEFVs are defined, as shown in Table 9.1.

The electric cargo bike looks like a real bike and is therefore agile, with a maximum payload of 350 kg. The bikes are suitable for delivering small volumes, such as food deliveries, mail and parcel delivery services. The electric cargo moped is really a moped; cycling is not needed. The maximum

Table 9.1   *Three types of LEFVs*

	Electric cargo bike	Electric cargo moped	Small electric distribution vehicle
Loading capacity (kg)	50–350	100–500	200–750
Vehicle weight (kg)	20–170	50–600	300–1,000

Source: Balm et al. (2018).

payload is 500 kg. Small volumes of construction materials and more heavier loads (like a keg of beer) can be delivered with this vehicle. The small electric distribution vehicle looks most like a mini-van. It has a maximum payload of 750 kg. The vehicle is used for retail and residential streams, such as waste collection, street cleaning and catering services. Manoeuvring and parking is much easier in dense city areas compared to a van. However, it is less agile compared to the bike and the moped, but still well suited for use in dense areas as parking and manoeuvring are much easier. This definition does not (yet) include light electric autonomous vehicles.

The number of different types of LEFVs on the market has increased and the performance of the LEFVs in terms of loading capacity, range and usability has improved. Still, logistics professionals seem to hesitate in making the switch to using LEFVs. Fleet decision-makers and city logistics operators show doubts about using LEFVs, as there are still many small engineering companies optimizing the design of the LEFVs instead of providing a full professional service (e.g. 24-hour maintenance services) (Balm et al., 2018).

Therefore, the following research question is posed in this chapter:

What are the success and failure factors of the introduction of LEFVs in city logistics and what are the future perspectives on LEFVs in city logistics?

This question is answered by carrying out an ex-post analysis based on the Technological Innovation System (TIS) framework (see Langeland et al., Chapter 7 in this volume). After this short introduction of LEFVs in this section, we continue with a description of the TIS framework evaluation operationalized according to Langeland et al. The last section ends with a conclusion and future perspectives on LEFV usage in cities based on the applied TIS framework.

EVALUATION BASED ON THE TIS FRAMEWORK

Literature on ex-post analysis based on real cases with LEFV usage is limited. The application of the TIS framework can provide insight on the cumulative systematic change towards more LEFVs being used in city logistics processes. While the processes are obviously important from a business perspective, the policy perspective is also important given that the negative externalities influence the living environments in cities as well. The TIS framework (Langeland et al., Chapter 7 in this volume; see Figure 9.1) has an active policy dimension and engages in developing targeted tools and policy mixes that can help to remove barriers. The system elements consist of actors (individuals, organizations and networks), institutions (habits, routines, norms and strategies), interactions (cooperative relationships) and infrastructures (physical, financial and knowledge). The main system functions are entrepreneurial activities, knowledge development and diffusion, influence on the direction of search, market formation, resources mobilization, legitimacy and development of positive externalities. The framework is represented in Figure 9.1 and will be elaborated on for our LEFV research experiences in Dutch cities and, where useful, extended to foreign experiences, to find the success and failure factors and the future perspectives on LEFVs.

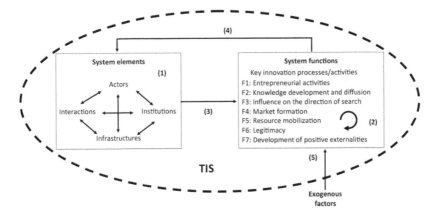

Figure 9.1   *The TIS framework (Langeland et al., Chapter 7 in this volume)*

System Elements

According to the TIS framework, actors, institutions, infrastructures and interactions need to be defined. The system elements are obtained from the literature and based on our practical research experiences in the European Union (EU; Cyclelogistics, 2014; Wrighton and Reiter, 2016; TRELab, 2019; Cairns and Sloman, 2019) and in the Netherlands (Balm et al., 2018).

Actors

The main actors in the application of LEFVs in the city logistics field are the municipalities, network organizations, logistics service providers, shippers, and receiving parties like consumers, storekeepers and constructors/servicemen (Lebeau et al., 2018).

Public actors
Municipalities consider LEFVs a sustainable solution for city logistics as they use less (urban) space, produce zero emissions and provide quick delivery in highly congested areas (Ploos van Amstel et al., 2018; Logistiek010, 2021). Still, they have worries about the safety issues related to the interaction of LEFVs with motorized and non-motorized traffic. Also, the potential market share is not fully clear to them.

Network organizations lobby for the use of new and cleaner vehicles and therefore can be stimulating agencies advocating the environmental benefits of the LEFVs to other stakeholders.

Suppliers of LEFVs
The supplier market of LEFVs (Original Equipment Manufacturers, or OEMs) is maturing. Within each type of LEFV, certain vehicles are dominating, like the Urban Arrow (and lookalikes) in the non-motorized bikes type, the Stint (and lookalikes) in the motorized bikes type and the Goupil/E-Seval (and lookalikes) in the motorized segment up to 750 kg. The market is still under development, and worldwide many different types of LEFVs are being developed that are dedicated to niche markets. Also, big OEMs (like Pon) are extending their market to LEFVs, which provides this market with strong backup facilities; for instance, the moment a LEFV breaks down, a new LEFV will be provided. At the same time these OEMs facilitate the development of LEFV lease/rental services, making the usage of LEFVs for logistics service providers easier as they do not need to make huge investments in their fleet.

LEFV users
Different professional users can be identified including logistics service providers, and constructors and servicemen (craftsmen).

The *logistics service providers*, and even retailers, consider LEFVs a potential solution for their future processes. However, they don't know whether the LEFVs are fit for their current processes. The planning, sorting, loading and invoicing of deliveries is currently geared towards the use of delivery vans and trucks. Efficient use of LEFVs requires a different view of logistics operations and customer segments. This is due to the smaller payload and delivery radius of LEFVs. Fleet managers in particular experience a lot of uncertainty as the suppliers of LEFVs are not yet sufficiently mature when it comes to their field service. Unforeseen issues are arising in practice during operations. Therefore, the service attitude of the suppliers in relation to operations, maintenance and repair is crucial.

Constructors and servicemen account for 35 per cent of the commercial movements in cities (Ploos van Amstel et al., 2020). The city is their area of operation, where they must have their materials and tools to do their job. Looking at the list of challenges, requirements and wishes, as a service company, a company should first define a service logistics strategy based on which customers they want to serve with which services. This service logistics strategy is in turn determined by the service and market segments that are chosen and the service agreements that are made with customers. The choice of vehicle is a result of this strategy. The business issues involved in the roadmap to zero emissions can be divided into those that are long term (strategical), medium term (tactical) and day-to-day (operational) (Ploos van Amstel et al., 2020).

Shippers
Shippers want their products transported in a fast, reliable manner, and at low cost. They wonder whether a carrier using LEFVs can guarantee the same service at the same cost.

Receivers
For storekeepers (B2B) and consumers (B2C), the reliability of delivery (level of service) is important. Storekeepers need to have personnel available to receive the goods, or to be available to receive a parcel, and may have limited space for receiving deliveries. Most storekeepers therefore want to have a minimum of deliveries during the day, as this distracts them from sales operations and can cause hindrance in the street. At the same time, they also appreciate if the delivery doesn't harm the living environment too much. In that sense, they have often a positive attitude towards the usage of LEFVs if it doesn't cost extra.

The consumer also likes to know when their ordered goods are going to be delivered. The consumer has become an important client of a city logistics system as the rise of e-commerce has been relentless (Statista, 2021).

Behavioural analyses (de Oliveira et al., 2017; Buldeo Rai et al., 2019) have shown that the means by which goods are delivered is not yet a decisive factor in people's choices in relation to home delivery. These studies show that fast and free delivery are still the most dominant attributes of choice. This is somewhat paradoxical as most citizens want deliveries to their homes to be safe, noise-free, healthy and less space consuming.

Another group of *receivers*, namely office, government and educational buildings, represent important clients because of their need for the supply of facility goods and services. When procuring facility goods, buyers can ask suppliers for smart and zero-emission deliveries. The facility service departments of public organizations generate many transport movements in cities. In public procurement, there is potential for smarter and cleaner urban distribution; public organizations can act as game-changers. Recent initiatives across Europe show that more public organizations are gaining knowledge of the transport volumes resulting from their purchases and the potential for improvement. For this improvement to be realized, a change is needed in the way public organizations make decisions about the selection of suppliers, incoterms, delivery service plans and performance indicators.

Together with their suppliers and logistics service providers, the Amsterdam University of Applied Sciences (AUAS) and the University of Amsterdam (UvA) are working on a sustainable supply of their premises. The impact has been fewer kilometres travelled by vehicles, more electric vehicles and more efficient handling of incoming goods. However, getting to the benefits requires a change in the way purchasers behave on a strategic, tactical and operational level. The result is an approach to sustainable logistics that can be scaled up and taken to other large cities.

Institutions

Following the TIS framework, the main institutions that can be identified in relationship to LEFVs in the Netherlands are the Amsterdam University of Applied Sciences (AUAS), the Rotterdam University of Applied Sciences (RUAS) and the HAN University of Applied Sciences. Together, these institutions initiated the two-year LEFV-LOGIC project that ran between 2016 and 2018. This project played a crucial role in knowledge development in city logistics by looking at the usage of LEFVs. The institutions set up new research initiatives and organized learning communities, among them new initiatives like the LEFV-LOGIC project (Ploos van Amstel et al., 2018).

Next, Transport Logistiek Nederland (representing the carriers) and evofenedex (representing the shippers) are relevant industry organizations helping their members (transport companies and shippers) to get acquainted with new technological developments in their field of operations.

The Dutch Safety Board is also an important actor as in some ways the LEFVs constitute a new type of vehicle that is operating on our public roads, bicycle lanes and sometimes even in pedestrian areas. The consequences of the access to roads and bike lanes should be carefully evaluated to provide the vehicles with a clear legal status.

Slob (2021) uses the concepts of institutional logic to capture the various values of actors involved in the development and implementation of LEFVs in the Dutch mobility sector. By means of qualitative research, he conducted interviews with these actors, which resulted in the distinguishing of three institutional logic types: the automobile logic, the society logic and the regulation logic.

- In the automobile logic people are used to car driving, which hinders LEFV implementation. The policy focus is aimed at car manufacturers and the fuel industry. LEFVs are often labelled as unsafe due to outdated rules and regulations, and the infrastructure system is built around car use with little room for LEFVs. Regulators pay little attention to LEFVs as a solution to social goals (e.g. inclusivity, less congestion, liveable streets) other than sustainability. Implementation of LEFVs is hindered due to unclear roles and poor communication between the Ministry of Transport and the Netherlands Vehicle Authority (RDW).
- The research describes how the infrastructure and regulation mechanisms are influenced by the car-dependent automobile logic, and how this makes it difficult for LEFVs to substitute the car. It is observed that actors aiming at the implementation of LEFVs have a variety of social values like safety, liveability and sustainability, institutionalized in the society logic.
- The regulation logic should balance between, on the one hand, securing the safety and performance of the current infrastructure system and, on the other hand, transitioning in line with the urgent call for a more holistic view on mobility. The holistic view of the society logic creates tensions with both the automobile logic with its economic focus and the regulation logic with its primary focus on safety.

Infrastructures

The LEFV is a vehicle and as such makes use of roads, bike paths and sometimes even pavements for (un)loading activities. Urban infrastructure and traffic rules are not yet prepared for an increase in the number of LEFVs. There is uncertainty over which part of the streetscape LEFVs will be allowed to use to drive, load and unload, and there is a shortage of parking facilities. Further speed limits on the road, the construction of bicycle streets, and the installation of loading and unloading spaces for LEFVs offer opportunities for better integration of LEFVs in traffic.

The growth of LEFVs in cities (especially in the Netherlands) leads to conflicts with regular bike traffic, causing dangerous situations on the bicycle lanes. The Dutch Fietsersbond (cyclists' union) prefers not to have cargo bikes on the bike lanes. They don't think the bike lane should be the 'drain' of the traffic system. Like the high-speed lanes for fast e-bikes, new ideas for having separate freight cargo bike lanes and for avoiding rush hours are often brought up for consideration by local authorities. Power grids will play a role in charging the vehicles. Joint efforts by the municipalities and the energy providers, involving good thinking and planning, are needed for the suitable locating and dimensioning of charging stations for all the electric vehicles in the future, including the LEFVs. Whether this should be public or private spaces, combined or not combined with (un)loading areas, is an interesting research question.

As the range of the LEFVs is limited, it is necessary to make use of (micro-) hubs. The hub is a fundamental infrastructure element for the functioning of a city logistics concept based on LEFVs. The usage of hubs touches again on constraints as the hubs take space (space is scarce and expensive) in cities; hubs can create congestion and need space for operations (including outside the building). Still, in practice there are some companies who are operating from a hub (for instance Fietskoeriers, Cycloon). The search for a hub location is like the search for a new house. Patience is needed and a long period of searching should be calculated for, especially for so-called micro-hubs in the inner city. A study by Ploos van Amstel et al. (2021) illustrates potential hub locations and their relative advantages (Table 9.2).

Interactions
Considering the TIS framework, interactions are here interpreted as interactions among the stakeholders involved. The interactions are mainly based on the development of LEFV experiences/knowledge in festivals, projects, Living Labs and daily practice.

The International Cargo Bike Festival (ICBF) is a yearly event organized by exhibitors and many passionate LEFV believers/users to bring the LEFV to the attention of municipalities and SMEs. The exhibitors include many small companies, not only cargo bike producers but also a range of related organizations and start-ups. More than forty countries participate in this event, with demonstrations of new prototypes as well as established LEFVs, knowledge dissemination through presentations, and exchange of experiences with LEFV usages. This yearly festival is a key event in motivating people to switch over to LEFVs. Although attention to this festival is growing, an important note should be raised here, namely that this event is mainly focused on LEFV lovers. The general public still don't know much about receiving goods through delivery by LEFVs.

Table 9.2   *Overview of potential hub locations*

Different options for hub locations for zero emission city logistics and their respective advantages:

Location	Advantages
1. At the employee's home	– Optional private use of zero-emission vehicle – The battery can possibly be charged at home – No additional transfer required at a hub in the city
2. At a secure parking facility in a public space or at a parking garage	– Option for sharing system – Can be set up through an (area-oriented) public–private partnership
3. At the company's own premises	– Option for sharing system and easy to realize – Visible for employees – Can be combined with central delivery of goods
4. At the premises of a friendly competitor (other service company)	– Many possibilities for locations and option for sharing system – Visible for employees – Can be combined with central delivery of goods
5. At a wholesaler where the company buys goods	– Many possibilities for locations and creates value for wholesaler – Can be combined with central delivery of goods – Possibility to create parking spaces with charging infrastructure
6. In a semi-mobile container in a parking lot or company site	– Can be moved depending on the amount of work in an area – Location for stock and delivery of materials in the hub – Suitable as an experiment after which the location may/may not become permanent

	7. At a public transport junction (train, metro)	– Accessible for employees who can travel with public transport – Possibility to connect to existing concepts (such as public transport/bicycle) – Promising areas for low-traffic/zero-emission transport
	8. At a logistics service provider (city hub)	– Making use of existing logistics facilities – Can be combined with central delivery of goods – Possibility to create parking spaces with charging infrastructure

Source: Ploos van Amstel et al. (2021).

A key project on the experiences with LEFVs is the LEFV-LOGIC project. The study has shown the potential for a market share of 10 to 15 per cent of delivery vehicle movements. The sectors with most potential in city logistics are food, construction, services, non-food retail, and post and parcel delivery. LEFVs demand a different logistics concept and transportation costs are determined largely by personnel costs. LEFVs can be beneficially deployed if the delivery can be performed faster than with a conventional vehicle. LEFV vehicle technology is still at an early stage, and LEFVs are not yet mass produced. There is currently a very limited offering when it comes to cooling capabilities and to standardized load carriers (containerization). In the case of small electric distribution vehicles, the electric delivery van is becoming increasingly competitive in terms of cost, speed, load capacity and deployability. Purchase subsidies, experiments with LEFVs and realization of policy objectives (such as emission-free or car-free cities) help to bring about a behavioural change among businesses. LEFVs have been successfully deployed in market segments where low weights and volumes are transported, in which operational excellence is key, or where the use of LEFVs contributes to a distinctive social and innovative value proposition. Recipients of goods or services themselves feel no urgency to pursue supply by LEFV by vendors and carriers but do respond positively if it happens. Based on these project experiences, many new initiatives have started and show positive results in terms of both business potential and sustainability (see also *F1: Entrepreneurial activities*).

Living Labs form a practice-based methodology for evaluating new solutions in the city logistics sector, such as LEFVs. To understand the full added value of a new solution in city logistics the evaluation should be done with representatives from all stakeholders. The Living Lab methodology starts with the multi-stakeholder commitment in the project, although each stakeholder can have its own stakes. In a controlled environment, the stakeholders are involved by providing them performance measurements and asking them about their opinions on these performances. Also, different experiments can be realized in a short time, which provide new insights for all stakeholders (Nesterova and Quak, 2016).

Conclusion on system elements
Following the Paris Agreement, by setting up the goals of CO_2 reduction, there is a strong consensus in cities that the environmental situation and related liveability in the cities need to be improved. The liveability is at risk both from traffic congestion caused by the lack of vehicle space and accidents and from emissions. A greater awareness leads to new thinking about delivering concepts with new elements. The LEFVs in combination with (micro-)hubs could serve some niche markets in city logistics like food, construction, services,

non-food retail, and post and parcel delivery. There is a positive attitude among the first suppliers and users of LEFVs. Companies are open to learning from and sharing experiences and are willing to exploit the possibilities in practice through the setting up of joint projects and Living Labs. Based on the dynamic interrelationships between the stakeholders involved, the adaptation of LEFVs in city logistics will become essential in making cities more sustainable.

System Functions

Following the TIS framework (Langeland et al., Chapter 7 in this volume), the system functions are clarified for the LEFV innovation. In accordance with the framework, entrepreneurial activities, knowledge development and diffusion, influence on the direction of search, market formation, resource mobilization, legitimacy and development of positive externalities (Bergek et al., 2010) are analysed for the LEFV innovation in this section.

F1: Entrepreneurial activities
Entrepreneurial activities can be divided in three categories: (1) general promotion of LEFVs (the ICBF has already been discussed as an interactions element, see above), (2) development of the vehicles and (3) development of the service supplier side. Examples illustrate the presence of active entrepreneurs as the prime indication of the performance of this innovative solution.

Supplier side: development of the vehicles
Since 2011 there has been a growth in the supply and use of light electric vehicles. The growth is evident not only in the numbers, but also in the diversity of types of LEFVs. Several Dutch companies, including Urban Arrow, Easy Go Electric and Stint Urban Mobility, started developing light electric solutions for passenger transport before 2010, following which they also began to see market potential in freight transport and started to standardize the cargo bikes. Internationally, companies like Goupil, Citkar and Velove are obtaining more and more market share as suppliers of LEFVs. Despite this growth, especially for the larger LEFVs, there are not many suppliers (as the Cargohopper and Picnic examples below show).

CARGOHOPPER
One of the first light electric vehicles was the Cargohopper. It was an adapted golf cart.

In the Cargohopper project, a private logistics company developed and tested a sustainable solution for inner-city deliveries using electric transportation. The solution was first piloted in the city of Utrecht, and was then replicated in Amsterdam in collaboration with the municipality (van Winden and van den Buuse, 2017). The project demonstrated that firm-level internal knowledge transfer is important for replication; effective management of ambidexterity is needed to move from exploration to exploitation; prospects of economies of scale can be a prerequisite for a scalable business model; and local-level regulations can be a driver of sustainable innovation.

Transport company Transmission initiated the project. Transmission, with various establishments across the Netherlands, developed the idea for the Cargohopper as a response to the growing number of Dutch cities that had introduced bans on large diesel trucks to limit pollution and congestion. The Cargohopper solution consists of two components: a LEFV and a smart distribution system. The LEFV has the features of a 'road train' with separate carriages, and delivers shipments to businesses in the city's central area where no diesel trucks are allowed. In a distribution centre (located at a facility just outside the inner city), shipments are processed, bundled and loaded onto the electric freight vehicle. These shipments are bundled by address into separate carriages, allowing efficient delivery to businesses based on the proximity of delivery addresses in the same area. The project stopped in 2017 after technical issues with the vehicle meant it didn't pass the rules set in legislation and because of a lack of support by the supplier, typical problems that face an early adopter.

PICNIC

Another example comes from Picnic. Picnic is an online supermarket currently operational in the Netherlands and Germany. Groceries are ordered from an app and delivered to the consumers using a light electric last-mile delivery vehicle. This vehicle is currently used for densely populated residential areas and works well enough to support the current number of Picnic deliveries.

However, the current last-mile delivery vehicle can't access all households. This is because its speed is limited to 50 km/h and it can only carry a limited volume of cargo. To expand and reach new households, Picnic is looking for a new last-mile delivery vehicle.

Having started a joint venture with VDL and TNO, Picnic is looking to design and build their very own last-mile delivery vehicle that is purpose-built for their needs. They are looking to become bigger, faster and stronger on the roads. Increasing the vehicle speed to 80 km/h and carrying more cargo allows them to reach the households that are currently out of geographic scope. Having a vehicle that can reach those extra households would significantly increase their customer base and consequently market share in the food market.

With the opportunity to build a purpose-specific vehicle, they can also control the aesthetics of the vehicle. Picnic relies heavily on their brand image and identity as a means of differentiation from their competitors. Therefore, translating their brand assets to the vehicle will give them a stronger brand presence in the consumer environment. Having identified the raison d'être of the vehicle – reaching new households in less densely populated areas – research was carried out to find different design cues and assets that could be leveraged in the exterior design of the new vehicle.

Accompanying these was a design vision also derived through research and by talking to the relevant stakeholders: 'Design a next-generation company icon, to remain a local hero.' Using this input, the sketch phase was conducted with a funnel approach, creating a broad spectrum of different designs and options that were in accordance with the design vision guided by the research. Through the method of elimination based on the principles of Quality Function Deployment (Cohen, 1995), and with the input from the different stakeholders, the sketch phase reached a point of maturity, which consequently yielded a final design that satisfied the requirements derived from the research and embodied the Picnic aesthetic while remaining functional for the runners.

Velove

Velove developed the Armadillo. An Armadillo uses only 6 per cent of the electricity of a small electric van, for the same transport work, and is more productive. In 2019 they won the first International Cargo Bike of the Year award in the category small electric vehicles. The Armadillos form the basis for successful last-mile solutions in Berlin, Oslo and Gothenburg. Velove also designed their vehicles to be suitable for containerized processes. At the same time leasing and rental solutions have been set up to make the market more accessible. The Armadillos are now operational in 20 different countries. To grow into a more mature business with a maximization of the uptime of the vehicles, Velove also organized two maintenance partner events to exchange knowledge and experiences on maintenance services (Erlandsson, 2020).

Supplier side: development of the service supplier side
From the service supplier side one can observe a tremendous growth of companies delivering products by LEFVs. An illustrative example of such a company is Pedal Me, an e-cargo bike logistics and pedicab company in London.

Pedal Me

The company was founded in 2017. It normally operates within five miles of Central London, using bikes built by Urban Arrow. The company also offers cargo deliveries. Pedal Me proved in a trial run that the company's riders delivered construction materials from Wood Green to Whitechapel faster than a van. The contractor plans to continue receiving deliveries by bike. During the COVID-19 lockdown in April and May 2020, Pedal Me partnered with Lambeth Council to deliver care packages to the individuals and families most in need. This was the largest operation conducted by Pedal Me to date, and perhaps the single largest e-cargo bike logistics operation in the UK. In total, the Pedal Me fleet covered over 20,000 km to distribute nearly 10,000 packages, moving around 150,000 kg across the borough of Lambeth. Recently they have opened a 6,500 sq. ft warehouse in Zone 1, the Central London zone, to support their freight operations (Pedal Me, 2021).

Conclusions – F1: Entrepreneurial activities
From the entrepreneurial activities and the lessons learned from the LEFV-LOGIC study (Ploos van Amstel et al., 2018) it could be concluded that the frontrunners of LEFVs have had an open mind to collaborating with suppliers to create a purpose-built vehicle for their specific needs (Picnic, DHL; see also *F5: Resource mobilization*). They are willing to adjust their designs according to the demands of their users and are open to experimenting. From the entrepreneurial activities, we learn that the most important issues are related to how to reduce the total cost of ownership compared to conventional vehicles/electric vans while at the same time continuing to guarantee a good quality service with everyday, intensive usage. The interaction with other traffic can sometimes lead to unsafe situations on the roads, which means that alongside the issue of specialization of LEFVs in specific niche markets, road safety and EU homologation are also important issues. Having a driving licence, use of a helmet, and a minimum age for driving are aspects where conformity is needed, with a minimum set of regulatory, technical and safety requirements being met by all countries. For suppliers of LEFVs, it is a challenge to meet the growing demand for these vehicles both in terms of producing LEFVs in sufficient numbers and in terms of providing the necessary maintenance. Many initiatives that started out with tests followed by experiments (Living Lab settings) have now reached the stage of real-world applications. Daily practice shows that there is still a poor set of suppliers, little cooperation among suppliers and little communalities in their vehicle designs (leading to unique maintenance instead of joint maintenance), and that full-service concepts are not yet fully provided, which can lead to uncertainty in the operations of the logistics service providers. Considering these factors, it can be said that the LEFV product is not sufficiently mature to make it a reliable proposition for its potential users. At this point it can be reflected that the maturity of the service levels diminishes as the size of the vehicles increases.

F2: Knowledge development and diffusion
Knowledge development in the LEFV sector is based on the setting up of (inter)national (feasibility) projects, Living Labs as practice-based research, and international cargo bike events. Some examples of knowledge activities are described in this section.

Cargo bike events

Earlier, the International Cargo Bike Festival (ICBF) was discussed. Here, the project RIPPL (Register of Initiatives in Pedal Powered Logistics), a knowledge-sharing project which highlights examples of businesses, organizations, projects and initiatives using bikes, trikes or pedal power, is mentioned. The project also identifies trends in cycle logistics and aims to share best practices and innovation from across the world. RIPPL is an initiative from Jos Sluijsmans and researcher Tom Parr. In the past, RIPPL has been funded by Topsector Logistiek, through the Dutch sustainable mobility organization Connekt. Although based in the Netherlands, Sluijsmans and Parr are aiming at exchanging knowledge and experiences throughout the world by setting up their own magazine and organizing events.

Scientific research

Although the Living Lab initiatives provide new knowledge and experiences with LEFVs, only a few successful cases for LEFVs have been found in scientific research (Schliwa et al., 2015; Lenz and Riehle, 2013). Simulation approaches and ex-ante analyses (Melo and Baptista, 2017; Gruber et al., 2014; Tipagornwong and Figliozzi, 2014; Arnold et al., 2018; Zhang et al., 2018; Gruber and Narayanan, 2019; Sheth et al., 2019; Naumov, 2021; Caggiani et al., 2021; Llorca and Moeckel, 2021) are most commonly applied in practical research. Fiori and Marzano (2018) developed an EFVs energy consumption model. The estimated model was validated by collecting real-world data from 144 observed trips for pickup/delivery made by five EFVs operating in the city of Rome.

In the literature only one real-life case can be found (Browne et al., 2011). Browne et al. performed an in-depth case study of Gnewt Cargo in London. In transport as well as environmental and financial terms, the trial of LEFVs was proven successful from the company's perspective, as the total distance travelled and the CO_2 emissions per parcel delivered dropped by 14 per cent and 55 per cent respectively using LEFVs. At the end of the project they decided to continue their delivery operations with the LEFVs.

LEFV-LOGIC: a national project

In the Netherlands, a research project with over 25 participants in the Netherlands developed insights into the types and logistics usage of LEFVs. In 2018, Ploos van Amstel et al. (2018), as part of the LEFV-LOGIC project, investigated for which types of goods the LEFVs are most promising within the framework of city logistics. Four crucial criteria have been identified for LEFV usage: small and light shipments, high network density, time-critical shipments and sufficient opportunities for growth and innovation. In line with these findings, they came up with the sectors mail, parcel and local retail deliveries, and smaller shipments in food, construction and service logistics that meet all the criteria. Moolenburgh et al. (2020), as part of the same project, performed case-based research. Several experiments were set up in different towns in the Netherlands to test LEFVs and collect knowledge on the practical experiences of their usage. Stakeholder consultation was performed to obtain feedback on LEFV usage. Also, the LEFVs were monitored with GPS loggers and cameras to obtain real-time measurements. Ten companies volunteered to take part in these experiments to experience using LEFVs.

This project shows that LEFVs are suitable for a wide range of users and applications, from independent entrepreneurs with a briefcase who want to transport a letter or small parcel to logistics service providers who transport roll containers. The expected fields of application for LEFVs (Ploos van Amstel et al., 2018) are proven to be viable in practice. The costs of the LEFVs can be up to 20–30 per cent less than those of the traditional delivery van or lorry. The use of LEFVs for short journeys in (inner) cities yields time savings due to the presence of cycle paths and one-way roads. The surveys show that bicycle routes in cities are on average 15 to 20 percent shorter than car routes. Together with the advantage of loading and unloading on footpaths, delivery times can be up to 30 per cent faster.

To deploy LEFVs efficiently, adjustments must be made in how logistics are planned, for example by clustering orders (even more) geographically and using planning software with routes suitable for LEFVs. This requires sufficient shipment density, or short distances between the stops. As the range of the vehicles is limited, all logistics concepts using LEFVs need to have a collection/consolidation point that is nearby.

The position of LEFVs in traffic, including the rules for the use of cycle lanes and pedestrian areas, is not unambiguous and requires further investigation. The integration of the vehicles into urban traffic networks is a necessity. Examples include the design of comfortable and safe routes, such as bicycle streets, and the creation of loading and unloading areas. Experimenting with LEFVs leads to greater awareness, knowledge and behavioural change. For instance, weather conditions can have a strong influence on the maintenance of the cargo bikes. Driving a LEFV takes some time to get used to in the beginning, but is perceived as simple. Drivers of LEFVs receive positive reactions from customers and the public.

In contrast to electric delivery vans, many LEFVs, particularly those that are more bicycle-like, have the advantage that the range is less dependent on interim charging. With limited use of LEFVs, businesses do not experience any barriers when charging. With an expansion of electric vehicles in the fleet, smart charging offers a solution to balance out any peaks and troughs in energy demand.

Living Labs

As follow up on the feasibility studies, Living Labs have been developed to cover the lack of practical knowledge on city logistics concepts with the use of LEFVs. Each year students of the Rotterdam Applied University of Sciences run a city hub with LEFVs. The hub is located on the Noord-West business park in Rotterdam, in-house at a logistics partner; ride and route planning software from RoutiGo is used, a dedicated software for cycling; and three electric vehicles are deployed: an Urban Arrow XL (cargo bike), a Sevic Cargo 500 (LEFV) and sometimes, as backup, a Nissan e-NV200 (which is not LEFV). The main services requested are the collection of returns and delivery of packages. Furthermore, light Value Added Logistics (VAL) activities are supplied and short-term stock storage services are offered. During the pilot, 3,108 inner-city kilometres were travelled, 726 stops were made, 4,278 parcels were delivered and at least 889 kg of CO_2 has been saved (van Duin et al., 2022). Transport space, vehicles, warehouse space, office space, software, personnel and knowledge are shared with logistics partners, and the available overcapacity pushed back. Ultimately, it turned out that a hub can be profitable, but at the same time it was realized that entrepreneurship, networking, good marketing and a wide range of services and collaborations are indispensable. The hub (working as a Living Lab) not only became visible in the city (in the COVID-19 period there was strong increase in customers of this service), but perhaps more importantly it has been shown that it is possible.

Conclusions – F2: Knowledge development and diffusion

Besides the most important knowledge that gets shared at events and comes out from experiences with LEFV projects, it can be concluded that most knowledge remains at local level, especially for the Living Lab experiences. At a national scale, universities (of applied sciences) have been working together more in long-term (eight years) city logistics programmes. This supports the development of scientific and practice knowledge by ensuring the knowledge is both verified and sound.

It is good to have a dedicated International Festival for Cargo Biking in order to keep the knowledge more focused. Other conferences such as the Transportation Research Board (TRB), the World Conference on Transport Research (WCTR) and the International Conference in City Logistics (ICCL) have themes on city logistics but still the attention to LEFVs remains fragmented and only accessible to scientific researchers. When it comes to knowledge sharing, the exchange of user experiences and best practices now seems to be most hotly in demand.

F3: Influence on the direction of search

This function (F3) encompasses mainly visions, targets and regulations set by governments and/or industries. For the LEFVs this is evidently the Green Deal for Zero-Emission City Logistics (Green Deal ZES, the Dutch climate agreement introducing the establishment of zero-emission zones for logistics in the centres of the 30 largest Dutch cities, as a direct follow-up on the Paris Agreement) and the resulting new plans for many cities to ban car traffic from the inner cities.

In 2014 the Green Deal ZES was signed by a list of Dutch governmental bodies, businesses and institutions. The parties signed up to the Green Deal ZES want to supply inner cities efficiently and emission-free by 2025. It was agreed that the period up to 2020 would mainly be used to 'experiment' with small, often local projects, to explore and learn as much as possible, and to share the acquired knowledge and experience with each other; pilots could fail, however, and therefore were continued rather than stopped, mainly to see where things went wrong and how things could be improved. These projects were called 'Living Labs'.

The Green Deal ZES has a term of ten years. In November 2014 there were 50 participating parties: governmental bodies, entrepreneurs (transporters and shippers), sector parties, interest representatives and knowledge institutes. Since then, more than 200 parties have joined, all experiences have been evaluated, and the projects are now being scaled up towards 2025, the year that, as far as possible, city logistics must be zero emission in at least 30 Dutch city centres. Supply to shops in those areas may only take place with zero-emission powered vehicles (for exact plans and details of the planned transition period, see GovNed, 2021a).

The Green Deal ZES is one of the six hundred measures in the Dutch climate agreement. To make this all happen a City Logistics Implementation Agenda has been developed (GovNed, 2021a) providing guidelines on how to set up regional cooperation and allow scope for local customization in preparation for the introduction of zero-emission zones.

Car-free/liveable cities

Some cities and neighbourhoods are beginning to rethink where cars can go – and redesigning streets to prioritize other uses, from public transportation to parks. It's happening around the world, including to major streets in cities like San Francisco and New York, but is occurring at the largest scale in many European cities.

The 'City as a City Launch' (Municipality of Rotterdam) is the provocative title of an initiative to redesign the inner city by allowing more space for bikers and pedestrians. This development can also be observed in many other Dutch cities, but in a less strict a way as is happening in Rotterdam. It is likely they will follow the Amsterdam agenda, as Amsterdam doesn't plan to fully ban cars. Rather, the key step in Amsterdam is removing 11,200 parking spots by 2025 and then using that space for wider pavements and bike lanes, trees and bike parking. Some streets will be narrowed or blocked off, and the city will issue fewer parking permits. It also plans to redesign roads for better cycling and add bike parking at metro stations, and may experiment with free public transit at rush hour. The few cars that are left will soon be electric or otherwise emissions-free. The same development is happening in other European cities: Milan is giving its squares back to residents, Paris has extended a 30 km/h speed limit to most of its city streets, and Barcelona is restructuring its city with fewer intermediate roads and has issued 12,000 free annual public transport tickets to former car owners.

Conclusions – F3: Influence on the direction of search
It is obvious that the Green Deal ZES and local car-reduction policies (including freight vehicles) represent a strong push towards LEFVs. At the same time a new type of space shortage is occurring at the edges of inner cities, where (micro-)hubs should be developed to facilitate the transhipment from trucks/vans to LEFV (see also '(micro-)hubs' in *F5: Resource mobilization*).

F4: Market formation
Market formation started with suppliers offering cargo bikes or other LEFVs for sale. Dutch companies today work together to offer 'full service' concepts for LEFVs that include, for instance, financing and maintenance. An example is DOCKR Mobility. DOCKR (a Pon Mobility company) helps entrepreneurs to navigate the inner city. It achieves that result by providing flexible contracts and electric vehicles with roomy cargo compartments, according to the one-stop-shop concept: insurance, periodic maintenance, and replacement transport are included as standard. Lease contracts can be terminated within one month, creating greater flexibility. Urban Arrow has a relationship with DOCKR as the latter leases out a big assortment of Arrows, and European lease company Paribas Arval offers all kinds of cargo bikes, including Arrows, as an option to fleet managers. On a small scale, similar initiatives are found elsewhere in the world, which sees the market for renting/selling LEFVs slowly developing.

Conclusions – F4: Market formation
A market for the supply of LEFVs is emerging. Many suppliers are coming to the market and the market is becoming more accessible with the introduction of rental contracts. This gives the companies options to experiment and learn with the LEFV usage for a period instead of having to make large fleet investments. A point of concern is the service and maintenance options. Due to the

high volume of delivery tasks, fall-back options need to be available in times of disruption.

F5: Resource mobilization

Resource mobilization is discussed here from the LEFVs' own development perspective, a sharing perspective, their linkage with hubs, and their linkage with the development of Physical Internet (PI).

LEFV development

Two developments are associated with the practical usage of LEFVs. The first development is the mobilization of the LEFVs themselves. Nowadays there are many (proto)types of LEFVs available on the market. Due to the need to overcome the limited loading capacity of normal-sized cargo bikes, the market is now booming for XXL cargo bikes, which can have a loading capacity up to a box of 1,350 litres and weight of 150 kg. As well as electric-driven engines, hydrogen engines are also becoming available on the market.

Sharing LEFVs

On the demand side one can observe that large companies purchase LEFVs exclusively for their own operations, because they often want to have full use of the LEFVs, are able to make the investments, and take the uncertainty related to innovations for granted. SMEs, however, cannot do this. Sharing is a new trend in logistics in general (Gesing, 2017), and sharing of LEFVs, where several parties share one or more LEFVs, can offer SMEs a solution. Sharing LEFVs may represent a good opportunity, as it could become attractive, as demonstrated by transport-sharing concepts for private individuals such as bicycle sharing from GoAbout or shared scooters from Felyx. Research (van den Band and Roosendaal, 2020) shows that the business community is willing to accept the concept of sharing LEFVs, because it can reduce transport movements, reduce costs and can provide additional, flexible capacity to one's own transport fleet. However, there are some challenges that need tackling, such as the need for mutual trust and a good platform that supports the concept and maintenance.

(Micro)Hubs

One should understand that the mobilization of LEFVs is not a standalone development but has links with the development of (micro)hubs. The LEFVs often form an integral part of the last-mile solution, where goods are traditionally transported to hubs located at the edges of the city centres (often just outside the zero-emission zones) and then transshipped from trucks/trailers/vans to electric vans and LEFVs which deliver the goods to the final destinations in the inner city or vice versa. Although the research related to hub location problems (Farahani et al., 2013) is extensive, in practice the value of these facility location models is limited as in many cases the space available for hubs is limited or is not available at the calculated best solution in terms of (distribution) cost. Still, the hunt for good hub locations is increasing while real-estate prices are rising excessively, leading to scarcity of affordable space in the cities and sometimes in neighbouring regional districts. With respect to the electricity supply, this is not such a big issue for LEFVs as for other electric vehicles as the recharging of significant numbers of LEFV batteries at the same time doesn't cause a big peak in (power) demand. Still, not much space can be found for installing hubs in the inner cities.

Physical Internet
Another trend that is important to mention here is the growth of the Physical Internet (PI). The PI concept is inspired by the digital Internet and is an open, collaborative, and standardized transportation concept in which containerized freight moves through a network from hub to hub in a self-organizing manner. It is hypothesized that the PI will significantly increase the efficiency, transparency, scalability and robustness of the transportation network (Ballot et al., 2014; Montreuil et al., 2010). Crainic and Montreuil (2016) have translated the PI vision to the context of city logistics. Combining the fields of city logistics and the PI, they have come up with the notion of the 'Hyperconnected City', in which a scalable distribution structure is developed based on standardization of loading units (PI containers) and information exchange, and where LEFVs could play an important role in the urban last mile.

An illustration of the principles of PI development is the initiative DHL took to integrate their networks based on standardization. DHL has been using bicycles for 20 years and was previously involved with bike designs and modifications, from which the Parcycle and Cubicycle emerged. DHL sees the bike as the new industry standard for delivering parcels. DHL Express is taking the next step in rolling out a fully containerized last-mile delivery process, starting in the Netherlands. The containerized delivery process enables the cost-efficient and secure transfer of goods between terminals, motor vehicles and specialized last-mile delivery vehicles. The containers are transported by motor vehicles to handover points in city centres. At the handover points, the containers are quickly and safely transferred to last-mile delivery vehicles, and the couriers have all the information about the contents in their hand units. The last-mile delivery vehicles used are Velove Armadillos (called Cubicycles within DHL) and electric vans. Which type of vehicle is chosen depends on the last-mile delivery zone. Parcels are sorted and handled only once in the delivery process, at the sorting terminal. This makes it economically viable for the last-mile delivery vehicle to refill during the workday, as the distance to return the empty container and pick up more goods can be minimized, and the transfer of the goods is fast and secure.

Conclusions – F5: Resource mobilization

It can be concluded that resource mobilization is developing well for the LEFV industry. Several different types of LEFVs are coming to the market, serving all kinds of city logistics niche markets including, among others, conditioned deliveries, service deliveries, facility deliveries and parcel deliveries. At the same time, PI can support future integration into the logistics processes in co-creation with the drivers by looking at standardization of LEFVs so that multipurpose vehicles and boxes can be utilized in different niche markets. For SMEs, opportunities for sharing LEFVs are arising through service platforms. The link with hubs development is evident as many times the final delivery from the hubs is carried out by a LEFV.

F6: Legitimacy

Industry associations and companies in a few countries have started to work on cargo bike standards in the last few years, but the need for a consistent European approach has really come together in 2020 under the umbrella of CIE's partnership with the European Cycle Logistics Federation (ECFL, 2021). With this umbrella view, coordinated by a joint Expert Group, we can see that there is little consistency in the national approaches, and even worse, some of the proposals could seriously damage the growth of some cargo bike and commercial vehicle services.

In November 2020, the expert group made its first proposals on how cargo bikes could be regulated to an EU review of mobility devices, an achievement celebrated as a major milestone in the development of the sector. Together with the proposals were the results of a comprehensive survey on cargo bikes in commercial use, the first of its kind at the international level. According to industry partners, this showed that cargo bikes are possibly the safest bikes in the world, because after millions of kilometres of use in commercial fleets not a single fatal accident had been reported, and there were few injuries to riders or the public. That allowed the new proposals for cargo bike regulation to be compared confidently to existing regulation for bikes and e-bikes.

Also, the industry organization LEVA-EU plays an important role at EU level. To promote the market uptake and deployment of light electric vehicles (LEVs), LEVA-EU guides LEV companies through the maze of rules and regulations governing the vehicles. A better understanding of these rules and regulations facilitates the market access and development of LEV businesses. In turn, the EU market benefits from a more varied and high-quality LEV offer. Rules and regulations are sometimes outdated or inaccurate and there-fore create legal bottlenecks for the LEV community. LEVA-EU has a direct line with EU decision-makers to negotiate improvement of the rules. It is also in contact with national decision-makers to exchange information on best practice. Furthermore, LEVA-EU works pro-actively and gives its members a voice in European LEV advocacy. It monitors all EU legislation and policy-making that may be relevant to LEV companies. In consultation with its members, it selects those issues that are most relevant to the LEV business and develops and implements a European advocacy strategy.

The current infrastructure system and urban planning are mainly organized around the use of cars. The transition towards a more sustainable mobility system is also mainly directed towards the replacement of fossil fuel cars by electric cars. However, with the development of many other types of LEFVs, it becomes clear that the infrastructure system cannot support all these different types of vehicles. Formerly, in the Netherlands, the main roads were used for cars, and bikes used the cycling lanes. Currently, there is a debate on which lanes LEFVs should use.

Former Minister of Infrastructure and Water Management Cora van Nieuwenhuizen recently presented a new framework for LEFVs (GovNed, 2021b). Among other things, this states that cargo bikes may have a maximum total weight of 425 kg, may be up to three metres long and may be driven by persons of 18 years and older. For access to bikeways the Minister proposes a maximum speed of 25 km/h and maximum width of one metre.

What regulations should be in place for LEFVs:

- *Infrastructural changes such as speed limits, car-free roads and road widening.* This will give all road users more space and the city's infrastructure will be equipped for a further increase in sustainable (freight) vehicles in the city centre.
- *Unambiguous technical requirements.* These ensure that all vehicles have the same safety standard. Currently, there are no technical specifications that LEFVs must meet. As a result, many cargo bikes are built with vulnerable components from consumer bikes.
- *Inspection of LEFVs by an official independent body such as the RDW.* This organization is objective and operates using European vehicle regulations, and it has the necessary relevant experience from the automotive sector.

Conclusions – F6: Legitimacy

Today it can be seen that the legislation on LEFVs is tightening and that it is dominated by safety concerns. It is obvious that this is happening as the usage of LEFVs is growing and the number of accidents involving them is concurrently increasing. Most Dutch people can remember the horrifying accident in 2018 involving a Stint, which at that time, as well as being a very popular vehicle for parcel delivery, was also used for passenger transport. The accident occurred at a level crossing, where a train collided with a Stint that was transporting young children to kindergarten. Four children sadly passed away.

F7: Development of positive externalities

According to a new analysis from the European Cyclists' Federation (ECF), there are almost three hundred tax-incentive and purchase-premium schemes for cycling offered by national, regional and local authorities across Europe to make it attractive to cycle more and drive less, to reduce transport CO_2 emissions and to provide important growth stimuli for the European bicycle industry.

The usage of electric cargo bikes in particular, as a category of LEFVs, can contribute to bettering the health of drivers. The use of cycling as a means of prevention against welfare diseases such as diabetes and obesity is gaining ground. However, it is still hardly known that cycling has other positive effects too, that is, it is reducing the risk of depression as it induces an increase in various neurotransmitters including dopamine (provides satisfaction and reward), serotonin (a happiness hormone) and endorphins (an anti-stress hormone) (MensLine, 2021).

CONCLUSIONS AND FUTURE PERSPECTIVES

In answering the research question '*What are the success and failure factors of the introduction of LEFVs in city logistics and what are the future perspectives on LEFVs in city logistics?*' the TIS framework has provided us insights on the usage of LEFVs in city logistics concepts.

The framework provides a complete overview of factors that have influence on the development of LEFVs. The distinction between the system elements, which are more related to the stakeholder dynamics, and the system functions is a valuable insight. However, working with the framework didn't provide us explicit insight on whether the factors are success or failure factors. At the same time, it is not fully clear whether a factor mentioned in the TIS framework is an internal strength or weakness, or whether it can be seen as an exogenous opportunity or threat. It doesn't provide any steering or support towards a further valorization/implementation of the LEFV innovation.

For this reason, an additional framework is suggested here to find strategies on how to proceed with the outcomes of the TIS framework by categorizing them into strengths (S), weaknesses (W), opportunities (O) and threats (T).

Here a special form of SWOT analysis is suggested, that is, the confrontation matrix of Kearns (1992). The confrontation matrix contrasts opportunities

Table 9.3   *SWOT strategies matrix*

	Strengths (S)	Weaknesses (W)
Opportunities (O)	*Challenge* Use the strengths to better exploit the opportunities	*Defend/improve* Use the opportunities and challenges to improve/cover the weaknesses
Threats (T)	*Protect* Areas where there are issues and a choice must be made to invest, disinvest or work together	*Keep the damage* Minimize the weaknesses and prevent the threats

Source: Kearns (1992).

and threats with strengths and weaknesses. This creates a matrix with four cells (Table 9.3) that can have the following meanings for research:

• Investigate many promising opportunities to make large innovation steps possible with the strengths (OS);
• Investigate whether opportunities can contribute to an improvement of weaknesses (OW);
• Investigate to what extent the threats have a negative impact on the strengths of the system (TS);
• Investigate to what extent the threats are threatening to the weaknesses and/ or can be converted into improvements or be made to disappear completely (TW).

The SWOT analysis can be used as a kind of compass for navigating the research opportunities and offers the possibility of continuously identifying where possible points of interest lie in the research field. The SWOT analysis diagram on the usability of LEFVs is presented in Table 9.4.

As can be gleaned from Table 9.4, due to their strengths, LEFVs offer good value propositions to stakeholders. At the same time, the list of opportunities forms a solid basis to extend the current usage of LEFVs. The weaknesses might come from the real-estate markets as the LEFVs require hubs for their operations. Space for hubs will be scarce and expensive in the next decade. The growing use of LEFVs asks for a safe position on the roads and more professional services from their suppliers. The threats come from the legal position on the road and the fierce competition from e-vans. Most of the threats can be covered by the LEFVs' strengths.

As can be noticed from Table 9.4, some additional factors are provided. Applying the TIS framework rigidly doesn't give space for these factors. Most of these factors are logistics characteristics of the LEFVs which are not explicitly addressed in the TIS framework.

Table 9.4   *SWOT analysis on usability of LEFVs in city logistics*

STRENGTHS	WEAKNESSES
TIS Functions	*TIS System Elements*
• F3: Influence on the direction of search. Fit with current views of liveable cities.	• System Infrastructures: bicycle lanes (unsafe interference with other cyclists).
• F4: Market formation. The supply of LEFVs is emerging.	• System Infrastructures: hubs (need for many locations).
• F6: Legitimacy. Implementation of LEFVs: (safety) position on the road due to influential groups (EFCL and national governments).	*TIS Functions*
	• F4: Market formation. No all-in-one solutions are widely available for fleet managers. DOCKR is still an exception in the market, with Pon Mobility as a supportive OEM.
Additional Factors	
• Perception is based on 'fun' factor: people love cargo bikes.	*Additional Factors*
• Zero emission, less space, easy loading/unloading.	• Payload.
• Fit for purpose for different segments of city logistics.	• Restricted radius.
	• Only suited for specific products (small, low weight/volume).
	• LEFVs are not linked to upstream processes: (micro)hubs, containerization, difficulty in scaling, IoT (Physical Internet).

OPPORTUNITIES/ENABLERS	THREATS/BARRIERS
TIS System Elements	*TIS Functions*
• System Actors: OEMs; logistics service providers having a hub; receivers; office, government and educational buildings procuring facility goods; demand for zero-emission deliveries.	• F6: Legitimacy. Implementation of LEFVs: current safety position on the road. Legislation needs to be further developed.
• System Interactions: Living Labs.	*Additional Factors*
TIS Functions	• Emerging autonomous delivery systems.
• F1: Entrepreneurial activities. (1) General promotion of LEFVS, (2) development of the vehicles and (3) development of the service supplier side.	• Human talent: not everybody wants to ride a LEFV.
• F2: Knowledge development and diffusion. Based on the setting up of (inter) national (feasibility) projects, Living Labs as more practice-based research, and international cargo bike events.	• Competition developing better delivery vans.
• F3: Influence on the direction of search. Green Deal ZES and fewer vehicles in the city: LEFVs are a real solution.	• High total cost of ownership.
• F5 Resource mobilization. LEFV development, sharing concepts, hubs and Physical Internet (PI).	• Not ready for many large-scale implementations.
• F7: Development of positive externalities. Health issues.	
Additional factors	
• Awareness: understanding the usefulness of LEFVs in different segments of city logistics.	

To conclude, the TIS framework as an ex-post analysis tool makes a lot of things clear in terms of system elements (the stakeholder environment) and system functions. However, to apply the TIS framework for deeper analysis in order to come up with strategies for innovation acceptance, the additional usage of a SWOT analysis is useful to position the elements and functions better. In particular, the discussion on the position of the elements and functions helps in understanding the implementation of the innovation and facilitates thinking on new strategies for implementation.

Looking again at the initial estimate of a potential market share of 10 to 15 per cent (Ploos van Amstel et al., 2018), a more optimistic estimate has been recently presented by Verlinghieri et al. (2021), who suggest that up to 51 per cent of all freight journeys in cities could be replaced by cargo bike. For other cities in the world, the findings from the TIS framework and the SWOT analysis are likely to be more or less identical. Together with Sweden and Denmark, the Netherlands has a leading position in terms of LEFVs usage, but one can observe a growing usage of LEFVs in cities throughout the world. In our opinion, more growth than the initial estimate of 15 per cent is certainly possible, however 51 per cent of all freight journeys is not a realistic estimate considering all the influencing factors derived from the TIS framework.

REFERENCES

Arnold, F., Cardenas, I., Sörensen, K. and Dewulf, W. (2018). Simulation of B2C e-commerce distribution in Antwerp using cargo bikes and delivery points. *European Transport Research Review*, *10*(1), 1–13.

Ballot, E., Montreuil, B. and Meller, R.D. (2014). *The Physical Internet: The Network of Logistics Networks*. Paris: La Documentation française.

Balm, S., Moolenburgh, E., Anand, N. and Ploos van Amstel, W. (2018). The potential of light electric vehicles for specific freight flows: Insights from the Netherlands. *City Logistics*, *2*, 241–260.

Bergek, A., Hekkert, M. and Jacobsson, S. (2010). Functions in innovation systems: A framework for analysing energy system dynamics and identifying goals for system building activities by entrepreneurs and policy makers. RIDE/IMIT Working Paper No. 84426-008.

Boer, E.D., Kok, R., Ploos van Amstel, W., Quak, H.J. and Wagter, H. (2017). *Outlook City Logistics 2017*. Topsector Logistiek, Delft.

Browne, M., Allen, J. and Leonardi, J. (2011). Evaluating the use of an urban consolidation centre and electric vehicles in central London. *IATSS Research*, *35*(1), 1–6.

Bruntlett, C. and Bruntlett, M. (2021). *Curbing Traffic: The Human Case for Fewer Cars in Our Lives*. Washington, DC: Island Press.

Buldeo Rai, H., Verlinde, S. and Macharis, C. (2019). The 'next day, free delivery' myth unravelled: Possibilities for sustainable last mile transport in an omnichannel environment. *International Journal of Retail Distribution Management*, *47*, 39–54.

Caggiani, L., Colovic, A., Prencipe, L.P. and Ottomanelli, M. (2021). A green logistics solution for last-mile deliveries considering e-vans and e-cargo bikes. *Transportation Research Procedia*, *52*, 75–82.

Cairns, S., and Sloman, L. (2019). *Potential for e-Cargo Bikes to Reduce Congestion and Pollution from Vans in Cities*. Transport for Quality of Life Ltd. https://www.bicycleassociation. org. uk/wpcontent/uploads/2019/07/Potential-for-e-cargo-bikes-to-reduce-congestion-and-pollution-from-vans-FINAL.pdf.

Cohen, L. (1995). *Quality Function Deployment: How to Make QFD Work for You*. Reading, MA: Addison-Wesley.

Crainic, T.G., and Montreuil, B. (2016). Physical Internet enabled Hyperconnected City logistics. *Transportation Research Procedia, 12*, 383–398.

Cyclelogistics (2014). https://www.cyclelogistics.eu/, visited 21 October 2021.

Dablanc, L. (2011). City distribution, a key element of the urban economy: Guidelines for practitioners. In Macharis, C. and Melo, S. (eds), *City Distribution and Urban Freight Transport: Multiple Perspectives*. Cheltenham, UK and Northampton, MA, USA: Edward Elgar Publishing, pp. 13–36.

Deloison, T., Hannon, E., Huber, A., Heid, B., Klink, C., Sahay, R. and Wolff, C. (2020). *The Future of the Last-Mile Ecosystem: Transition Roadmaps for Public- and Private-Sector Players*. World Economic Forum, Geneva, Switzerland.

DfT (2021). *Provisional Van Statistics 2019–20*. Department for Transport UK.

ECFL (2021). *European Cycle Logistics Federation*. https://eclf.bike/, visited 28 September 2021.

Erlandsson, J. (2020). *Velove 2019 Highlights*. https://www.velove.se/news/velove-2019-highlights, visited 17 August 2021.

Farahani, R.Z., Hekmatfar, M., Arabani, A.B. and Nikbakhsh, E. (2013). Hub location problems: A review of models, classification, solution techniques, and applications. *Computers and Industrial Engineering, 64*(4), 1096–1109.

Fiori, C., and Marzano, V. (2018). Modelling energy consumption of electric freight vehicles in urban pickup/delivery operations: Analysis and estimation on a real-world dataset. *Transportation Research Part D: Transport and Environment, 65*, 658–673.

FMI (2021). *Analysis and Review: Cargo Bike Market by Propulsion – Conventional and Electric for 2021–2031*. https://www.futuremarketinsights.com/reports/cargo-bike-market, visited 24 August 2021.

Gesing, B. (2017). *Sharing Economy Logistics: Rethinking Logistics with Access over Ownership*. DHL Trend Research, Troisdorf.

GovNed (2021a). *New Agreements on Urban Deliveries without CO_2 Emission*. Government of the Netherlands. https://www.government.nl/latest/news/2021/02/11/new-agreements-on-urban-deliveries-without-co2-emission, visited 28 September 2021.

GovNed (2021b). *Kamerbrief over kader Lichte Elektrische Voertuigen*. https://www.rijksoverheid.nl/documenten/kamerstukken/2021/07/13/kader-lichte-elektrische-voertuigen, visited 28 September 2021.

Gruber, J., Kihm, A. and Lenz, B. (2014). A new vehicle for urban freight? An ex-ante evaluation of electric cargo bikes in courier services. *Research in Transportation Business and Management, 11*, 53–62.

Gruber, J., and Narayanan, S. (2019). Travel time differences between cargo cycles and cars in commercial transport operations. *Transportation Research Record, 2673*(8), 623–637.

Kearns, K.P. (1992). From comparative advantage to damage control: Clarifying strategic issues using SWOT analysis. *Nonprofit Management and Leadership, 3*(1), 3–22.

Lebeau, P., Macharis, C., Van Mierlo, J. and Janjevic, M. (2018). Improving policy support in city logistics: The contributions of a multi-actor multi-criteria analysis. *Case Studies on Transport Policy, 6*(4), 554–563.

Lenz, B., and Riehle, E. (2013). Bikes for urban freight? Experience in Europe. *Transportation Research Record, 2379*(1), 39–45.

Llorca, C., and Moeckel, R. (2021). Assessment of the potential of cargo bikes and electrification for last-mile parcel delivery by means of simulation of urban freight flows. *European Transport Research Review, 13*(1), 1–14.

Logistiek010 (2021). LEVV als vervanger van de Rotterdamse bestelauto? https://logistiek010.nl/artikel/levv-als-vervanger-van-de-rotterdamse-bestelauto/, visited 20 October 2021.

Melo, S., and Baptista, P. (2017). Evaluating the impacts of using cargo cycles on urban logistics: Integrating traffic, environmental and operational boundaries. *European Transport Research Review, 9*(2), 30.

MensLine (2021). *Cycling – The Exercise for Positive Mental Health.* https://mensline.org.au/mens-mental-health/cycling-the-exercise-for-positive-mental-health/, visited 28 September 2021.

Montreuil, B., Meller, R.D. and Ballot, E. (2010). Towards a Physical Internet: The impact on logistics facilities and material handling systems design and innovation. In Gue, K. (ed.), *Progress in Material Handling Research.* Charlotte, NC: Material Handling Industry of America, pp. 305–327.

Moolenburgh, E.A., van Duin, J.H.R., Balm, S., van Altenburg, M. and Ploos van Amstel, W. (2020). Logistics concepts for light electric freight vehicles: A multiple case study from the Netherlands. *Transportation Research Procedia, 46*, 301–308.

Naumov, V. (2021). Substantiation of loading hub location for electric cargo bikes servicing city areas with restricted traffic. *Energies, 14*, 839.

Nesterova, N., and Quak, H. (2016). A city logistics living lab: A methodological approach. *Transportation Research Procedia, 16*, 403–417.

Oliveira, L.K. de, Morganti, E., Dablanc, L. and Oliveira, R.L.M. de (2017). Analysis of the potential demand of automated delivery stations for e-commerce deliveries in Belo Horizonte, Brazil. *Research in Transportation Economics, 65*, 34–43.

Pedal Me (2021). https://pedalme.co.uk/, visited 17 August 2021.

Ploos van Amstel, W., Balm, S., Tamis, M., Dieker, M., Smit, M., Nijhuis, W. and Englebert, T. (2020). *Zero-Emission Service Logistics in Cities.* Amsterdam University of Applied Sciences.

Ploos van Amstel, W., Balm, S., Tamis, M., Dieker, M., Smit, M., Nijhuis, W. and Englebert, T. (2021). Gas op elektrisch: Servicelogistiek zero emissie de stad in. Publicatiereeks HvA Faculteit Techniek, No. 17. Onderzoeksprogramma Urban Technology, Faculteit Techniek, Hogeschool van Amsterdam.

Ploos van Amstel, W., Balm, S., Warmerdam, J., Boerema, M., Altenburg, M., Rieck, F. and Peters, T. (2018). *City Logistics: Light and Electric: LEFV-LOGIC: Research on Light Electric Freight Vehicles.* Amsterdam University of Applied Sciences, Faculty of Technology.

Schliwa, G., Armitage, R., Aziz, S., Evans, J. and Rhoades, J. (2015). Sustainable city logistics – making cargo cycles viable for urban freight transport. *Research in Transportation Business and Management, 15*, 50–57.

Sheth, M., Butrina, P., Goodchild, A. and McCormack, E. (2019). Measuring delivery route cost trade-offs between electric-assist cargo bicycles and delivery trucks in dense urban areas. *European Transport Research Review, 11*(1), 1–12.

Slob, A.W. (2021). How policymakers can overcome competing values in the pursuit of solutions for societal problems – light electric vehicles in the Dutch mobility sector from an institutional logis perspective. Master's thesis, Utrecht University, Faculty of Geosciences.

Statista (2021). *E-commerce Share of Total Global Retail Sales from 2015 to 2024.* https://www.statista.com/statistics/534123/e-commerce-share-of-retail-sales -worldwide/, visited January 2021.

Tipagornwong, C., and Figliozzi, M. (2014). Analysis of competitiveness of freight tricycle delivery services in urban areas. *Transportation Research Record, 2410*(1), 76–84.

TRELab (2019). *Rome Logistics Living Lab – Cargo Bike.* http://www.trelab.it/2019/04/18/rome-logistics-living-lab-cargo-bike/, visited 13 November 2019.

Van den Band, N., and Roosendaal, B. (2020). Tijd voor DEELLEVV's: een verkennend onderzoek. *Logistiek+, Tijdschrift voor Toegepaste Logistiek*, (10), 110–125.

Van Duin, J.H.R., van den Band, N. and Moolenbergh, E. (2022). Real time learning with light electric freight vehicles for urban freight distribution: A living lab approach. *Logistiek+, Tijdschrift voor Toegepaste Logistiek*, (12), 38–55.

Van Winden, W., and van den Buuse, D. (2017). Smart city pilot projects: Exploring the dimensions and conditions of scaling up. *Journal of Urban Technology, 24*(4), 51–72.

Verlinghieri, E., Itova, I., Collignon, N. and Aldred, R. (2021). The promise of low carbon freight benefits of cargo bikes in London. White paper, Pedal Me, August 2021.

WEF (2020). *The Future of the Last-Mile Ecosystem.* World Economic Forum, Geneva, Switzerland.

Wrighton, S., and Reiter, K. (2016). CycleLogistics – moving Europe forward! *Transportation Research Procedia, 12*, 950–958.

Zhang, L., Matteis, T., Thaller, C. and Liedtke, G. (2018). Simulation-based assessment of cargo bicycle and pick-up point in urban parcel delivery. *Procedia Computer Science, 130*, 18–25.

10. Automated driving on the path to enlightenment?

Maaike Snelder, Gonçalo Homem de Almeida Correia and Bart van Arem

INTRODUCTION

The development of automated cars started about a century ago as a curiosity, almost part of some kind of science fiction depiction of the future. Naturally underlying the innovation was the objective to save time that could be used in more enjoyable ways. At that time improving traffic, safety, and energy consumption were not among the priorities of car makers.

The first radio-controlled driverless car was introduced to the public on the McCook Air Force test base in Dayton, Ohio, USA on 5 August 1921 (Kröger, 2016).

> In the summer of 1925, Houdina's driverless car, called the American Wonder, traveled along Broadway in New York City—trailed by an operator in another vehicle—and down Fifth Avenue through heavy traffic. It turned corners, sped up, slowed down and honked its horn. Unfortunately, the demonstration ended when the American Wonder crashed into another vehicle filled with photographers documenting the event. (Engelking, 2017, p. 1)

This first failure stresses the importance of safety for the people within the vehicle and other road users, which cannot be solved without proper technology. This technology would only become available decades later (US Department of Transportation, 1994).

Technology such as advanced driver assistance systems (ADASs) like blind-spot monitoring, adaptive headlights systems, obstacle and collision warning, lane-keeping support, emergency braking systems, and eco-driving support hit the market for private cars later in the 20th century. In 2014 the deployment rate averaged over 28 European countries was 2.7–12.6 percent for five safety-related ADASs and 23 percent for eco-driving support (Kyriakdis et al., 2015). In that period there was a lot of optimism about automated driving. It was expected that by 2021 the majority of vehicles would drive automatically (Underwood et al., 1991).

From 2010 onward more advanced automated driving projects gained momentum. In 2019 Tesla vehicles equipped with autopilot had already 6.9 percent of the market in the Netherlands (Autoweek, 2021). Waymo, formerly the Google self-driving car project, has introduced a ride-hailing service in Phoenix, Arizona (Waymo, 2021). Their vehicles can drive autonomously on certain roads under certain conditions. For freight transport, Waymo has introduced automated trucks, which promise to provide continuous uninterrupted driving along with flexible scheduling and routing at lower operating costs.

It is important to understand that automation is not a binary property of vehicles. The Society of Automotive Engineers provides a taxonomy with detailed definitions for six levels of driving automation, ranging from no driving automation (level 0) to full driving automation (level 5) (SAE International, 2021). When vehicles are fully automated, a driver is no longer needed. For people without a driver license, that opens a whole new world of opportunity, because all of a sudden they can use automated cars for their trips. Former car drivers become passengers who can do all kinds of activities within a vehicle that they couldn't do before. This innovation in transport might have a huge impact on daily activity and travel patterns. The impact of lower levels of automation will be less disruptive, but still substantial. Each level of automation can be considered an innovation in transport because at every level the automation increases safety and driving comfort and has potential other implications as described in more detail in this chapter. At levels 1–2, the driving automation system provides the driver with longitudinal and lateral control, that is, adaptive cruise control (ACC) as well as lane-keeping and parking support. However, at these levels, the driver is still responsible for monitoring the environment. At level 3, the automated driving system (ADS) monitors the environment and executes driving tasks in certain operational design domains (ODDs), allowing the drivers to avert their attention from driving tasks while being ready to take back control in case of a failure when approaching difficult driving conditions. Level 4 is expected to handle the fail-safe situation autonomously; however, the ODD is still limited. This implies that levels 3–4 might need dedicated infrastructure or roads with other specific infrastructure requirements. Finally, at level 5, the vehicles can drive safely anywhere at any time. However, level 5 vehicles are not to be expected in the near future due to their demanding safety requirements in any ODD (Shladover, 2016).

Besides automation, connectivity plays an important role. When vehicles can communicate with each other (V2V) and with the infrastructure (V2I and I2V), they can drive at much shorter time headways, which increases the road capacity and reduces energy use. With respect to I2V communication, five levels for infrastructure support for automated driving (ISAD) have been recently specified, ranging from A to E (Carreras et al., 2018). E represents no infrastructure support, D represents static digital information like digital map data complemented by physical reference points, C represents dynamic digital information provided to automated vehicles (AVs) in digital form, like dynamic road signs and dynamic information about warnings, incidents, and weather conditions, B represents cooperative perception, which means that the infrastructure is capable of perceiving microscopic traffic situations and of communicating to vehicles (I2V), and A represents the highest infrastructure support level where AVs are guided by the infrastructure to optimize traffic flow by sending out gap and lane change advice messages.

For freight transport truck platooning has gained momentum since 2014. Janssen et al. (2015) wrote a white paper that explained why truck platooning is the future of freight transportation, also emphasizing that companies benefit from lower fuel consumption and improvements in (driver) productivity while society benefits from fewer accidents, less congested roads, and lower carbon emissions. Since then, the automotive industry and numerous start-ups have been involved in many successful real-world trials. Since 2018 the European project called ENSEMBLE has been working on inter-brand platooning technology, which is an important next step for truck platooning. In September 2021 a platoon consisting of seven trucks, one from each leading European manufacturer, is expected to drive from a logistics hub to Barcelona harbor (ENSEMBLE, 2021).

In public transportation, automation technologies were introduced in the early 1980s, with the first automated metro in Europe starting operation in 1983. In 1999 the Rivium ParkShuttle in the Netherlands brought automated public vehicles onto the roads with the very first automated shuttle (Transdev, 2021). Hagenzieker et al. (2020) presented an overview of 118 pilots with automated shuttles, that is, vehicles with predominantly low speeds, low capacities, and short operation routes, across Europe. They conclude that the vast majority of automated bus system pilots operate on an on-demand basis and as an access and egress mode for main facilities and/or public transport lines. Most pilots still have a steward on board, due to legislation and technological challenges as well as passengers requesting them, raising concerns regarding (e.g. economic) efficiency. For a more complete view of the implications of automation for public transport the reader may consult Correia (2021).

Based on the above, it can be concluded that many developments have taken place in the past century in the field of automation of cars, trucks, and public transport. This is especially true for the lower levels of automation in combination with electrification. An important next step, and a potential game changer, is the transfer of control from the driver to the vehicle. However, this is challenging as various circumstances fall outside of the ODD of many partially automated vehicles (Calvert et al., 2020).

With respect to the deployment of vehicles with higher forms of automation there are still many uncertainties, which lead to questions about expected societal impacts and success and failure factors for automated driving. To address these questions, this chapter describes the expected societal impacts of automated cars levels 3, 4, and 5, truck platoons, and automated shuttles, and derives success and failure factors for these innovations in transport based on findings from the STAD (Spatial and Transport Impacts of Automated Driving) research project and associated work by the authors. The chapter starts out from the STAD project because it is one of the few projects about the societal impacts of automated driving. Particular topics that were taken into account were 1) travel and location choice behavior, 2) freight and logistics applications, 3) infrastructure service networks, 4) urban design and traffic safety, 5) spatial structure and economy, 6) integrated model for the impacts of automated driving, and 7) use cases and demonstrators. A complete overview of the project can be found in van Arem (2021). Note that other aspects like legislation and technological developments are also crucial for the success of automated driving, but were outside the scope of the STAD project. Therefore, this chapter describes the impact of automated driving that can be expected once the legislative and technological challenges have been resolved.

SOCIETAL IMPLICATIONS OF AUTOMATED DRIVING

In this chapter, the conceptual ripple model presented by Milakis et al. (2017a) is used to describe the societal implications of automated driving (see Figure 10.1). The ripple model represents a metaphor of how the impacts of driving automation propagate over time from changes in traffic and travel characteristics (first ripple or first-order) to spatial implications such as infrastructure and location choice (second ripple or second-order) and ultimately to economic and societal changes (third ripple or third-order). The ripple model of automated driving should not be taken too strictly, because feedback loops can occur and sometimes the time delay between successive effects in different ripples can be negligible. The general idea is that with an increase in penetration rates of higher levels of AVs the expected societal impacts in all ripples will be higher. The paper of Milakis et al. (2017a) includes a literature review that describes what was known about the impacts mentioned in the ripple model up to 2017. This section enriches these insights with the findings of the STAD research project and associated research work and derives the most important success and failure factors for automated driving based on these findings. In Figure 10.1 the ripple model is reproduced with the identification of where each of the STAD research topics fit. The next subsections will present the results organized per each of the ripple levels (categories marked in bold).

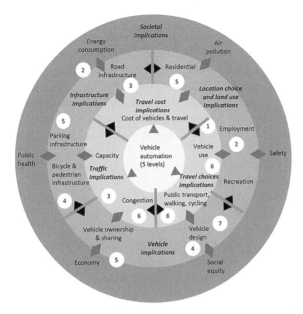

STAD research
1. Travel and location choice behavior
2. Freight & Logistics applications
3. Infrastructure service networks
4. Urban design and traffic safety
5. Spatial structure and economy
6. Integrated model for the impact of
   automated driving
7. Use cases and demonstrators

Figure 10.1   Mapping of STAD research project onto the ripple model of
Milakis et al. (2017a)

First-Order Implications of Automated Driving

Based on their literature review, Milakis et al. (2017a) concluded that the fixed costs of AVs will very likely be higher than for conventional vehicles due to the advanced hardware and software technology involved. The price difference could be gradually reduced to $3000 or even less with mass production and the technological advances of AVs. Regarding the generalized transport costs, these are expected to decrease because of 1) a reduction in costs per kilometer, 2) increased road capacity (> 10 percent when the penetration rate is higher than 40 percent and vehicles are connected), 3) reduced travel times, 4) reduced search time for parking, 5) a potential decrease in the value of time because of more travel comfort, 6) enhanced travel safety, 7) higher travel time reliability, and 8) the possibility to perform activities other than driving, like working, meeting, eating, or sleeping. It was noted that the enhanced road capacity and reduced generalized travel costs might increase vehicle travel demand (3–27 percent), vehicle kilometers driven, and congestion. Below, the findings from the STAD project and related research with respect to the first-order implications of automated driving are summarized, resulting in success and failure factors for automated driving. The focus was on travel choice implications and traffic implications. Travel costs implications have not been studied.

Travel choices implications
The research about travel and location choice behavior aimed to understand the impact of level 5 vehicle automation on activities and travel behavior. Based on focus group research among 27 commuters (Pudăne et al., 2019) and a stated activity-travel survey among 509 commuters in the Netherlands (Pudăne et al., 2021), it was concluded that AVs are expected to allow their users to engage in a broad range of non-driving activities while traveling. Pudăne et al. (2019) classified on-board activities in four quadrants according to their novelty and priority level: 1) current activities with high priority like work, study, eat, apply make-up; 2) new activities with high priority like prepare dinner, wash oneself, brush teeth, do administration; 3) current optional activities like read news, check phone, make phone calls; and 4) new optional activities like exercise, play games, watch movies. This classification is helpful in under-standing the potential re-arrangements of daily activities. It was found that the availability of on-board activities influences the (time-geographic) constraints of daily activities and may lead to complex re-arrangements of daily activity patterns especially for the high-income and higher-educated groups. This re-arrangement of activities may have a substantial impact on trip patterns during the day and therefore on congestion levels. Engaging in any activity during travel worsens congestion, at least when assuming that AVs do not increase bottleneck capacity (Pudăne, 2020). In a parallel study on the value of travel time (VOTT) inside an AV, Correia et al. (2019) ran a stated preference survey comparing a conventional vehicle with an AV with an office interior and an AV with a leisure interior in commuter trips. A sample of travelers in the Netherlands answered the survey, from which it was possible to estimate a lower VOTT in an AV prepared for work when compared to the current VOTT in a conventional car while the leisure vehicle was not perceived as more attractive. The authors attribute this difference to the difference of performing both activities in a car in relation to performing them at the usual place. In fact, the VOTT of working in a car can be demonstrated as being the result of the difference between the experience of working at a "normal" work-place and the experience of working while traveling. Following this logic, the

VOTT of the leisure car is the difference between having leisure at a "normal" place and the experience of having leisure inside a car. It can be concluded that both activities are seen as being very different, with the work vehicle providing a more equivalent experience to current work than the leisure vehicle, which seems to provide an experience that is far from what the respondents considered to be their current leisure experience (Pudãne and Correia, 2020). This research shows that the type of activity can lead to a very different experience, thus adding to the uncertainty of the effects of vehicle automation.

Traffic implications
Snelder et al. (2019) developed a model to explore the modal shift and traffic implications of connected and automated driving (level 3/4 and level 5) and shared mobility. It is an explorative iterative model that uses an elasticity model for destination choice, a multinomial logit model for mode choice, and a network fundamental diagram to assess traffic implications. The model uses literature-based assumptions in combination with findings from the other STAD research projects about the impact of AVs on capacity, VOTT, monetary cost per kilometer, and other user preferences captured by the mode-specific constant as an input. Because the impact of AVs on the costs, VOTT, and mode-specific constants are highly uncertain, a sensitivity analysis was carried out. The model was applied in a case study of the Dutch province of North-Holland, in which the potential impacts of automated and shared vehicles and mitigating interventions were explored. In this case study, four scenarios were explored, in which 100 percent of the vehicles have SAE-level 3/4 or 5 and people have a low or high willingness to share. A 100 percent penetration rate was chosen to get insights into maximum effects. The results show that a shift to automated private cars and automated taxis can be expected. This increases the accessibility of many regions for many people, including for those who are not allowed to drive. In the most extreme scenario, L5-no-sharing, the share of car trips including new modes increases from 41 percent to 68 percent. In the scenario with a high willingness to share (L5-sharing) the share of car trips including new modes increases to 62 percent. The increased mobility has negative effects on congestion. The results of the sensitivity analyses for scenario L5-sharing showed that varying the mode-specific user preferences (i.e. mode-specific constant) that were not explicitly included in the utility function had the largest impact. If the new concepts appear to be less attractive than we assumed, their total modal share might reduce from 62 percent to 44 percent. Charging the ownership costs to the user also makes a large difference. It reduces the modal share of the new modes to 46 percent. Owners of the automated (shared) taxis and vans will likely do this at least to some extent. A strong mix of interventions will be needed to keep delays at the same level as

in the reference scenario. This is especially the case in (very) highly urbanized areas. In other areas, the interventions can be more modest.

Success and failure factors related to the first-order implications of automated driving

Based on the expected first-order implications of automated driving, it can be concluded that the fact that people can spend their time in AVs in a better way is an important success factor for automated driving. A reduction in cost per kilometer adds to the success of AVs, whereas higher ownership costs are a potential failure factor. An increase in capacity, in the case of cooperative AVs, can also be a factor for success. A reduction in public transport trips and walking and cycling and an increase in car trips and vehicle kilometers driven can however be a failure factor, especially if the increase in traffic volumes is larger than the increase in capacity, resulting in more congestion.

Second-Order Implications of Automated Driving

Vehicle automation can have second-order implications on vehicle ownership and vehicle design, location choices and land use, and transport infrastructure. With respect to vehicle ownership, Milakis et al. (2017a) concluded that shared AVs could replace from about 67 percent up to over 90 percent of conventional vehicles, delivering equal mobility levels. Concerning land use, it was concluded that AVs could enhance accessibility citywide, especially in remote rural areas, triggering further urban expansion. AVs could also have a positive impact on the density of economic activity at the center of the cities. With respect to transport infrastructure, the focus has been on required parking spaces. It was concluded that parking demand for AVs could be shifted to peripheral zones. On the one hand, parking demand for shared AVs can be high in city centers, if the vehicle is not allowed to move without people in it. On the other hand, shared AVs could significantly reduce parking space requirements by up to over 90 percent. The overall reduction of the conventional vehicle fleet and parking spaces could vary according to the automated mode (vehicle-sharing, ridesharing, shared electric vehicle), the penetration rate of shared AVs, and the presence or absence of public transport. Below, the findings from the STAD project and related research with respect to second-order implications of automated driving are summarized, resulting in additional success and failure factors for automated driving.

Vehicle implications

According to the ripple model, automated driving may have implications on vehicle ownership and sharing and on the design of vehicles. Ostermeijer et al. (2019) developed a model to explore what the potential implications on vehicle ownership and on car demand are when households own private AVs of level 4/5 that will be parked at locations in the periphery, where parking costs are relatively low. First, they developed an approach to estimate implicit residential parking costs and then examined the effect of these costs on household car ownership. They applied their approach to the four largest metropolitan areas of the Netherlands and found that for city centers, annual residential parking costs are around €1000, or roughly 17 percent of car ownership costs, and are more than double the parking costs in the periphery. The disparity in parking costs explains around 30 percent of the difference in average car ownership rates between these areas and corresponds to price elasticity of car demand of about −0.7. They concluded that when private AVs of level 4/5 can be parked at locations in the periphery, this is expected to increase car demand by around 8 percent in the center and 5 percent in the urban ring, and that there would be no change in the periphery. When AVs are shared and therefore parking costs approach zero, car demand is predicted to increase by around 14 percent in the center, 11 percent in the urban ring, and 5 percent in the periphery.

The design of AVs might affect the crossing behavior of pedestrians and bicycles at intersections. To understand the impact of the design of AVs and urban intersections on crossing behavior, Nuñez Velasco (2021) used a virtual reality platform. The role of several characteristics of AVs, such as their physical appearance, whether there is a driver present in the vehicle, and the presence of external communication interfaces (i.e. screens mounted on AVs to communicate with other road users), were investigated. Concerning crossing intentions of pedestrians, it was concluded that a zebra crossing and larger gap size between the pedestrian and the vehicle increases the pedestrian's intention to cross. In contrast to what was expected, participants intended to cross less often when the speed of the vehicle was lower. Despite that the vehicle type affected the perceived risk, no significant difference was found in the crossing intention. However, pedestrians who did recognize the vehicle as an AV had, overall, lower intentions to cross (Nuñez Velasco et al., 2019). For cyclists, the gap size and the right of way were found to be the primary factors affecting the crossing intentions of the individuals. The vehicle type and vehicle speed did not have a significant effect on the crossing intentions. Cyclists' statements of whether they trusted AVs more or less as compared to conventional vehicles were found to be a stronger predictor of the crossing intentions compared to their trust in AVs by itself. Furthermore, those that reported being low risk-seeking cyclists had a higher intention to adapt their speed than those that reported being high risk-seeking cyclists. Overall, a positive relation was found between cycling speed adaptation and perceived behavioral control, and a negative relation between cycling speed adaptation and perceived risk, when interacting with an AV compared to a conventional vehicle (Nuñez Velasco et al., 2021).

Rad et al. (2020) used models to simulate on a screen different crossing situations in the context of having to catch a train in a serious gaming environment. Respondents needed to observe the situation and choose to cross or to wait for the next gap. In some experimental configurations, the AVs communicated their intention to continue or not to continue their trajectories using lights. The subjects of the experiment were also asked to fill in a questionnaire about usual behavior in traffic, as well as attitudes and risk perceptions toward crossing roads. The results of generalized linear mixed models applied to the data showed that besides the distance from the approaching vehicle and existence of a zebra crossing, pedestrians' crossing decisions are significantly affected by their age, their familiarity with AVs, the communication between the AV and themselves, and whether the approaching vehicle is an AV. Moreover, the introduction of latent factors as explanatory variables into the regression models indicated that individual-specific characteristics like willingness to take risks and violate traffic rules and trust in AVs can have additional explanatory power in the crossing decisions (Rad et al., 2020).

Location choice and land use implications
Legêne et al. (2020) and Hollestelle (2018) studied the spatial impacts of automated driving. Legêne et al. (2020) developed a geospatially disaggregated system dynamics model for the city of Copenhagen and focused on explaining the effects of vehicle automation on the city structure. The analysis led to two distinct scenarios. In one scenario, AVs lead to more vehicle use, which leads to more urban sprawl and more congestion as a consequence. In the other scenario, more shared use of cars leads to less traffic and more open space in the city through converting parking space and road infrastructure. Hollestelle (2018) combined a transportation model with an agent-based location choice model and a research-by-design approach. He concluded that urban centers in particular are vulnerable to induced travel demand, which threatens accessibility levels and facilitates the process of suburbanization. However, when this decrease in accessibility is compensated by increased travel comfort and spatial quality gains, the opposite might occur.

According to the ripple model, automated driving may also have an impact on employment and jobs in different areas. With respect to implications on employment, the focus of the STAD project has been on the impact of truck platooning on employment. With progressing technology, drivers may rest while being in the truck. One step further is that drivers are not required anymore in some of the trucks in a platoon. Hence, platooning technology has a significant impact on the jobs of truck drivers. Driver acceptance of this emerging technology is, therefore, an important factor in the implementation of platooning and, consequently, automated driving in general. Bhoopalam et al. (2021) concluded, based on focus groups, that drivers foresee that platooning will eventually become a reality and that it will have a negative impact on the quality of their work and their job satisfaction.

Infrastructure implications
The introduction of AVs might have an impact on road infrastructure, parking infrastructure, and bicycle and pedestrian infrastructure. With respect to road infrastructure, Madadi (2021) claims that reaching a high market penetration rate of fully automated vehicles will be a gradual process that will take several decades. Thus, for a long time, a heterogeneous mix of traffic with AVs of different automation levels and regular vehicles on the roads will be inevitable. During this transition period with mixed traffic, relying on driving automation technology alone without infrastructure support might compromise the potential safety and efficiency gains of AVs. A proper infrastructure can support AV functionalities, extend their ODD, and improve safety for all road users, while lack of proper infrastructure can negatively influence these factors. Madadi et al. (2021) developed an optimization model that determines which motorways and regional roads in a network can best be upgraded to a so-called connected AV-ready subnetwork where AVs and non-automated vehicles can drive in mixed traffic, dedicated lanes, or dedicated links (i.e. road sections). The optimization is based on a trade-off between infrastructure adjustment costs and network performance benefits because of lower total travel cost, a decrease in total travel time, and a minor increase in total travel distance (Madadi et al., 2020). For the upgrade to an AV-ready subnetwork a multi-stage optimization model was developed that not only determines which roads should be upgraded, but also when they should be upgraded. One of the main conclusions is that different network layouts for accommodating AVs in road networks are relevant for different market penetration rates of AVs. For lower market penetration rates, AV-ready subnetworks, which are suitable for AVs in mixed traffic, appear to be the most efficient configuration. However, starting from an around 30 percent market penetration rate, dedicated AV lanes become relevant, and can efficiently host the AVs. Road types play a crucial role in the choice of network configuration as well. Motorway on-ramps and off-ramps, single-lane roads, and major regional roads that include sections with a single lane have been shown to be appropriate for mixed traffic, while dedicated AV lanes are most suited for motorways. Finally, it was concluded that an effective AV-ready subnetwork including an appropriate selection of links (~20 percent) to be upgraded with infrastructure adjustments to accommodate AVs can deliver a large proportion of the benefits (~70 percent) obtainable from upgrading infrastructure on all links with significantly lower investment cost.

Truck platooning may also have implications on infrastructure investments. Bhoopalam et al. (2018) conducted a literature review about truck platoon planning and the impact of platooning on routing and network design. They concluded that parts of the network may be heavily used by platoons and require infrastructural changes such as reinforcement of roads, new lanes, and additional communication support in tunnels. Also, since the starting locations of trucks influence platooning opportunities, there is an incentive to move facilities such as warehouses and depots closer to each other to achieve economy of agglomeration.

Boersma et al. (2018a) used existing cases of automated driving to derive success and failure factors for automated driving with a specific focus on the implementation of automated shuttles in the Netherlands like the Rivium Parkshuttle in Rotterdam, the WEpod in Ede/Wageningen (Boersma et al., 2018c), and a shuttle in Appelscha (Boersma et al., 2018b), as well as on an automated mini bus in the Oku-Eigenji area in Japan (Boersma et al., 2019). An important conclusion is that the focus of most pilots has been on technical feasibility with short route lengths and low speeds. The pilots showed that it is technically feasible to operate without a steward on board. However, attention should be given to the space needed for the vehicle and remaining space for other road users. Furthermore, it is advisable to control intersections where automated shuttles cross other traffic. In order to make automated shuttles more accessible, future pilots should aim to roll out transit lines throughout larger (and denser) areas where there is actual demand for shuttles. Current legislation in many countries, requiring stewards on board and low operating speeds, is mentioned as an important factor that withholds automated bus systems and shuttles from practical implementation and utilization (Hagenzieker et al., 2020). At the same time, safety and monitoring are crucial for the success of automated driving (Santoni de Sio et al., 2022).

Success and failure factors related to the second-order implications of automated driving

Based on the expected second-order implications of automated driving it can be concluded that the fact that AVs of level 4/5 can park themselves at locations in the periphery, where parking costs are relatively low, is a success factor for automated driving, because it saves money for users and reduces local parking pressure. The fact that AVs enable people to live further away from their work locations is a success factor as well because it enables them to live at more preferred locations. At the same time, both can be a failure factor when they result in more congestion and urban sprawl. More shared use of vehicles is a success factor because it leads to less traffic and better spatial quality in cities. Investments in physical road infrastructure, including AV-ready subnetworks, dedicated lanes, controlled intersections, and zebra crossings, and investments in digital infrastructure, including monitoring, supervision, and even remote control, are other success factors for automated driving. Carefully selecting the right location for automated driving systems, and especially shuttles where there is actual demand for them, is important for the success of these systems. Current legislation in many countries with respect to stewards on board and operating speeds is a failure factor. Finally, drivers' acceptance of truck platooning is an important factor in its implementation and, consequently, automated driving in general.

Third-Order Implications of Automated Driving

The third ripple focuses on the societal implications of AVs. Milakis et al. (2017a) found that fuel savings can be achieved by various longitudinal, lateral (up to 31 percent), and intersection control (up to 45 percent) algorithms and optimization systems for AVs. Vehicle automation can lead to lower emissions of NOx, CO, and CO_2 and advanced driver assistance systems and higher levels of automation (level 3 or higher) can enhance traffic safety. A higher level of automation, cooperation, and penetration rate could lead to higher fuel savings and even lower emissions. The shared use of AVs could further reduce emissions. Impacts on long-term energy consumptions, long-term air pollution, social equity, the economy, and public health were still uncertain in 2017. The research in the STAD project with respect to the third ripple was limited to the impact of truck platooning on fuel cost savings, as explained below.

Energy consumption
Truck platooning has an impact on fuel costs because of short headways between vehicles as well as on associated emissions. Bhoopalam et al. (2020) developed an optimization technology to match trucks into platoons. They showed that when platoons of two or three trucks are formed, > 98 percent of the vehicles can be matched in a platoon, > 89 percent of the kilometers driven can be as part of a platoon, < 3 percent of the vehicles have to take a detour, and 8.8–9.5 percent fuel savings can be achieved.

Success and failure factors related to the third-order implications of automated driving
It can be concluded that fuel savings and lower emissions are a success factor for automated driving. Other third-order implications of automated driving have not been studied extensively. Impacts on safety, social equity, public health, and the economy still need to be studied.

Overview of Success and Failure Factors for Automated Driving

The previous section illustrated that automated driving can have both positive and negative first-, second-, and third-order impacts and that the impacts depend on many different success and failure factors. This section combines

all the insights to give a coherent set of success and failure factors for automated driving (see Figure 10.2). It should be noted that there is a difference
between automated driving technologies that can still progress within our
current mobility system (including the legal and policy context) and automated
driving functions like sustained level 3 and level 4 automation and beyond that
require that fundamental challenges with respect to human factors, technology,
infrastructure, and legislation are addressed at the same time. Yet, this does not
mean that these higher levels of automation are not feasible. The first car was
considered dangerous a century ago, and ACC, automatic braking, and autopilot systems were all considered to face strong barriers (of different kinds)
before their introduction. And still they were introduced.

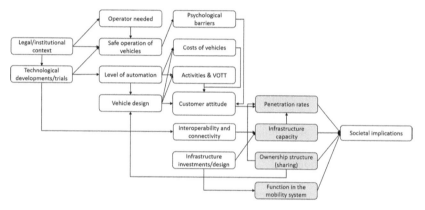

Figure 10.2   *Success and failure factors and relationships for automated
                  driving*

The previous section illustrated that the penetration rates, the infrastructure
capacity, ownership structure (sharing), and the function in the mobility
system determine to a large extent how large and how positive or negative the
societal implications will be. These factors (indicated in gray) depend in their
turn on many other factors as described below.

Penetration rates will increase when customers have a positive attitude
toward AVs and are willing to buy or use (in the case of shared vehicles) the
vehicles. Note that there is not one average customer, but the attitude of customers depends on their socio-demographic and economic characteristics. The
willingness to buy or use an AV depends on the ownership and usage costs of
the vehicles, on the VOTT, and on vehicle design. When people can do more
activities in a vehicle, they will be more inclined to use AVs. The costs of the
vehicles, the VOTT, and the vehicle design depend on the level of automation and technical developments and trials or pilots to test the technological

feasibility of AVs. Customer attitude also depends on psychological barriers to using and interacting with AVs. Safe operation of vehicles, including a trustworthy combination of human/machine-operated driving, monitoring, and service tasks and including safe interaction with pedestrians and cyclists, might reduce these psychological barriers. Legislation to force safe operation and to allow automated driving on the roads is a precondition for the success of automated driving.

The impact of automation on *infrastructure capacity* is also important for the success of automated driving. As explained before, the capacity might slightly reduce when AVs are not connected to each other and/or the infrastructure. However, when they are connected, capacity is expected to increase, which will have a positive impact on travel times. Reduced travel times will encourage more people to buy or use AVs, which results in higher penetration rates. Investments in physical and digital infrastructure can thus increase the success of automated driving. STAD research showed that it is not necessary to invest in all roads at once. Investing in a carefully selected selection of roads can already result in large benefits.

The *ownership structure* of AVs, or the extent to which people are willing to share vehicles, has a huge impact on the size of the vehicle fleet and required space for parking. If people are not only willing to share vehicles, but are willing to share rides, this is expected to lead to a substantial reduction in congestion.

Finally, the *function of AVs in the mobility system* is important for the success of automated driving. For example, automated shuttles for first and last mile transport compete with other modes of transport like (shared) bikes and public transport. Therefore, it is important to carefully select locations where demand is expected to be high enough to operate with a positive benefit–cost ratio. Another important notion is that when AVs become "too" attractive, they might cause a modal shift from walking, cycling, and public transport to AVs, which potentially has negative impacts on congestion, livability, and public health. This calls for a clear vision on the role of AVs within the mobility system, taking societal goals into account.

It is argued by business experts that AVs are in the "trough of disillusionment" according to Gartner's hype cycle (Hull, 2021). According to this hype cycle (a methodology developed and used by consultancy firm Gartner to explain how a technology or application evolves over time) the AV technology has passed the periods "Technology Trigger" and "Peak of Inflated Expectations" and is now in this trough. The question is whether this technology will get past the trough and start on its way on the "Slope of Enlightenment" toward the "Plateau of Productivity". Innovation theories (see Part I of this book) and the results of this chapter may give some food for thought on this question. First, as this chapter shows, it is clear that realizing an AVs system involves the initiatives of, and adaptations by, many actors, like firms, users, institutions, and governments. The AV innovation, therefore, fits perfectly one of the conclusions of Greenacre's (2012) review on modern innovation theories: "innovation is a dynamic, systemic process, arising out of the interplay between actors and institutions, and involving both knowledge flows and market interactions in a context of inherent uncertainties." So, evolving in a context of uncertainty is inherent to all innovations, but perhaps too much uncertainty at a certain point in time slows down the innovation process so that it falls into a trough. This chapter clearly conveys the current broad level of knowledge uncertainty related to AVs. There is uncertainty in the potential social implications of the technology, in user acceptance, in profitability for firms producing or using the technology, and so forth. Second, to speculate on passing through the trough to the Slope of Enlightenment the literature on the dynamics of transitions in socio-technic systems may be helpful. In a paper on the transition pathways from horse-drawn carriages to automobiles (1860–1930) in the United States, Geels (2005) showed that particular niches played a crucial role in the final wider adoption of the gasoline human-driven automobiles. So, to speculate if the AV technology will get past the trough, it is perhaps best to analyze in the near future if there is growth of this technology in a particular niche (or in niches), such as in automated shuttle buses applications.

CONCLUSIONS AND RECOMMENDATIONS

In this chapter the societal impacts of automated cars (level 3, 4, and 5), truck platoons, and automated shuttles have been discussed based on the ripple model (Milakis, 2017b) and these findings have been used to derive success and failure factors for automated driving. First- and second-order implications have been extensively studied. More research however is needed to fully understand the third-order implications of automated driving with respect to air pollution, safety, social equity, the economy, and public health.

With respect to methods and models it can be concluded that the "ripple model" of automated driving (Milakis, 2017b) has proven to be a useful model to analyze the societal impacts of automated driving. Current "classic" tools and methods have been considered too rigid and unsuitable to assess the impacts of automated driving given all the interactions in the ripple model. Several new approaches are emerging, such as the ones that have been developed within the STAD research project. Virtual reality and simulation models have been developed to analyze interactions of AVs with pedestrians and cyclists and to study changes in activity and travel patterns. Exploratory traffic and transport models have been developed to assess the impact of automated driving on modal split and congestion, taking uncertainties that are related to the introduction of AVs into account. Truck platoon planning algorithms have been developed to assess the impact of truck platooning. An optimization model has been developed for adaptive planning of road networks for automated driving. Finally, spatial models have been developed to assess the spatial implications of automated driving. These developments are an important step toward a brand-new generation of models and tools that are needed to assess the societal implications of automated driving. Besides new models, also new data collection methods are required as input to the models and to calibrate and validate them.

This chapter showed that in the short term "autopilot" systems can increase traffic safety and they have an effect on VOTT and can therefore cause a modal shift from walking, cycling, and public transport to the car. Investments in physical and digital infrastructure for automated driving quickly pay off. Upgrading 20 percent of the motorways and regional roads will allow AVs to drive 70 percent of the kilometers in automatic mode (Madadi et al., 2021). At low penetration rates, investments in AV-ready subnetworks for mixed traffic are recommended. At higher penetration rates (> 30 percent), dedicated lanes for AVs can yield additional benefits. For freight transport, truck platooning is expected to bring benefits for companies from lower fuel consumption and improvements in (driver) productivity while providing society benefits from fewer accidents, less congested roads, and lower carbon emissions. STAD research showed that for platoons of two trucks, the energy savings and improvement of the traffic flow outweigh the time needed to form the platoon. Automatic shuttles are seen as a promising solution for first and last mile transport. They can operate safely at low speeds on many roads. The behavior of the shuttles should be adapted to the expectations of cyclists and pedestrians. When crossing the road, pedestrians pay attention to speed, zebra crossings, and gap size between the pedestrian and the vehicle. The vehicle type and level of automation appeared to be less relevant. Communication of automatic vehicles with cyclists and pedestrians can make crossing easier.

In the longer term, automated driving can lead to changes in activity planning, travel patterns, and destination choice. When level 4/5 shared vehicles hit the market, less parking space and more drop on/drop off places will be needed. The freed-up parking space would lead to an increase in prosperity for non-car users because this space can be used for other purposes.

The societal impacts of automated driving are highly uncertain and will depend to a large extent on the penetration rates of AVs, infrastructure capacity, ownership structure (sharing), and the function of the AVs in the mobility system. Legislation and technological developments and trials/pilots are a precondition for the success of automated driving.

Overall, it can be concluded that it is still uncertain whether or not the higher levels of AVs will reach the "Slope of Enlightenment". Fundamental challenges with respect to human factors, technology, infrastructure, and legislation need to be addressed at the same time. The fact that automated driving can contribute to many societal goals increases the chance that this technology will become a success. However, based on the findings of this chapter it can also be concluded that they are not a "silver bullet" to create accessible and livable cities and regions. Therefore, it is recommended to focus on all travel modes including cycling and public transport and to invest in automation, connectivity, sharing, and electrification. A multimodal vision on future mobility systems along with quadruple helix stakeholder engagement will be needed to decide where and when to invest in different solutions.

REFERENCES

Autoweek (2021). Verkoopcijfers Tesla. https://www.autoweek.nl/verkoopcijfers/tesla/.

Bhoopalam, A., Agatz, N., and Zuidwijk, R. (2018). Planning of truck platoons: A literature review and directions for future research. *Transportation Research Part B: Methodological, 107*, 212–228.

Bhoopalam, A., Agatz, N., and Zuidwijk, R. (2020). Spatial and temporal synchronization of truck platoons. *ERIM Report Series Research in Management Erasmus Research Institute of Management.*

Bhoopalam, A., van den Berg, R., Agatz, N., and Chorus, C. (2021). The long road to automated trucking: Insights from driver focus groups. *ERIM Report Series Research in Management Erasmus Research Institute of Management.*

Boersma, R., Mica, D., van Arem, B., and Rieck, F. (2018a). Driverless electric vehicles at Businessparkt Rivium near Rotterdam (the Netherlands): From operation on dedicated track since 2005 to public roads in 2020. Paper presented at the 31st International Electric Vehicles Symposium and Exhibition and International Electric Vehicle Technology Conference, Kobe, Japan, 1–3 October.

Boersma, R., van Arem, B., and Rieck, F. (2018b). Application of driverless electric automated shuttles for public transport in villages: The case of Appelscha. *World Electric Vehicle Journal, 9*(1), 15. https://doi.org/10.3390/wevj9010015.

Boersma, R., van Arem, B., and Rieck, F. (2018c). *Casestudy WEpod: Een onderzoek naar de inzet van automatisch vervoer in Ede/Wageningen.* Delft/Rotterdam: Spatial and Transport Impacts of Automated Driving (STAD).

Boersma, R., van Arem, B., and Rieck, F. (2019). Assessment of automated mini-bus operation in the Oku-Eigenji area in Japan. Paper presented at the 13th Intelligent Transport Systems European Congress, Brainport, Eindhoven, the Netherlands, 3–6 June.

Calvert, S., van Arem, B., Heikoop, D., Hagenzieker, M., Mecacci, G., and Santoni de Sio, F. (2020). Gaps in the control of automated vehicles on roads. *IEEE Intelligent Transportation Systems Magazine, 13*(4), 146–153.

Carreras, A., Daura, X., Erhart, J., and Ruehrup, S. (2018). Road infrastructure support levels for automated driving. Paper presented at the 25th ITS World Congress, Copenhagen, Denmark, 17–21 September.

Correia, G. (2021). Planning for public transport with automated vehicles. In R. Vickerman (ed.), *International Encyclopedia of Transportation*, 198–205. Elsevier. https://doi.org/10.1016/B978-0-08-102671-7.10640-2.

Correia, G., Looff, E., van Cranenburgh, S., Snelder, M., and van Aerm, B. (2019). On the impact of vehicle automation on the value of travel time while performing work and leisure activities in a car: Theoretical insights and results from a stated preference survey. *Transportation Research Part A: Policy and Practice, 119*, 359–382. https://doi.org/10.1016/j.tra.2018.11.016.

Engelking, C. (2017). The "driverless" car era began more than 90 Years ago. *Discover magazine*, 13 December. https://www.discovermagazine.com/technology/the-driverless-car-era-began-more-than-90-years-ago.

ENSEMBLE (2021). Platooning together. https://platooningensemble.eu/.

Geels, F.W. (2005). The dynamics of transitions in socio-technical systems: A multi-level analysis of the transition pathway from horse-drawn carriages to automobiles (1860–1930). *Technology Analysis and Strategic Management, 17*(4). https://doi.org/10.1080/09537320500357319.

Greenacre, P. (2012). Innovation theory: A review of the literature. ICEPT Working Paper ICEPT/WP/2012/011.

Hagenzieker, M., Boersma, R., Nuñez Velasco, J., Öztürker, M., Zubin, I., and Heikoop, D. (2020). *Automated Buses in Europe: An Inventory of Pilots, version: 1.0.* TU Delft.

Hollestelle, M. (2018). Automated driving: Driving urban development? An integrated modelling and research-by-design approach on the spatial impacts of automated driving. Master's thesis, Delf University of Technology.

Hull, D. (2021). Hyperdrive daily: Self-driving's trough of disillusionment continues. *Bloomberg.com*, 14 April. https://www.bloomberg.com/news/newsletters/2021-04-14/hyperdrive-daily-self-driving-s-trough-of-disillusionment-continues.

Janssen, G., Zwijnenberg, J., Blankers, I., and de Kruijff, J. (2015). Truck platooning: Driving the future of transportation. White paper, TNO Voorwaarden, Delft, the Netherlands.

Kröger, F. (2016). Automated driving in its social, historical and cultural contexts. In M. Maurer, C. Gerdes, B. Lenz, and H. Winner (eds.) *Autonomous Driving: Technical, Legal and Social Aspects*, 41–68. Springer. https://doi.org/10.1007/978-3-662-48847-8.

Kyriakdis, M., Van de Weijer, C., van Arem, B., and Happee, R. (2015). The deployment of advanced driver assistance systems in Europe. *SSRN Electronic Journal*. https://doi.org/10.2139/ssrn.2559034.

Legêne, M., Auping, W., Correia, G., and van Arem, B. (2020). Spatial impact of automated driving in urban areas. *Journal of Simulation, 14*(4), 295–303. https://doi.org/10.1080/17477778.2020.1806747.

Madadi, B. (2021). Design and optimization of road networks for automated vehicles. Doctoral thesis, Delft University of Technology. https://doi.org/10.4233/uuid:c21d4943-b848-4e77-b5b8-a5423f751dbd.

Madadi, B., van Nes, R., Snelder, M., and van Arem, B. (2020). A bi-level model to optimize road networks for a mixture of manual and automated driving: An evolutionary local search algorithm. *Computer-Aided Civil and Infrastructure Engineering, 35*(1), 80–96.

Madadi, B., van Nes, R., Snelder, M., and van Arem, B. (2021). Optimizing road networks for automated vehicles with dedicated links, dedicated lanes, and mixed-traffic subnetworks. *Journal of Advanced Transportation, 2021*, 8853583. https://doi.org/10.1155/2021/8853583.

Milakis, D., Snelder, M., van Arem, B., van Wee, B., and Homen de Almeida Correia, G. (2017a). Development and transport implications of automated vehicles in the Netherlands: Scenarios for 2030 and 2050. *European Journal of Transport and Infrastructure Research, 17*(1), 63–85. https://doi.org/10.18757/ejtir.2017.17.1.3180.

Milakis, D., van Arem, B., and van Wee, B. (2017b). Policy and society related implications of automated driving: A review of literature and directions for future research. *Journal of Intelligent Transportation Systems, 27*(4), 324–348. https://doi.org/10.1080/15472450.2017.1291351.

Nuñez Velasco, J. (2021). Should I stop or should I cross, interactions between vulnerable road users and automated vehicles. Doctoral thesis, Delft University of Technology. https://doi.org/10.4233/uuid:f9c3ef7d-66df-4f59-8eae-28cfb3b4499e.

Nuñez Velasco, J., de Vries, A., Farah, H., van Arem, B., and Hagenzieker, M. (2021). Cyclists' crossing intentions when interacting with automated vehicles: A virtual reality study. *Information, 12*(1), 7.

Nuñez Velasco, J., Farah, H., van Arem, B., and Hagenzieker, M. (2019). Studying pedestrians' crossing behavior when interacting with automated vehicles using virtual reality. *Transportation Research Part F: Traffic Psychology and Behaviour, 66*, 1–14.

Ostermeijer, F., Koster, H., and van Ommeren, J. (2019). Residential parking costs and car ownership: Implications for parking policy and automoated vehicles. *Regional Science and Urban Economics, 77*, 276–288. https://doi.org/10.1016/j.regsciurbeco.2019.05.005.

Pudăne, B. (2020). Departure time choice and bottleneck congestion with automated vehicles: Role of on-board activities. *European Journal of Transport and Infrastructure Research, 20*(4), 306–334. https://doi.org/10.18757/ejtir.2020.20.4.4801.

Pudăne, B., and Correia, G. (2020). On the impact of vehicle automation on the value of travel time while performing work and leisure activities in a car: Theoretical insights and results from a stated preference survey—a comment. *Transportation Research Part A: Policy and Practice, 132*, 324–328. https://doi.org/10.1016/j.tra.2019.11.019.

Pudăne, B., Rataj, M., Molin, E., Mouter, N., van Cranenburgh, S., and Chorus, c. (2019). How will automated vehicles shape users' daily activities? Insights from focus groups with commuters in the Netherlands. *Transportation Research Part D: Transport and Environment, 71*, 222–235. https://doi.org/10.1016/j.trd.2018.11.014.

Pudăne, B., van Cranenburgh, S., and Chorus, C. (2021). A day in the life with an automated vehicle: Empirical analysis of data from an interactive stated activity-travel survey. *Journal of Choice Modelling, 39*, 100286. https://doi.org/10.1016/j.jocm.2021.100286.

Rad, S., Correia, G., and Hagenzieker, M. (2020). Pedestrians' road crossing behaviour in front of automated vehicles: Results from a pedestrian simulation experiment using agent-based modelling. *Transportation Research Part F: Traffic Psychology and Behaviour, 69*, 101–119. https://doi.org/10.1016/j.trf.2020.01.014.

SAE International (2021). *Taxonomy and Definitions for Terms Related to Driving Automation Systems for On-Road Motor Vehicles.*

Santoni de Sio, F., Mecacci, G., Calvert S., Heikoop, D., Hagenzieker, M., and van Arem, B. (2022). Realising meaningful human control over automated driving systems: A multidisciplinary approach. *Minds & Machines*. https://doi.org/10.1007/s11023-022-09608-8.

Shladover, S. (2016). The truth about "self-driving" cars. *Scientific American, 314*, 52–57. https://doi.org/10.1038/scientificamerican0616-52.

Snelder, M., Wilmink, I., van der Gun, J., Bergveld, H., Hoseini, P., and van Arem, B. (2019). Mobility impacts of automated driving and shared mobility: Explorative model and case study of the province of north Holland. *European Journal of Transport and Infrastructure Research, 19*(4). https://doi.org/10.18757/ejtir.2019.19.4.4282.

Transdev (2021). ParkShuttle Riviium. https://www.transdev.nl/nl/onze-routes/vervoersgebieden/parkshuttle-rivium.

Underwood, S., Chen, K., and Erwin, R. (1991). Future of intelligent vehicle-highway systems: A delphi forcast of markets and sociotechnological determinants. *Transportation Research Record, 1305*, 291–304.

US Department of Transportation (1994). *Precursor Systems Analyses of Automated Highway Systems—AHS Comparable Systems Analysis*. Federal Highway Administration Publication No. FHWA-RD-95-120.

van Arem, B. (2021). About STAD. www.stad.tudelft.nl.

Waymo (2021). Company. https://waymo.com/company/#story.

11. Assessing policies to scale up carsharing[1]

Karla Münzel, Marlous Arentshorst, Wouter Boon and Koen Frenken

1.   INTRODUCTION

Our current transportation system is heavily based on the use of private cars. This leads to a range of problems including the emission of greenhouse gases, air and noise pollutants, the depletion of resources, congestion, and inefficient land use (European Environment Agency (EEA), 2018; Eurostat, 2016). It is commonly recognized that the unsustainability of the transport system cannot be solved by technological innovations alone. Instead, we also need to change the way we use the different mobility options available (OECD and ITF, 2017).

Carsharing, a service where consumers share access to cars, is a mobility innovation that is based on the use of underutilized assets and on access instead of ownership. Carsharing can be an efficient way for consumers to access a car when needing one without the costs and the hassle of owning one. Carsharing is found to have a positive influence on multiple urban problems, such as reducing the number of cars and parking spots needed, the number of kilometers driven by users, emissions, and congestion, as well as increasing access for underserved groups (Chen and Kockelman, 2016; Giesel and Nobis, 2016; Nijland and van Meerkerk, 2017; Schreier et al., 2018). Carsharing can thus have a positive societal impact through acting as a means in achieving multiple societal goals, such as reducing emissions, improving livability in cities, and increasing equitable access to mobility. Because of this potential contribution to societal goals, policy makers are interested in scaling up carsharing and learning about supportive policy measures. Carsharing is an example of a socio-institutional innovation involving new business models, new user practices, and new government policies, as part of a transition towards a more sustainable transport system (Pel, Chapter 2 in this volume).

While carsharing schemes date back at least to the 1980s in Western Europe, they only play a minor role in present mobility systems. Most people favor ownership of (one or more) cars to cover their individual transport needs, which means that the car regime is still dominated by private ownership. The notion of regime, in short, refers to "established practices and associated rules" that apply in a particular socio-technical system (Geels, 2011; Pel, Chapter 2 in this volume). In the case of the mobility system, the current regime is characterized by private car ownership, the practice of car commuting, the car as a status symbol, and the supporting infrastructure of roads, parking lots, and facilities accessible by car (Meelen et al., 2019). In this respect, carsharing is still a niche product serving the needs of particular users, mostly inhabitants of large cities with higher education and higher incomes and who are often environmentally motivated (Burkhardt and Millard-Ball, 2006; Dill et al., 2016; Namazu and Dowlatabadi, 2018). The niche is only to a limited extent "protected" by municipalities through support measures, primarily by providing cheap or free parking spaces for shared cars.

While carsharing has been extensively studied empirically (for a review, see Münzel et al., 2019), scant attention has been paid to the question of which policies governments could adopt to further promote carsharing. Exceptions are the studies of Shaheen et al. (2004) and Enoch and Taylor (2006), who presented early reviews of support measures for carsharing, while Akyelken et al. (2018) recently showed empirical evidence of the need for policy measures. Other articles reported on policies supporting carsharing only as a sidenote (e.g. Millard-Ball et al., 2006; Prettenthaler and Steininger, 1999; Shaheen et al., 2006). Because of the potential benefits carsharing can offer, policy makers, businesses, and environmental organizations alike aim to upscale carsharing, which supports the transition to a more sustainable system based on access to, instead of ownership of, mobility. Public policy can – among others – contribute to such transition and act as a success factor in scaling up carsharing. It has remained unclear, however, which policy measures can upscale carsharing adoption and which policy measures are unsupportive or even act as failure factors for the scaling up of carsharing. Here, we provide a review of measures recommended and validate their effectiveness and feasibility with multiple stakeholders for the case of the Netherlands. We further identify which measures are perceived as not achievable or controversial, what barriers limit the implementation of measures, and which roles need to be taken up by different stakeholders.

2. CONTEXT – CARSHARING IN THE NETHERLANDS

An interesting, innovative, albeit not successful project in Amsterdam can be considered the forefather of modern carsharing. In 1972 the Witkar ("white car") project was launched, featuring small electric cars that could be taken instantaneously from charging stations in the city center on one-way trips.

The idea of small shared cars as a sustainable alternative to the private car was born, although this particular project was ahead of its time and failed a few years later, partly because it lacked municipal support (KiM Netherlands Institute for Transport Policy Analysis, 2015; Nijland and van Meerkerk, 2017). The first successful carsharing organizations started in the late 1980s in Switzerland and Germany and offered cars in a business-to-consumer (B2C) roundtrip business model, where the organizations each owned a fleet of cars that could be booked per hour or per day and taken from specific parking spots or stations and had to be returned to the same stations (Münzel et al., 2018; Truffer, 2003). This type of organization also arrived in the Netherlands at the beginning of the 1990s, and carsharing programs were stimulated by the Ministry for Transport as it considered that carsharing could help achieve the goals of reducing car use and vehicle emissions (Ministerie van Verkeer en Waterstaat, 1988). In 1995 the government together with a number of mobility providers also supported the establishment of the "Stichting voor Gedeeld Autogebruik" (Foundation for Shared Car Use) to develop concepts for accessing a car without owning one (Enoch and Taylor, 2006). B2C roundtrip carsharing grew slowly but continuously, provided by a handful of organizations that operated nationally or only locally. In 2010 another type of carsharing was introduced in the Netherlands where the cars available to be rented belong to private car owners and the organization offers only the platform and insurance. This peer-to-peer (P2P) business model was able to achieve quick growth since private car owners have basically zero marginal costs in putting up their car on the platform (Meelen et al., 2019). A third type of carsharing that has been offered since 2015 in Amsterdam only is B2C free-floating carsharing, where cars can be found, picked up, and left anywhere in the operating area without having to return it to the same station.

Taken together these three forms (B2C roundtrip, P2P, B2C free-floating) amounted to around 41,000 shared cars and around 400,000 users of carsharing services in the Netherlands in spring 2018. The more recent data on 2020 suggest that this number has further increased to 65,000, mainly due to the increase in P2P sharing (CROW, 2021). Growth in carsharing supply is mainly taking place in the largest cities and can be observed to be most strong for the P2P carsharing type. Figure 11.1 shows the growth of carsharing supply in the Netherlands up until 2018, for which public data are available. These numbers should be regarded in relation to the 8.4 million cars currently on Dutch roads and 11.2 million driver's license holders (CBS, 2019a, 2019b). Carsharing in the Netherlands is thus still a niche market and users can be seen as belonging to the early adopter category (Rogers, 2003), and comparable to other West European countries (Münzel et al., 2019).

In 2015 a Green Deal between governmental authorities, companies, and environmental organizations was set up with the aim to stimulate the scaling

Number of shared cars

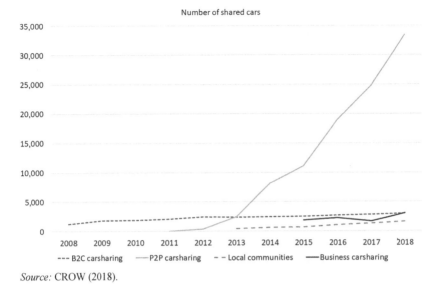

Source: CROW (2018).

Figure 11.1   The growth in the supply of shared cars in the Netherlands

up of carsharing and reach 100,000 carsharing cars by 2018 (Rijksoverheid, 2015), and it was renewed in 2018 with the ambition to reach 700,000 carsharing users by 2021 (Rijksoverheid, 2018). The Deal has the goal to encourage companies, governments, and citizens to meet their mobility needs in a way that makes maximum use of the opportunities that carsharing concepts offer and to reduce the number of parking spaces, making more space available for greenery, recreation, and clean mobility modalities. Two publications recommending policy measures followed from working groups of Green Deal participants during the first Green Deal phase. They were both called "Rode Loper" (red carpet) and intend to give recommendations to local (Autodelen.info, 2018, De Rode Loper voor autodelen) and national authorities (Autodelen.info, 2019; De Rode Loper voor de Rijksoverheid) on how to produce regulations and policy conditions that stimulate the scaling up of carsharing. They cover the topics of removing barriers that are preventing carsharing organizations from scaling up their services, like improving processes; promotion and communication campaigns towards consumers and companies; decreasing the attractiveness of private car ownership; and parking policies. They also give suggestions for monitoring and research on the developments of the market. At present mostly the largest cities of the country have dedicated policies to support carsharing and make it an important part of the mobility system. Two examples with dedicated policy plans for the support of carsharing are the cities of Utrecht and Amsterdam, which also have the largest supply of shared cars per capita (Münzel et al., 2019). Both cities made plans to formulate a vision on carsharing seen in the wider domain of urban planning and climate action, incorporating the integration of carsharing in new developments, the stimulation of innovative initiatives, and measures to improve communication with carsharing organizations and inhabitants (Gemeente Amsterdam, 2019; Gemeente Utrecht, 2015). These two publications developed by Green Deal participants as well as the plans published by the cities signify new initiatives to implement policy measures that support the scaling up of carsharing. These recent activities by a specific group of actors (Green Deal participants) raise questions on how a larger circle of stakeholders evaluates them and how the measures can be shaped and implemented on a larger scale in practice.

3.   METHOD

To gain insight into and understanding of measures to make carsharing an important part of the mobility system – their perceived impact, feasibility, and related barriers – we identified measures from the literature and policy reports and presented a shortlist of these measures for reflection by various relevant stakeholders during a workshop. Together with the workshop participants, we aimed to identify those measures perceived to be the most impacting and feasible and those perceived as not achievable or controversial, including barriers that limit implementation, and the envisioned roles of different stakeholders to realize the measures.

We started with a *review of the literature* to identify measures for introducing and scaling up carsharing by governmental and market actors. We consulted scientific articles, international and national policy reports that described policy measures for scaling up carsharing, and articles and reports that described barriers to the scaling up of carsharing, including the proposed solutions to overcome these barriers. We searched using Google and Google Scholar with the keywords "carsharing AND policy/policies" and "carsharing AND support", as well as searching on websites of carsharing associations (in the Netherlands, Germany, United Kingdom, Belgium), for documents on policies and support measures. Furthermore, through snowballing, we followed relevant publications in the reference lists. In total ten articles and eight reports were identified that described measures. The measures were subsequently categorized based on the sector it was envisioned would implement, or was already implementing, the measures (government or automotive sector). Based on the categories identified, a shortlist of four categories with 42 measures to introduce and scale up carsharing was made, which was subsequently validated by both us and an independent carsharing policy expert. As a result of

Table 11.1   *Expertise and working field of consulted experts*

Position	Actor group
Workshop	
Senior advisor	Government (national level, infrastructure)
Department head	Government (national level, infrastructure)
Policy advisor	Government (national level, economy/climate)
Senior researcher	Government (national level, planning agency)
Policy officer	Government (provincial level)
Policymaking intern	Government (provincial level)
Director	Automotive sector (carsharing)
Director	Automotive sector (carsharing)
Communication manager	Automotive sector (carsharing)
Product manager	Automotive sector (rental, carsharing)
Communication and corporate social responsibility manager	Automotive sector (leasing)
Product marketing manager	Automotive sector (leasing)
Section manager	Automotive sector (industry association)
Program leader	Environmental organization
Advisor/architect	Consultancy (planning, citizen involvement)
Interviews	
Senior policy officer	Government (Dutch municipality network)
Project manager and policy advisor	Government (local level)
Transportation planner	Government (local level)
Senior advisor public affairs	Consumers (touring club)
Consultant	Consumer network (private carsharing)

the input obtained, eventually, 21 measures in four categories were included on the shortlist. The reviewed and validated measures are presented in Section 4.

To gain an understanding of the perceived impact and feasibility of the identified measures in the context of the Netherlands, we presented the short-list for reflection by various relevant stakeholders during a workshop. Experts from governmental institutions (local, regional, and national level) and the automotive sector (including carsharing, dealer association, rental, and leasing organizations) as well as knowledge experts (from universities and consultancies, environmental organizations) were invited to discuss measures to upscale carsharing. In total 15 experts participated in the workshop (Table 11.1).

Workshop participants were divided into four groups of three to four participants based on their working field in order to ensure a variety of actor groups.

The workshop consisted of three stages. It started with an introduction of the project, after which, as a first stage, the participants engaged in a discussion, led by a facilitator, about which measures were perceived as needed to make carsharing part of the Dutch mobility system. The shortlist of 21 measures, divided into the four categories, was presented to the participants. As part of the discussion, participants were asked to place the identified measures in a timeline that contained a "main road" to visualize those measures perceived as necessary, and a "side road" to visualize those measures perceived as less important or slowly contributing to the aim of scaling up carsharing. With this, insight was obtained on the order, the envisioned effect, and the potential coherence of different measures. Participants were also asked to identify and discuss measures that were missing and to remove those measures that were considered unrealistic, not desirable, or not contributing to the aim of scaling up carsharing. As a conclusion to the first stage, all the groups presented their outcomes. In stage two, participants were asked to prioritize three measures taken from the previous exercise that they perceived as most important, and these were explored in depth with a focus on barriers related to the measures, solutions to overcome these, and the identification of actor(s) perceived as initiators and executors of the measure. As a third step of the workshop, the four groups presented their outcomes to each other and participants were invited to contribute to the outcomes of other groups by going into dialogue with each other. Potential options for future collaboration were explored and questions raised about the barriers or tasks of each other's organizations.

To ensure that all key measures were identified and addressed, five organizations who were unable to join the workshop, but whose input was considered relevant and necessary, reflected upon the results of the workshop in separate feedback interviews (see Table 11.1). A summary of the workshop discussion was presented together with the shortlist of discussed measures. The interviewees were asked to reflect on the measures identified as desirable, undesirable, and controversial and were asked to state their perspectives on these, as well as add measures they thought to be important. In addition, barriers for taking on new roles and implementing measures as an organization were discussed in depth.

Data Analysis

Notes were made during and after the workshop by the facilitators for further analysis and a summary of the outcomes of the workshop was sent to the participants for checking. The workshop notes, as well as the additional insights gained from the feedback interviews, were included in the analysis dataset. Using thematic and open coding, we identified, coded, described, and categorized topics in the data obtained. This resulted in the identification of different aggregation levels of the scaling-up challenge of carsharing, which, in turn, formed the basis for further interpretation and definition of the articulated barriers and solutions to translate these barriers into opportunities for policy makers, innovators, and others to support and facilitate the scaling up of carsharing to become part of the mobility system.

4.   POLICY MEASURES

Carsharing can contribute to reaching multiple societal goals, like decreasing emissions, increasing livability, and increasing equity, and the scaling up of carsharing is therefore seen as a valuable goal by policy makers. Various success and failure factors for the scaling up of carsharing and its contribution to reaching societal goals have been identified in earlier studies. There is a need to identify the most promising measures that enable the scaling up of carsharing and relate the measures to reaching societal goals.

Previous literature on carsharing has reviewed measures to support carsharing directly or indirectly. The first type of policy supports the niche of carsharing while the latter type of policy weakens the dominant regime of private car ownership.

We first treat *niche-supporting measures*. The measures help the niche practice of carsharing develop and expand through protecting it, directly helping it expand, or taking away barriers. Most measures discussed are (to be) taken by local authorities, while some are (to be) applied at a higher level or by other stakeholders. The literature showcases examples of measures being taken since the 1990s in North America and Europe.

Parking: The review of the literature identifies that niche support through measures on the topic of parking are most prominent. Measures on this topic include the provision of parking spots for free or for reasonable prices, in attractive locations (close to transit hubs, preferably on-street for increased accessibility and visibility and in front of public land instead of private property for increased acceptance), and without long bureaucratic processes (Akyelken et al., 2018; Enoch and Taylor, 2006; Kent and Dowling, 2016; KiM Netherlands Institute for Transport Policy Analysis, 2015; Le Vine, 2012; Loose, 2009c; Millard-Ball et al., 2006; Prettenthaler and Steininger, 1999; Shaheen et al., 2006, 2004; Stars project, 2019; Steininger et al., 1996; Vanhee, 2010), as well as measures stimulating developers to include carsharing in the plans for new buildings or areas (Akyelken et al., 2018; Enoch and Taylor, 2006; Loose, 2009c; Millard-Ball et al., 2006; Shaheen et al., 2006; Stars project, 2019; Vanhee, 2010).

Start-up support: Another measure supporting the carsharing niche is direct help in the start-up phase of a carsharing organization through, for example, start-up grants that help overcome the high initial costs of setting up a carsharing service or organizational help for initiatives (Enoch and Taylor, 2006; Shaheen et al., 2006, 2004). A more indirect but powerful support measure municipalities can take in the emergence phase is using the carsharing service themselves for municipal employees, which provides the carsharing organization with a base demand and a number of bookings (Enoch and Taylor, 2006; Loose, 2009c; Millard-Ball et al., 2006; Shaheen et al., 2006; Stars project, 2019).

Information and promotion: Using carsharing as a municipality is also a measure to promote carsharing to the inhabitants of a city. This and other measures on communication, promotion, and information are also reported in the literature. Providing information to both inhabitants and businesses about carsharing and making carsharing visible, as well as joint marketing efforts of, for example, a municipality or public transit operators and carsharing organizations, are considered important measures in supporting the carsharing niche (Akyelken et al., 2018; Enoch and Taylor, 2006; KiM Netherlands Institute for Transport Policy Analysis, 2015; Loose, 2009b; Prettenthaler and Steininger, 1999; Shaheen et al., 2004; Vanhee, 2010). Some studies report the importance of setting up organizations for information dissemination on the local or the national level, that can support awareness campaigns (Enoch and Taylor, 2006; Loose, 2009b). Political support that puts carsharing on the political agenda is likewise identified as a support measure (Akyelken et al., 2018; Enoch and Taylor, 2006; KiM Netherlands Institute for Transport Policy Analysis, 2015; Loose, 2009b; Vanhee, 2010).

User incentives: Measures supporting the niche can also be aimed at the consumers and can be financial in nature, especially targeting specific groups, or give special rights to users of carsharing vehicles like special lane access or access to restricted zones in a city (Enoch and Taylor, 2006; Loose, 2009c; Shaheen et al., 2006, 2004).

Integration with public transit: Previous studies have put forward measures to integrate carsharing into public transport provision (Akyelken et al., 2018; Enoch and Taylor, 2006; KiM Netherlands Institute for Transport Policy Analysis, 2015; Loose, 2009a, 2009c; Röhr and Rovigo, 2017; Shaheen et al., 2010, 2006). Measures on this topic include providing parking spots at public transport locations and offering combined access passes. Some more recent studies indicate the importance of including carsharing in an integrated Mobility-as-a-Service (MaaS) offer (Akyelken et al., 2018; Stars project, 2019).

Legal measures: Legal measures can be taken to support and protect the carsharing niche. A call for recognizing carsharing as a unique mode of transport in legal frameworks has been reported (Autodelen.info, 2019; Stars project, 2019). Some studies also mention measures of exempting carsharing from specific taxes (Akyelken et al., 2018; Enoch and Taylor, 2006; Shaheen et al., 2006, 2004).

Planning: Studies report the importance of integrating carsharing into the planning of authorities. This applies to transport strategies but also the wider domain of urban planning as well as planning for reaching climate targets or increasing social cohesion. When changes relating to parking, parking standards, and carsharing are part of a larger mobility plan, lower numbers of parking spots are more acceptable to residents (Enoch and Taylor, 2006; Kent and Dowling, 2016; Millard-Ball et al., 2006; Stars project, 2019; Vanhee, 2010).

Although most studies focus on the measures taken to support the carsharing niche that are described above, some also report on measures that bring *changes to the car regime*.

Vision: Integrating carsharing into the planning and into the development visions of authorities on the transportation system and the built environment can also be seen as a measure to change the car-focused transportation regime.

Parking: Two measures often mentioned are making changes to parking norms and thus decreasing the number of required parking spots in new building projects or reducing the number of issued parking permits for residents (Akyelken et al., 2018; Enoch and Taylor, 2006; KiM Netherlands Institute for Transport Policy Analysis, 2015; Loose, 2009c; Millard-Ball et al., 2006; Shaheen et al., 2006; Stars project, 2019; Vanhee, 2010). Taking away parking spaces that are already built seems to be a much harder and contested measure to be taken by municipalities, although increased parking pressure can have a positive effect on carsharing (Akyelken et al., 2018). But acceptance can be increased if alternatives are available, such as offering carsharing and removing parking spots only where alternatives such as public transport are available (Enoch and Taylor, 2006). Furthermore, good communication to residents about changes is important (Autodelen.info, 2018; Vanhee, 2010), as well as giving back to the community through placing something at the location of the former parking spot that increases livability (e.g. greenery, play area) (Autodelen.info, 2018). Acceptance is also increased if requests for carsharing parking and removal of parking spots come from community groups (Autodelen.info, 2018). This process can be stimulated.

Taxation: Another measure with the power to change the regime concerns taxation. Studies report the possible impact that higher taxes on car ownership (variable and fixed costs) and removing tax incentives for company cars and their use can have (Akyelken et al., 2018; KiM Netherlands Institute for Transport Policy Analysis, 2015; Prettenthaler and Steininger, 1999; Shaheen et al., 2006; Stars project, 2019).

Table 11.2 gives an overview of the measures reported in the literature. We also added a column indicating whether local or national authorities are to initiate the policy measure.

These measures collected from the literature review were used to compile a shortlist of measures that were then presented in the workshop. The measures were categorized into three levels (local, national, all levels of government) following the categories of the reviewed measures in Table 11.2. Next to policy measures, measures taken by industry players can also influence the scaling up of carsharing and are included in discussions within the Green Deal network. We therefore decided to include measures to be taken by industry players on the shortlist discussed during the workshop. Five industry-led measures were developed and included in the shortlist, based on the results of an expert interview with a coordinating member of the Green Deal network and on our own expertise of the carsharing field. Next to conducting several academic studies on carsharing, we have followed and actively participated in a wide range of stakeholder and policy events over the last four years. These events focus on an increase of cooperation and collaboration of mobility providers to aggregate services and thus increase user convenience, as well as on sharing data needed by authorities and researchers to improve planning. Table 11.3 gives an overview of the measures presented in the workshop.

5.   RESULTS

The main finding holds that only a few out of the 21 measures are unequivocally perceived as both effective and feasible. Many possible measures are not paid much attention or are judged to be of little importance or as having limited impact. Some other measures are contested by the participants, who assess feasibility and desirability differently. The following section describes the insights gained from the workshop combined with the results from the additional interviews on the evaluation of the 21 measures presented. Furthermore, the roles of different stakeholders in implementing measures that support the scaling up of carsharing are discussed.

Positively Evaluated Measures

Three types of measures are identified as most important to support the scaling up of carsharing: 1) the implementation of encouraging and supporting measures by municipalities, 2) the development and implementation of visions on sustainable mobility systems and urban planning that include carsharing as an integral part, and 3) a national support or coordination hub.

First, the stakeholders agree that municipalities need to start taking action and implement measures that encourage carsharing by users and support pro-

Table 11.2 *Measures supporting carsharing as reported in the literature*

	General measure area	Specific applied measures	Who?
Niche support	Parking policies	Providing parking spots for free/reasonable prices	L
		Making processes for requesting spots easy and fast	L
		Stimulating developers to include carsharing parking spots	L
	Start-up help for carsharing providers/ citizen initiatives	Start-up grant provision	A
		Organizational help and facility provision	L
		Providing base demand through using it as a municipality	L
	Providing information about carsharing and raising awareness	Providing information to citizens and businesses	L
		Making carsharing (spots) visible through signs etc.	L
		Joint marketing with providers	L
		Setting up an organization for information dissemination that leads awareness campaigns	N
	Political support	Putting carsharing on political agenda	A
	Incentives for consumers	Financial incentives (e.g. vouchers for specific citizen groups)	L
		Special rights for carsharing users (access to special lanes or to restricted areas)	L
	Integration with public transport	Providing parking spots at public transport locations	L
		Offering combined access passes	A
		Joint marketing campaigns	A
		Including carsharing planning in MaaS plans	A
	Legal measures	Recognizing carsharing as a separate mode in legal frameworks	N
		Exemption from specific taxes	N
	Integration into planning	Integration of carsharing into transportation and urban planning	A
Regime change	Integration into visions	Integration of carsharing into future visions/agendas on the transportation system and urban development	A
	Parking policies	Changing parking norms	L
		Reducing the number of parking spots	L
	Taxation	Increasing taxes on car ownership	N
		Decreasing tax incentives for company cars in use by individual employees	N

Note: L = local authorities; N = national authorities; A = authorities at all levels.

Table 11.3   *Shortlist of measures presented in the workshop*

Implementation level	Measures
Government (general)	• Drawing up a vision for the mobility system and area development that includes carsharing
	• Defining shared mobility as a fully-fledged category in traffic law
	• Invest in modalities that will be used more (public transport and bicycle) as a result of carsharing
	• Allowing or initiating experiments in the field of car use
	• Open data rules: when issuing carsharing concessions, companies are obliged to share data and work together
	• Subsidies for the car industry to invest in carsharing
Government (national level)	• Tax measure: making the "company car" less attractive
	• Tax measure: higher tax on car ownership
	• Set up a national/provincial coordination center for carsharing
	– National campaigns
	– Support for provinces and municipalities
	– Standardization process for interoperability
	– Harmonization of policies
	• Financing of research on carsharing
	• Subsidies for pilots or promotion campaigns
Government (local level)	• Adjust parking policy in favor of shared cars (lower parking standards, make permits more expensive, increase paid parking areas, remove parking spaces, reasonable rates for carsharing companies)
	• Information provision to residents and companies (structural communication, visibility)
	• Improve communication with carsharing companies and speed up processes
	• Start-up assistance for carsharing companies (e.g. by using carsharing as a municipality)
	• Help for citizens' initiatives that commit to/set up carsharing
Industry	• Cooperation in the form of an umbrella booking platform
	• Collaboration with other mobility providers for data standards for aggregation of available services
	• Collaboration between carsharing companies in marketing campaigns (and use of the universal logo)
	• Increase cooperation with public transport
	• Make usage data available to government/research institutions

viders of carsharing. The considered package of measures includes those on parking, communication with carsharing providers, and communication and information provision towards citizens and businesses. These measures are in line with those recommended in the "Rode Loper" (red carpet) documents and findings of previous research. Supporting carsharing by providing parking for carsharing cars at a reasonable price and through facilitating charging infrastructure for shared electric vehicles are envisioned as uncontroversial and desirable measures. At the same time, making concrete changes to the existing regime of private car ownership through, for example, taking away parking spots or raising parking prices, is perceived to be more difficult for municipalities as they are facing criticism from citizens who feel that their "right" to an affordable parking spot is being taken away. Raising, for example, the cost of residential parking permits to trigger giving up a car is seen as a realistic option only in large cities where parking congestion is high and ample alternatives to private car ownership are available. All stakeholders also consider information provision key for successful support of carsharing. Some of them emphasize that information provision and clear communication with providers is valuable and should be implemented, but that promoting specific forms of carsharing or specific providers over others by municipalities is problematic. According to them, municipalities should provide a level playing field. Other participants raise the potential of municipalities collaborating with carsharing organizations to set up promotional and information campaigns around the advantages and possibilities of carsharing. In discussing the measures to be taken up by municipalities around information provision and parking, participants conclude that the strongest measures are already known on a national level and in the group of Green Deal carsharing members, but that this knowledge is not widely known at the municipal level and implementation is as a result limited to a few of the largest cities in the Netherlands. Other municipalities are envisioned to lack knowledge and awareness as well as the capacity to implement carsharing measures. Participants voice a need for convincing carsharing arguments and practical tools or guidelines in order for these other municipalities to implement carsharing. It remains an open question as to which actor can play the role of disseminating knowledge about measures and of offering assistance in introducing new policy measures that can promote the upscaling of carsharing.

Second, the workshop participants and interviewees articulate the need for the development and implementation of visions on sustainable mobility systems and urban planning that include carsharing as an integral part. According to them, these visions can be linked to plans for achieving environmental and climate goals. Through connecting carsharing to climate goals, it can be brought to municipalities that have not yet considered it as something worth stimulating. Examples of provinces and municipalities that have started to integrate carsharing into future visions and planning or set up action plans on carsharing were discussed and the potential impact of integrating this measure at all levels judged to be high.

Third, setting up a national support or coordination hub is considered an important measure to help the upscaling of carsharing, and relates to the first-mentioned measure and the need for inter-authority learning. A national authority could coordinate the provision of information about the options, benefits, availability, and ease of use of carsharing to citizens, companies, and local authorities. Knowledge about impactful measures has to be disseminated to lower-level authorities and help with implementation needs to be provided. In order to bring carsharing to the attention of various different stakeholders, a national campaign needs to inform them about the benefits for a city or neighborhood, such as saving space and increasing the quality of life, as well as for individuals. Furthermore, attention can be drawn to potential benefits for different target groups and regions, for example for lower income groups or smaller municipalities. In addition, the provinces are mentioned as possibly being able to take a key role in integrating carsharing into planning and diffusing knowledge on best practices and effective policy measures to municipalities. Provinces can connect carsharing to regional mobility schemes and to measures they currently need to set up in collaboration with municipalities to reach climate goals.

Negatively Evaluated Measures

Some measures are perceived to be not feasible or not desirable. Discussions on changes in the national taxation regime that would increase taxes on ownership of cars are instantly discarded by the participants as being not possible in the current political climate. Such a measure would run against an emerging consensus to raise higher taxes on car use (road pricing with a fixed price per driven kilometer) while in exchange lowering the taxes of car ownership. Equally, participants from the automotive sector and the Dutch touring club, representing large numbers of citizens, do not envision higher taxation on car ownership as desirable. Because of this strict "no-go" statement made by several participants, discussions during the workshop did not go further into the topic, but in two feedback interviews an interesting argument came up, emphasizing that any new taxes on cars can be seen to be positive as they draw the attention of users to the high costs of cars and can trigger thinking of alternatives. Subsidies for the car industry to invest in carsharing are also clearly evaluated negatively. Participants are of the opinion that the car industry is not in need of subsidies from the government in order to be able to invest in new business models and services.

Unnecessary and Unimportant Measures

Multiple measures are perceived to be unnecessary or not very important by the participants and as a result were not discussed in depth. Indicating carsharing as a new category in transport law as well as increased cooperation with public transport stay mostly undiscussed as a result. Investing in other modalities like public transit and cycling infrastructure is evaluated to be only of importance later on, once carsharing has grown more. The measure of providing carsharing companies with start-up assistance through, for example, municipalities becoming launching customers of carsharing is partly evaluated positively but is perceived to be not that important and as not having a large effect. During the workshop, no attention was given to the measure of allowing or initiating experiments in the field of car use, although in the later interviews experiments such as MaaS pilots or shared electric vehicles in neighborhoods, set up by a province or municipality, are named as being positive examples of putting alternatives to car ownership on the agenda.

Controversial Measures

Not all measures are unanimously assessed or valued positively by participants. For some of the measures discussed certain stakeholders envision barriers, while others question the effects or desirability of the measures. Sharing of usage data by mobility providers both with each other and with authorities or research institutes is such a controversial measure. Providers are hesitant to share privacy-sensitive or competition-sensitive data. Authorities and research institutes, on the other hand, require usage data to be able to assess the usage and effects of offered services, so as to be better able to integrate carsharing into planning. Furthermore, they see an aggregation of supply on one platform as a powerful tool to raise the interest in and usability of carsharing for consumers. Providers articulate being more open to such an aggregated umbrella platform once a functioning MaaS platform is in place and the benefits for companies and the role of governmental regulation are clear.

In the group of participating stakeholders there is disagreement about the need for more research on the topic of carsharing. Some participants believe that research into the impacts of carsharing and its different forms is important, so that they can be clearly mapped and carsharing can thereby be better integrated into policymaking and planning decisions. For an accurate impact study, however, carsharing companies should be willing to provide usage data to independent researchers. Other participants do not see the need for such a study and are of the opinion that enough is known about the positive impact of carsharing and about barriers for scaling up. A question that thus arises for further discussion is on the subject of determining which data can be possibly shared by providers and what is needed for authorities to take effective action.

Although it is seen as necessary by all stakeholders for municipalities and provinces to be better informed and helped in implementing measures supporting carsharing, opinions differ on whether a national coordination center or authority for carsharing is necessary and if a national authority should be setting up a campaign promoting carsharing. Some feel that an "official" national coordination center would make processes unnecessarily complex, or that national authorities should not be running "marketing campaigns" for carsharing providers. Others see a need and potential for having one national coordination center that can support provinces and municipal authorities, spread knowledge, standardize processes, and harmonize policies.

In sum, this study shows that, according to the participating stakeholders, information about the advantages of carsharing is key. These advantages need to be made clear to various stakeholder groups. Private consumer groups, as well as companies, can be convinced by the different advantages carsharing can offer, be it, for example, cost-efficiency, ease, access opportunities, or increasing space availability. Local and regional authorities need to be convinced of the advantages of carsharing and the benefits a dense network of carsharing can offer a municipality or region, benefits such as decreasing space scarcity in the larger cities or increasing mobility access in more rural areas or for specific inhabitant groups. Here, the question remains who needs to take on the role of informing these different stakeholder groups. Some argue that solely the carsharing providers should be promoting and marketing their services, while many participants are of the opinion that authorities also have an interest in scaling up carsharing and should thus get more involved in informing about and promoting carsharing.

6.   SUMMARY

All stakeholders participating in the workshop or interviews agreed that carsharing can contribute to reaching the climate targets which the Dutch government has set itself following the Paris Agreement. In addition, carsharing can increase livability in crowded cities through being part of a transition of the mobility system. Because of this contribution of carsharing towards reaching societal goals, authorities on all levels should have an interest in supporting its scaling up. This study shows that there are a number of success and failure factors for scaling up carsharing. Table 11.4 summarizes them. The success and failure factors address challenges at the niche as well as the regime level, challenges that can be addressed with the right policy measures.

Measures supporting the carsharing niche and those more directly changing the car ownership regime were both discussed. However, the focus of the participating stakeholders clearly is on measures supporting the carsharing niche, which they evaluated as feasible and desirable. Measures challenging the established regime of private car ownership are perceived as being impossible or at least controversial. As some of the participants represented regime actors (e.g. the Dutch touring club), these evaluations can be seen as typical regime reactions. Furthermore, the workshop showed that measures to be taken by

Table 11.4   *Success and failure factors for scaling up carsharing*

Success factors	Failure factors
Support for carsharing by municipalities: Dedicated encouraging and supporting measures for the supply and use of carsharing have a positive effect on the scaling up of carsharing (e.g. measures on parking, communication with carsharing providers, communication and information provision towards citizens and businesses).	*Slow and difficult processes of authorities in communication with providers:* Scaling up is slowed down by unstandardized and lengthy processes that carsharing providers face when, for example, applying for parking permissions.
Provision of parking spaces for carsharing cars: Providing parking for carsharing cars is often named as one of the most important measures a municipality can take to support carsharing development. Carsharing parking should preferably be on-street, well visible, and close to public transport stops. Providing it with no or low costs can offer additional support.	*High attractiveness of car ownership:* A high attractiveness of car ownership through low costs, availability of good infrastructure, and high convenience levels decreases the chances of consumers reducing ownership and using carsharing.
Unattractive parking situation: Low parking norms (e.g. in new development/ redevelopment of areas), removal of parking spots, and high parking prices can decrease the attractiveness of car ownership and use and stimulate the use of carsharing.	*Low knowledge base on benefits of carsharing:* A low information level about carsharing, its benefits, supply, and operation mode in the general public leaves use of carsharing at a low level. A low knowledge base within national and local authorities on the contribution of carsharing to reaching societal goals reduces use of supportive measures and integration of carsharing into visions and plans.
High costs of car ownership: Higher taxation on car ownership can potentially lead to decrease in car ownership and increased use of carsharing.	*Unmonitored development:* A lack of knowledge on the development of carsharing (supply, use, effects) disables policy makers from implementing concrete supportive measures.
Multimodal mobility vision: Municipalities having a multimodal mobility vision for the future that is not focused on car mobility make carsharing an integral part of the solution for the mobility system and solving urban and climate challenges.	*Ambiguity in governmental roles:* Missing clarity on which governmental body should take up the role of spreading information on successful policies and processes to local authorities.
	Low collaboration: A low level of collaboration between different stakeholders slows down the development of the carsharing market.

Success factors	Failure factors
Support for local authorities: Information provision of recommended policies and governmental actions that stimulate carsharing to local government, including measures like improving governmental processes, promotion and communications campaigns, decreasing attractiveness of car ownership, parking policies, monitoring and researching developments. Having a national support or coordination hub to spread knowledge and best practices, and to standardize processes (through practical tools and guidelines), can be helpful in this.	*"Unlevel" playing field*: Uneven support of different (types of) providers of carsharing can lead to an unlevel playing field and lack of competition on the market.
Good communication about policy changes: Good communication towards citizens and businesses about changes in policies and a larger vision make policy change more feasible and socially and politically accepted (e.g. communication about benefits, changed parking situations, and alternatives).	
Experiments: Experiments in the field of mobility services can increase knowledge on effects and increase acceptance by citizens and businesses.	
Integrated mobility services: The integration of services can improve the service quality and attractiveness of mobility services for users, and the sharing of data by mobility service providers with each other and with authorities and research institutes can improve knowledge on effects and developments.	

authorities were discussed more freely, while measures to be taken by stake-holders from the industry, like collaborating with other providers or sharing data with authorities, although potential success factors, were controversial because of competitive pressures among providers.

Measures to be taken by authorities and perceived to have the highest impact are changing parking policies, actively promoting carsharing, and integrating carsharing into planning around transportation and urban develop-ment. Stakeholders also perceived information provision about the advantages of carsharing to citizens, companies, and local authorities as a promising policy. These measures have been identified as important success factors. Most measures that all stakeholders agree on can be taken at the municipal level, but municipalities are in need of help in order to take on this substantial role. A bridging function needs to be fulfilled between, on the one side, the knowledge available at the national level and in the Green Deal community and, on the other side, the lack of knowledge available at the majority of municipalities. That bridging of the gap in knowledge and ambition between the national government and municipalities is difficult and can potentially act as a failure factor. This has also been mentioned by Akyelken et al. (2018) for the case of the Tel Aviv region and the Israeli government. However, there are also controversies about how far governmental support for specific solutions or providers should go. Actors from the current regime perceive strong support from authorities and promotional activities for carsharing to be possibly unfair and undesirable. When taking on the goal of building a level playing field for different providers, policy makers need to be aware of the failure factor of an unlevel playing field and use measures that are not supporting one solution over the other but that rather take away advantageous regulation supporting the old regime of private car ownership instead of supporting the niche of car-sharing with new measures. This could also prevent new regulations becoming outdated quickly in such a dynamic market and would be in line with the warning from Le Vine (2012) and KiM Netherlands Institute for Transport Policy Analysis (2015) that policy makers, especially those at the local level, should stay flexible in policy use and strive for diverse options.

It can be concluded that supporting the carsharing niche offers options for accelerating the upscaling of carsharing. The niche can be supported by improving processes for carsharing companies, by developing parking policies that facilitate carsharing, and by encouraging and supporting neighborhood initiatives in setting up carsharing solutions. Large-scale changes at the regime level, for example through substantial tax increases, are difficult or not feasi-ble, but at a smaller, local level, changes in the regime, for example through changes in parking policy, can have a major impact.

7.   DISCUSSION

The current car regime is based on private car ownership and its supporting infrastructures. In addition, private car ownership is embedded in our social and cultural system and has symbolic power next to being a convenient transport mode (Truffer, 2003). The current regime has led and still leads to negative consequences on, for example, climate, livability, and equity. Carsharing can act as a means to achieve positive impacts on multiple societal goals. Given the regime in place, changes are incremental and geared to optimize the current system, with the capabilities and resources of incumbent players being used. More radical change is restricted since the established rules, structures, and culture lead to slow changes in regulations, norms, and practices. Carsharing offers an alternative to the regime of private car ownership. It makes use of existing regime infrastructure but builds on new behavioral practices, cultures, and business models. The socio-technical system of the regime is relatively stable, but larger societal trends, such as growing urbanization, the growing awareness of climate change, growing digitization, and the growing service economy, can influence the system and open a window of opportunity for a niche innovation like carsharing to break through and move into (or replace) the dominant regime (Geels, 2002, 2004; Loorbach, 2007; Pel, Chapter 2 in this volume). Nonetheless, changes in policies and new supportive measures are necessary for carsharing and other new mobility forms to scale up.

The insights from the various stakeholders show that large-scale changes at the regime level (like changes in taxation) are more difficult to implement or lack (political) feasibility, while smaller regime changes, often at the local level, can also have a substantial stimulating impact (e.g. changes in parking policy). Slower, but continuous changes to the regime in a small local setting can create the right "protection" in order for the innovation to be successful. Changes in parking policies seem to be of major importance for attracting new consumers to carsharing and giving up private car ownership.

The current measures that support the carsharing niche are an example of "Strategic Niche Management" (Kemp et al., 1998), as the measures are providing a "protective space" for carsharing providers. This is most literally exemplified by the dedicated parking spots for carsharing operators, while other measures including subsidies and the integration of carsharing with public transport services provide further niche support. However, while these measures contribute to the further upscaling of carsharing, it is unlikely that carsharing on its own will fundamentally change the car regime. Rather, carsharing policies in the Netherlands follow a "fit and conform" empowerment logic (Smith and Raven, 2012), rendering the niche innovation competitive given otherwise largely unchanged selection environments. The development of carsharing as an alternative for private car ownership has not prompted any bold measures to phase out private car ownership.

Our study makes clear that policy makers and other stakeholders refrain from policies that discourage private car ownership, even though this is a potential failure factor for the upscaling of carsharing and the positive impacts on reaching societal goals. Instead, they agree on measures that stimulate the niche of carsharing to further grow. There is thus backing for creating support for the carsharing niche, but little for breaking down the established regime. As a result, policy inconsistencies emerge where the regime logic hampers the further growth of carsharing. Problems because of these inconsistencies can be identified at local and national authorities as well as when looking at industry stakeholders. First, at the *local level* changing parking regulations exemplify the inconsistencies between niche support and regime change. While carsharing can be supported through providing parking spots in crowded locations, where it could then free up space because people decrease car ownership (Enoch and Taylor, 2006), municipal regulation limits the possibilities to actually remove parking spots. Taking away parking spots decreases earnings and this loss has to be compensated elsewhere. Second, the *national government* is using inconsistent policies: on the one hand they want to stimulate the use of shared mobility as they see its potential; on the other hand higher taxes on car use (road pricing) are being discussed in combination with a decrease in taxes on car ownership to compensate citizens and keep costs in balance. Such a reduction in the cost of car ownership obviously will slow down the scaling up of carsharing. Apparently, challenging the current regime by increasing taxes both on car use and on private car ownership is considered a political no-go, comparable to earlier findings by Akyelken et al. (2018). Only large cities with a green, progressive electorate have developed ambitious plans to change the car regime, including measures on parking, reducing car ownership and use, improving communication with carsharing providers, supporting innovative carsharing initiatives, and integrating carsharing into new developments (Gemeente Amsterdam, 2019; Gemeente Utrecht, 2015).

Apart from ambiguities in government policies, *industry stakeholders* active in the carsharing market also maintain inconsistent perspectives. B2C carsharing providers want to position themselves as the only truly sustainable and thus best solution for car ownership alternatives. They clearly search for support from authorities for the carsharing niche and for bringing changes to policies supporting the current regime of private car ownership. P2P carsharing providers, by contrast, operate more in line with the current car regime, as P2P sharing is based on people owning private cars that are rented out on the platform. Hence, while B2C providers emphasize the need to challenge private car ownership, P2P providers do not call for disruptive changes to the car regime. Outsider actors moving into the carsharing market, like car rental and leasing organizations, and organizations representing the current users of the regime, like a touring club, are also less inclined to challenge the current regime, as their organizations are well established in the current automotive industry and profit from a stable regulatory environment.

On a final note, our study makes clear that policies supporting carsharing should be discussed within the context of the multimodal mobility system as a whole. Several stakeholders consider the scaling up of carsharing as a means to an end rather than a goal in itself. The scaling up of carsharing, then, may be better seen as one out of multiple and complementary solutions towards decreasing car ownership and use, as also emphasized earlier by Millard-Ball et al. (2006). Indeed, stakeholders agree on the importance of a functioning multimodal mobility system in which carsharing is an integral part. This should be taken into account in the planning and visions for the mobility system and in wider urban planning. Accordingly, stakeholders' opinions converge towards considering MaaS as the desirable new paradigm in mobility. At the same time, however, changing the focus from carsharing to MaaS may well slow down sustainability improvements in the mobility system, as many technical and institutional challenges surrounding MaaS are still unresolved. As the upscaling of carsharing as one specific solution does not jeopardize a more comprehensive transition towards MaaS as such, the wish to move to MaaS does not constitute an argument to reduce support of carsharing.

Our study has some limitations. As it focuses on the Dutch context, our findings are only to a limited extent generalizable to other countries, as the regulatory situation, stakeholder composition, and political landscape have an important impact on which measures are perceived to be most impactful and feasible. Having said this, the current regime of private car ownership is a global regime, which suggests that our findings may well be relevant to other countries as well. The literature review also showed that policy measures in place or being discussed indeed overlap greatly between countries. Our method of hosting one workshop with a limited number of participants is also a limiting factor when interpreting and generalizing the results, as there might be more opinions and perspectives on measures and barriers perceived by other stakeholders. At the same time, the participants represented a wide variety of stakeholder groups and we supplemented their views with insights from five additional interviews. It has to be noted that some measures were not discussed in depth as they were directly pushed off the table as being not feasible or desirable. This limits the understanding of the possible effects they could have on the scaling up of carsharing. Similarly, the possibility to overcome barriers in upscaling through the combination of measures was not discussed. Furthermore, a workshop setting can lead to some participants dominating the discussion with their views and opinions. To avoid this from happening, we split up into groups, with each having a facilitator moderating the discussion.

Next to these limitations concerning the method, it has to be noted that we did not analyze the direct impact of the specific policy measures on societal goals and can therefore not draw conclusions on the effectiveness of single measures in impacting societal goals such as emission reduction or reduced urban space used for cars.

Future research on carsharing policy could include qualitative and quantitative analyses of the impact of the discussed policy measure. Also, similar workshops in other countries, or multi-country workshops, to compare which situations lead to different outcomes in terms of what measures are perceived as useful, could be valuable. Collecting these different perspectives can generate insights into best practices that many countries can profit from as well as reveal contextual factors that need to be taken into account in carsharing policies. Furthermore, future research could apply a wider focus to analyze measures supporting the larger transition of the mobility system instead of focusing on carsharing services. Finally, our research also makes clear that municipalities need more practical help in setting up and implementing measures supporting the growth of carsharing. Developing a tool or a template for an action plan that municipalities can easily fill in and adjust to their local context could be a valuable option to contribute to realizing a larger role for carsharing in the mobility system.

NOTE

1. We thank Martien Das for her help in setting up the expert workshop and all participants of the workshop and the experts we interviewed for their valuable insights. This work was supported by Dialogic, the Rathenau Institute, and NWO under the "Sustainable Business Models" program (No. 438-14-904). Declarations of interest: none.

REFERENCES

Akyelken, N., Givoni, M., Salo, M., Plepys, A., Judl, J., Anderton, K., and Koskela, S. (2018). The importance of institutions and policy settings for car sharing – evidence from the UK, Israel, Sweden and Finland. *European Journal of Transport and Infrastructure Research*, *18*(4), 340–359.

Autodelen.info. (2018). *Rode loper voor autodelen* [*Red Carpet for Carsharing*]. Retrieved from https://static1.squarespace.com/static/5818ae386a49632b8ef6ad95/t/5922f2203e0.

Autodelen.info. (2019). *Rode loper autodelen voor de Rijksoverheid* [*Red Carpet Carsharing for the National Government*]. Retrieved from https://static1.squarespace.com/static/5818ae386a49632b8ef6ad95/t/5aba0674562fa7e192075c71/152214079 2222/Rode+loper+voor+de+Rijksoverheid+autodelen+versie+3.pdf.

Burkhardt, J., and Millard-Ball, A. (2006). Who is attracted to carsharing? *Transportation Research Record: Journal of the Transportation Research Board*, *1986*, 98–105. https://doi.org/10.3141/1986-15.

CBS (2019a). Personenauto's. Aantal personenauto's neemt verder toe [Passenger cars. The number of passenger cars is increasing further]. Retrieved from https://www.cbs.nl/nl-nl/maatschappij/verkeer-en-vervoer/transport-en-mobiliteit/infra-ver voermiddelen/vervoermiddelen/categorie-vervoermiddelen/personenauto-s.

CBS (2019b). Rijbewijzen. 8 op de 10 volwassenen hebben een autorijbewijs [Driving licenses. 8 out of 10 adults have a car driving license]. Retrieved from https://www .cbs.nl/nl-nl/maatschappij/verkeer-en-vervoer/transport-en-mobiliteit/mobiliteit/per sonenmobiliteit/categorie-personenmobiliteit/rijbewijzen.

Chen, T.D., and Kockelman, K.M. (2016). Carsharing's life-cycle impacts on energy use and greenhouse gas emissions. *Transportation Research Part D: Transport and Environment*, *47*, 276–284. https://doi.org/10.1016/j.trd.2016.05.012.

CROW (2018). Aanbod deelauto's [Supply of shared cars]. Retrieved from https://www.crow.nl/dashboard-autodelen/home/aanbod/aanbod-resultaat.

CROW (2021). Aanbod deelauto's [Supply of shared cars]. Retrieved from https://www.crow.nl/dashboard-autodelen/home/aanbod/aanbod-resultaat.

Dill, J., Howland, S., and McNeil, N. (2016). A profile of peer-to-peer carsharing early adopters: Owners and renters. Paper presented at the Transportation Research Board 95th Annual Meeting, Washington DC, January 10–14. Retrieved from https://trid .trb.org/view/1394245.

Enoch, M.P., and Taylor, J. (2006). A worldwide review of support mechanisms for car clubs. *Transport Policy*, *13*(5), 434–443. https://doi.org/10.1016/J.TRANPOL .2006.04.001.

European Environment Agency (EEA). (2018). *Progress of EU Transport Sector towards Its Environment and Climate Objectives*. Publications Office. https://doi .org/10.2800/954310.

Eurostat. (2016). *Urban Europe: Statistics on Cities, Towns and Suburbs*. Publications Office. https://doi.org/10.2785/91120.

Geels, F.W. (2002). Technological transitions as evolutionary reconfiguration pro-cesses: A multi-level perspective and a case-study. *Research Policy*, *31*(8–9), 1257–1274. https://doi.org/10.1016/S0048-7333(02)00062-8.

Geels, F.W. (2004). From sectoral systems of innovation to socio-technical systems: Insights about dynamics and change from sociology and institutional theory. *Research Policy*, *33*(6–7), 897–920. https://doi.org/10.1016/J.RESPOL.2004.01 .015.

Geels, F.W. (2011). The multi-level perspective on sustainability transitions: Responses to seven criticisms. *Environmental Innovation and Societal Transitions*, *1*, 24–40. https://doi.org/10.1016/j.eist.2011.02.002.

Gemeente Amsterdam. (2019). *Agenda autodelen [Carsharing Agenda]*. Amsterdam. Retrieved from https://autodelen.info/publicaties/2019/2/20/gemeente-amsterdam -stelt-de-agenda-autodelen-vast.

Gemeente Utrecht. (2015). *Maatregelen stimuleren autodelen [Measures Stimulate Carsharing]*. Utrecht. Retrieved from https://api1.ibabs.eu/publicdownload.aspx ?site=utrecht&id=27367.

Giesel, F., and Nobis, C. (2016). The impact of carsharing on car ownership in German cities. *Transportation Research Procedia*, *19*, 215–224. https://doi.org/10.1016/j .trpro.2016.12.082.

Kemp, R., Schot, J.W., and Hoogma, R. (1998). Regime shifts to sustainability through processes of niche formation: The approach of strategic niche management. *Technology Analysis and Strategic Management*, *10*, 175–195.

Kent, J.L., and Dowling, R. (2016). "Over 1000 cars and no garage": How urban planning supports car(park) sharing. *Urban Policy and Research, 34*(3), 256–268. https://doi.org/10.1080/08111146.2015.1077806.

KiM Netherlands Institute for Transport Policy Analysis. (2015). *Carsharing in the Netherlands: Trends, User Characteristics and Mobility Effects*. The Hague. Retrieved from https://www.researchgate.net/publication/311774025_Carsharing _in_the_Netherlands_User_characteristics_and_mobility_effects.

Le Vine, S. (2012). *Car Rental 2.0: Car Club Innovations and Why They Matter*. London: RAC Foundation. Retrieved from http://www.racfoundation.org/assets/rac _foundation/content/downloadables/car_rental_2.0-le_vine_jun12.pdf.

Loorbach, D. (2007). Governance for sustainability. *Sustainability: Science, Practice and Policy, 3*(2), 1–4. https://doi.org/10.1080/15487733.2007.11907996.

Loose, W. (2009a). *Collaboration with Public Transport Operators. momo Car-Sharing Fact Sheet No. 1*. momo Car-Sharing. Retrieved from https://ec.europa.eu/energy/ intelligent/projects/sites/iee-projects/files/projects/documents/momo_car-sharing _f01_collaboration_with_public_transport_operators_en.pdf.

Loose, W. (2009b). *State Support for Car-Sharing. momo Car-Sharing Fact Sheet No. 5*. momo Car-Sharing. Retrieved from http://www.momo-cs.eu/index.php?obj=page &id=151&unid=b5ca99a6ae72ca8fc303558a1e4ee0d7.

Loose, W. (2009c). *Support by Local Councils. momo Car-Sharing Fact Sheet No. 7*. momo Car-Sharing.

Meelen, T., Frenken, K., and Hobrink, S. (2019). Weak spots for car-sharing in The Netherlands? The geography of socio-technical regimes and the adoption of niche innovations. *Energy Research and Social Science, 52*, 132–143. https://doi.org/10 .1016/J.ERSS.2019.01.023.

Millard-Ball, A., Murray, G., and ter Schure, J. (2006). Carsharing as parking management strategy. Paper presented at the Transportation Research Board 85th Annual Meeting, Washington, DC, January 22–26. Retrieved from https://trid.trb.org/view/ 776448.

Ministerie van Verkeer en Waterstaat. (1988). *Tweede Struktuurschema Verkeer en Vervoer, Tweede Kamer, vergaderjaar 1988–1989 [Second Structure Plan for Traffic and Transport, House of Representatives, 1988–1989 Session]*. The Hague: Ministerie van Verkeer en Waterstaat.

Münzel, K., Boon, W., Frenken, K., Blomme, J., and van der Linden, D. (2019). Explaining carsharing supply across Western European cities. *International Journal of Sustainable Transportation*, 1–12. https://doi.org/10.1080/15568318.2018.1542 756.

Münzel, K., Boon, W., Frenken, K., and Vaskelainen, T. (2018). Carsharing business models in Germany: Characteristics, success and future prospects. *Information Systems and E-Business Management, 16*(2), 271–291. https://doi.org/10.1007/ s10257-017-0355-x.

Namazu, M., and Dowlatabadi, H. (2018). Vehicle ownership reduction: A comparison of one-way and two-way carsharing systems. *Transport Policy, 64*(May), 38–50. https://doi.org/10.1016/j.tranpol.2017.11.001.

Nijland, H., and van Meerkerk, J. (2017). Mobility and environmental impacts of car sharing in the Netherlands. *Environmental Innovation and Societal Transitions, 23*, 84–91. https://doi.org/10.1016/j.eist.2017.02.001.

OECD and ITF. (2017). *ITF Transport Outlook 2017*. Paris: OECD Publishing.

Prettenthaler, F.E., and Steininger, K.W. (1999). From ownership to service use life-style: The potential of car sharing. *Ecological Economics*, *28*(3), 443–453. https://doi.org/10.1016/S0921-8009(98)00109-8.

Rijksoverheid. (2015). Over drie jaar honderdduizend deelauto's in Nederland [One hundred thousand shared cars in three years time]. Retrieved from https://www.rijksoverheid.nl/actueel/nieuws/2015/06/03/over-drie-jaar-honderdduizend-deelauto-s-in-nederland.

Rijksoverheid. (2018). Meer deelauto's voor betere bereikbaarheid en schonere lucht [More shared cars for better accessibility and cleaner air]. Retrieved from https://www.rijksoverheid.nl/actueel/nieuws/2018/10/04/meer-deelauto's-voor-betere-bereikbaarheid-en-schonere-lucht.

Rogers, E.M. (2003). *Diffusion of Innovation* (5th ed.). New York: Free Press.

Röhr, T., and Rovigo, M. (2017). Public service approach to car-sharing in mid-sized towns: The example of Belfort (France). *IET Intelligent Transport Systems*, *11*(7), 403–410. https://doi.org/10.1049/iet-its.2016.0259.

Schreier, H., Grimm, C., Kurz, U., Schwieger, B., Keßler, S., and Möser, G. (2018). *Analysis of the Impacts of Car-Sharing in Bremen, Germany*. team-red. Retrieved from https://northsearegion.eu/media/5724/analysis-of-the-impact-of-car-sharing-in-bremen-2018_team-red_final-report_english_compressed.pdf.

Shaheen, S., Cohen, A., and Martin, E. (2010). Carsharing parking policy: Review of North American practices and San Francisco, California, Bay Area case study. *Transportation Research Record: Journal of the Transportation Research Board*, *2187*(1), 146–156. https://doi.org/10.3141/2187-19.

Shaheen, S., Cohen, A., and Roberts, J. (2006). Carsharing in North America: Market growth, current developments, and future potential. *Transportation Research Record: Journal of the Transportation Research Board*, *1986*(1), 116–124. https://doi.org/10.3141/1986-17.

Shaheen, S., Schwartz, A., and Wipyewski, K. (2004). Policy considerations for carsharing and station cars: Monitoring growth, trends, and overall impacts. *Transportation Research Record: Journal of the Transportation Research Board*, *1887*(1), 128–136. https://doi.org/10.3141/1887-15.

Smith, A., and Raven, R. (2012). What is protective space? Reconsidering niches in transitions to sustainability. *Research Policy*, *41*(6), 1025–1036.

Stars project. (2019). *10 Recommendations to Help Policymakers Implement Car Sharing in Europe. Policy Brief*. Retrieved from http://stars-h2020.eu/wp-content/uploads/2019/01/STARS-Policy-Brief-4-pages.pdf.

Steininger, K., Vogl, C., and Zettl, R. (1996). Car-sharing organizations. *Transport Policy*, *3*(4), 177–185. https://doi.org/10.1016/S0967-070X(96)00024-8.

Truffer, B. (2003). User-led innovation processes: The development of professional car sharing by environmentally concerned citizens. *Innovation: The European Journal of Social Science Research*, *16*(2), 139–154. https://doi.org/10.1080/13511610304517.

Vanhee, J. (2010). *Momo Car-Sharing Deliverable 5.3. Guideline for Municipalities and Governments*. momo Car-Sharing. Retrieved from https://ec.europa.eu/energy/intelligent/projects/sites/iee-projects/files/projects/documents/momo_car-sharing_car_sharing_guidelines_for_public_authorities_en_en.pdf.

12. Mobility-as-a-Service: how governance is shaping an innovation and its outcomes

Wijnand Veeneman

INTRODUCTION

Mobility-as-a-Service (MaaS) is an innovation in which several new technologies are applied to integrate mobility options for the traveler, with the promise of seamless supply of mobility through a wide variety of modes (see for example Finger et al., 2015 and Smith, 2020). For the user, those mobility options are brought together in a single interface (Durand et al., 2018), mostly through an app. The mobility options included use a variety of infrastructures, vehicles, and services from different providers, all with their own strengths and weaknesses. In that fragmented landscape there is no single best way of travel in all situations for all travelers, making tailored advice and integrated provision of the services the big promise.

This chapter considers the expected societal impact of MaaS and the main factors driving that impact. The chapter points at how governance of MaaS implementation plays a major role in what that impact will be. MaaS proves to have a wide variety of forms in which it can be implemented. The form chosen, and consequently its societal impact and value, relies heavily on the way key actors in the field relate to the innovation. That relation is formed by the existing governance of mobility and the changes made to that governance to accommodate MaaS solutions.

To consider the social impacts and the factors driving them, this chapter will lean on the perspective of transitions theory, as presented in Chapter 2 of this book in which Pel shows the perspective on transport innovations shifting from a focus on ex-ante evaluation and implementation of technologies per se to one on institutional refocusing on societal value, with technologies being one of the shifting elements.

This chapter shows there has been a shift in the last few years in how MaaS is regarded and treated as an innovation, from a clear innovative concept about a service to help travelers deal with a fragmented landscape of mobility (fitting the earlier perspective that Pel puts forward in Chapter 2) towards a more messy and incremental system innovation (fitting his latter perspective), from a singular private innovation to be implemented (the earlier perspective) to an intricate effort from both governmental and private entities (the latter perspective), from a new service to a rethink of governance, and from a system improvement to a system innovation. Cases from three countries illustrate empirically how different governance contexts shape the innovation and how the innovation changes the governance, in line with that wider system perspective.

The chapter starts by explaining the context in which MaaS is being introduced, what MaaS entails, and the changing roles of private and public parties in the development of MaaS since its inception. It introduces two perspectives (narrow and broad) on MaaS and includes three illustrative cases showing how the broader perspective is recognizable in the development of MaaS in three European cities: Amsterdam, Birmingham, and Helsinki. It continues the discussion on broad and narrow applications with social implications, success and failure factors, and research challenges.

THE FRAGMENTED LANDSCAPE OF TRANSPORT SERVICES

To understand the potential of MaaS one has to understand the problems of the current mobility landscape with its variety of transport modes. These modes entail a wide variety of infrastructures, vehicles, and services that allow people (and goods) to go to places and back. This chapter will focus on person transport. Transport generally uses an infrastructure, an immovable technological asset that facilitates things to move. On most infrastructures the moving things are vehicles. These are the first distinguishing features of different modes: they have different infrastructures and different vehicles. Planes, trains, cars, bicycles, and their respective airfields, rail lines, roads, and paths, all represent different modes. Note that walking is the mode that doesn't need a vehicle, which makes it particularly convenient.

Second, there is the distinction between a vehicle being controlled by the traveler and a vehicle being controlled for the traveler. When that controlling for the traveler involves a transaction (the traveler paying the driver and the use of a vehicle), that can be considered a full transport service. In private or individual transport, the traveler controls the vehicle and passengers are generally not paying this driver. There is no transaction. In shared transport, the traveler is generally paying for vehicle control (like a train, bus, or taxi/ride hailing) or just vehicle use (like shared car, bike, or scooter). The word "public" is often used for a subset of shared modes in two ways. On the one

Table 12.1   *Different transport services and modes*

Transport service	Infra-structure	Vehicle	Driver	Locations and times*	Trips	Examples
Public	Dedicated	Service	Service	Scheduled lines	Combined	Train, metro, tram, some bus
	Shared	Service	Service	Scheduled lines	Combined	Bus, ferry, some tram
Shared	Shared	Driver owned	Driver traveler	Continuous	Combined	Ride sharing
	Shared	Driver owned	Service	Service hours	Dedicated	Ride hauling
	Shared	Service	Driver traveler	Continuous	Dedicated	Free-roaming bike or scooter sharing
	Shared	Service	Driver traveler	Continuous	Dedicated	Car and bike rental and docked sharing
	Shared	Service	Service	Service hours	Combined	Jitney
	Shared	Service	Service	Continuous	Dedicated	Taxi
Private	Shared	Driver owned	Driver traveler	Continuous	Dedicated	Car, motorbike, bicycle, scooter

Note: * Continuous is always dependent on possible erratic supply.
Source: Veeneman et al. (2020).

hand, "public" is used for transport services that are scheduled for combined use by travelers, like bus, tram, and metro. On the other hand, "public" is often used for all modes available to everyone who can afford them, including the shared services. Here we will use the first definition for public and discuss shared modes as a separate category (see Table 12.1).

Third, modes can be distinguished by the spatial scale in which they operate. A train on an urban street is generally called a tram, while a train connecting cities in a large country is a different mode altogether. A bus can take people downtown or take them across the continent. These services are generally operated by different operators, with different ticketing and pricing schemes. Obviously, it is exactly this wide variety of modes – with a wide variety of strengths and weaknesses, run by different operators, with various ticketing and pricing schemes – that is the basis for the promise of MaaS.

Historically, the mobility landscape could be seen mostly as a simpler dichotomy (see also Enoch, 2015 and Wong et al., 2020), namely public transport (provided with shared use of vehicles, controlled by others along fixed routes and schedules) and private transport (with privately owned vehicles on publicly owned infrastructures driven by travelers on any route made available by the network), with a limited role for taxis. Public and private transport modes both have vehicle- and infrastructure-related inefficiencies in terms of coordinating supply and demand. In terms of infrastructure, both public and private transport suffer peak use, leading to congestion, reduced throughput, and capacity loss. In terms of vehicles, in private transport, cars and bikes are most of the time parked, sitting unused, and in public transport, seats are hauled around often empty. The promise of shared modes is that they offer a way to reduce those inefficiencies related to vehicles. That partly explains their growth and the move away from the simple dichotomy to a more varied mobility landscape.

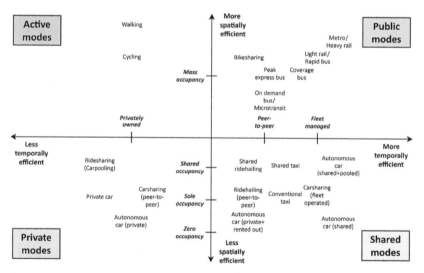

Source: Wong et al. (2020).

Figure 12.1   Different modes and their strengths

Figure 12.1 illustrates this further by showing how the private modes generally make less efficient use of space (capacity in passengers per square kilometer used by the system) and time (capacity in passengers per hour the system is available). The figure also shows how that dichotomy between private and public modes misses out another important category of modes beside public, private, and shared, namely active modes. Often, the travel chains of individual travelers use several types of modes, with links in low demand areas more efficiently provided with individual transport and links in high demand areas more efficiently provided through shared services: think about someone cycling to a station to take a train or taking the car to an airport to take a plane.

Shared modes have become a staple of the mobility landscape through platform technologies. They have allowed more intelligent coordination between supply and demand of capacity in vehicles in real time. Consider the way in which companies like Uber and Lyft scaled up the coordination that taxi providers traditionally had provided using phone and radio, based on the propagation of smartphones with internet connectivity, locational positioning technology, and payment possibilities, and their apps connecting to their services: the platform. That allowed for a far more efficient coordination of *sequentially shared **transport** services*, or taxis. Every smartphone owner can use Uber with the same app across the world. The same platform technologies also allowed for *sequentially shared **vehicle** services*, like bike and car sharing. In that context, those technologies allow for far more efficient linking between supply and demand; a single transport or vehicle service can now be used constantly by any smartphone owner in the world, because it allows all to find, reserve, pay for, and use a transport or vehicle service made available for sharing.

With a clear challenge of a fragmented and further fragmenting landscape of mobility services and the promise of platform technologies addressing the key issues, it does make sense that expectations on what effects an integrating platform like MaaS can bring are high. However, these expectations have been both positive and negative, which this chapter addresses later. First, the next section discusses how these platform technologies can be combined into MaaS.

MOBILITY-AS-A-SERVICE

The potential of a coordinated supply of mobility has improved significantly with the rise of information, communication, and sensor technologies that have become commodities in all modern smartphones. Planning travel, making capacity available at the right time and location, paying for the services – all these elements of mobility service provision can be supported by these technologies. First, these technologies allowed for real-time information on planned public transport services, moving from printed timetables for separate public transport services to real-time planning apps for all services in an area on any connected device. Second, these technologies allowed for the rise of the shared modes mentioned above, since they facilitated reserving, tracking, and unlocking cars, scooters, and bikes and identifying and localizing the users that want to use them. Third, these technologies now hold the promise to tie everything together, providing the mobility user a one-stop shop for all their mobility needs. This is where MaaS comes in.

Various authors give a wide range of definitions of MaaS. Cruz and Sarmento (2020) cite 15 different definitions. The first definition (Hietanen, 2014) focuses on a single interface for all mobility needs. Others add packaging of services (Cox, 2015; Holmberg et al., 2016), first general and later personalized (Atasoy et al., 2015), and integration of mobility services through a platform (Finger et al., 2015), for planning, booking, and paying (Smith, 2020). Smith and Hensher (2020) point at the fact that many early definitions use value-laden terms (seamless and user-centric) rather than functional (planning of multimodal trips) or structural (application and a platform) terms.

Here we will define MaaS as a substantial subset of the following functional elements (like Jittrapirom et al., 2017) with the specific aim (see also Durand et al., 2018) to *integrate different mobility services* for the traveler at various levels of integration (Sochor et al., 2018).

- Provision and traveling (the trip)
  - The MaaS user is supported with access to public services and provision of shared and individual services like a car or a taxi, and during travel with information on the trip and replanning in case of delays or cancelations.
- Planning and booking (the plan)
  - The MaaS user is supported from a single interface to plan multimodal trips and make reservations for those parts of the trip that are individually provided, like the reservation of a shared car or taxi.
- Remuneration and identification (the transaction)
  - The MaaS user is supported with identification for all other elements (access to platform and services, plan and book, individualized packaging) and an integrated payment system allowing the simplified use of the system.
- Platform and packaging (the offer)
  - The MaaS user is supported by an integrated platform for all the functions above, generally via an app, and can choose various plans or packages to provide for their mobility needs, with more inclusive plans or plans excluding some services or allowing for a limited amount of travel per period.

In the original and rather narrow perspective (Hietanen, 2014), MaaS is provided by a company using a (global) platform to integrate all (or at least most) of the above functions. The company becomes the intermediate player between the providers of various transport and vehicle services and the user, by integrating these services for that user. This was seen before in communication, with companies providing plans combining television, home and mobile telephone, and internet access, selling the variety of services in packages to users. In addition, it takes cues from platform companies like Booking.com, Takeaway.com, and Uber as they proved it is possible to become a global integrator. Uber, for example, provides taxi services from Sydney to Santiago and from Singapore to Sacramento (see also Finger et al., 2015). So, MaaS service providers offer packaged mobility to travelers anywhere as an intermediate between this traveler and a wide variety of mobility providers, like public transport operators, car and bike sharing companies, taxi companies, and more. And indeed, companies have sprung up with MaaS services (like Optimod, UbiGo, Tranzer, and Whim), integrating services through MaaS apps and mostly also providing packages of mobility services to travelers (see Cruz and Sarmento, 2020). In that sense the innovation has materialized in real services. This particular implementation can be seen as the market-driven and narrow form of MaaS.

This narrow form is now seen as only one of the possible forms in which MaaS can be provided, and a broader perspective has gained traction, allowing for more varied ways in which platform technologies can support integrating different mobility options. The broader perspective (see also Jittrapirom et al., 2017) also has a wider variety of incarnations all helping the user overcome the boundaries between modes, often also with a stronger role of government than in the narrow perspective. However, in that broader perspective it is far less clear what the expected end-state is, and consequently it is less clear how mature the innovation is.

MARKET AND GOVERNMENT IN MAAS

In recent years, the broader perspective has received more attention, as it has become clear that for the narrow version to function, a great deal of conditions have to be met that are not under the control or influence of the market. Governments in many regions and countries are stepping in. Their position towards MaaS in the narrow version is somewhat incongruous. On the one hand, they are looking for ways in which to incorporate MaaS, as it is a promising way to improve mobility options with all the benefits for the economy and society without expensive investments in infrastructure or the costs of running scheduled services in areas and at times of low demand. They expect MaaS to provide easier access to public and shared services for less experienced travelers, with positive effects on congestion and emissions, safety, and health. These governments are seeking improved integration, even if that does not provide the full set of MaaS functionality. On the other hand, MaaS could actually result in negative societal effects. If MaaS service providers provide more travelers with the means to use a car and the app often advises travelers to use that car-based option, the reduction of congestion and emissions and the improvement of safety and health, along with other public values governments might seek with mobility policies, might be thwarted (Becker et al., 2020). Governments understand that the promise of MaaS to integrate a fragmented landscape of services is attractive, but that the societal effects of MaaS very much depend on the way that the service is providing options and advice to the traveler. Further along we have a broader look at the possible effects of MaaS.

The possible downsides of MaaS have led governments to take an active role in its implementation in various ways. In the analysis below, we see regulation of transport providers opening up to MaaS. In addition, we see regulation of pricing of transport services bought in bulk by MaaS service providers for packaging in their offers to travelers. Moreover, we see experimenting with MaaS implementations, facilitation of MaaS service provision, and even the procuring of MaaS services from providers for a particular region, including the provision of shared services. The ideas on introducing MaaS seem to shift from a possible disruptive innovation by commercial market players to an innovation that is well embedded in the existing landscape of players (see also Geels, 2005). And this shift will have an effect on the societal consequences that MaaS will have. To understand how this shift has influenced the role of different players in MaaS and possible societal effects, the next section describes three real-world cases.

CASE DESCRIPTIONS: AMSTERDAM, BIRMINGHAM, HELSINKI[1]

With the early examples of MaaS, after the arrival on the market of MaaS Global and the first experiments in Helsinki, it became clear that MaaS implementation as a market initiative without government involvement would be challenging. The involvement of public transport operators, often under the control of public authorities, was needed for MaaS service providers to be able to include public transport services in their overall packages. Governments themselves, confronted with a growing number of shared mobility services and the high costs of improvements to public transport services and private transport infrastructure, sought ways to let MaaS do its work improving mobility without inducing a great deal of costs and by letting different services work within the mobility landscape in the way they would work best. And their expectation was that MaaS would deliver on integrating that ecosystem. However, governments were also confronted with the possibility that the implementation of MaaS focused on the interests of private players could have detrimental effects on public values, as mentioned before (see Hirschhorn et al., 2019).

With the governments developing their role related to MaaS, the views on its implementation seemed to have changed from a narrow private actor add-on to the existing mobility landscape, to a more tailored implementation within existing governance. In implementing MaaS, governance changes that would allow better integration of mobility options were also developed, like required sharing of information or regulated pricing of bulk sales of services. As stated in Chapter 2 of this book, this means new ways of initiating change. For example, in the Netherlands, the national government started various experiments with many stakeholders involved in specific MaaS implementations, and it worked in close collaboration with these stakeholders, directly engaging with them to understand and develop the wider mobility service, make the required institutional and governance changes, and evaluate real-world outcomes. Compare this to the Finnish case, where the national government's main contribution was to open the market for MaaS by way of new legal rules. Or the British case, where the national government focused mainly on informing the field with research and policy briefs.

This chapter presents the way in which three cities, in the context of their respective countries, introduced MaaS and highlights how the approach to the introduction was related to the existing governance and triggered new questions about that governance. And it looks at the way in which the more disruptive, niche-driven and narrow implementation seems to have given way to a more encapsulated, regime-driven and broad implementation.

Amsterdam in the Netherlands

Vervoerregio Amsterdam (VRA) is the transport authority in the Amsterdam region, and functions as a cooperation of the municipalities in that region. It is responsible for tendering public transport for the region, covering around 300 square kilometers, and separated into four different concessions. The concession for Amstelland Meerlanden (AML) was coming up for competitive tendering in 2018. In 2016 the transport authority recognized in their requirements the potential of MaaS for the concession. In the earlier concession in the area, the operator was relatively free to design the services. In the new concession, the authority was, on the one hand, taking more control over the planning of scheduled services but, on the other hand, giving a lot of freedom to the operator to include MaaS in the bid and eventually in the services provided for the region in the concession. MaaS was seen as a potentially valuable addition to the scheduled services, particularly in those situations where these scheduled services were less efficient or effective. The concession was won by Connexxion, a Dutch Transdev daughter company. The company added AML Flex, an electric car taxi service, to be booked through the wider concession specific app, phone, or website. To strengthen integration, the service could be paid for with the national smart card (OV-Chipkaart). In addition, a shared bike system was introduced in the concession. At implementation, it became clear that the authority and the operator disagreed on what patronage data would be made available to the authority. This triggered a rethink at the authority on the governance of data in this and other concessions and a reassessment of the roles of the operator and authority as service integrators.

There are a few elements of governance on a national level that are relevant for what is happening in this regional concession. Regional governments in the Netherlands have a strong role in public transport and, with a few exceptions, are responsible for tendering services out to private parties. In those tendered concessions, most regional authorities set general norms on public transport services (like minimum levels of service) for the operator designing the services, allowing the operators to define the services. However, some authorities take a more hands-on approach by defining the scheduled services themselves and only asking the operators to carry them out. Funding in the region of Amsterdam is provided by the national government, with a lot of freedom for the authority to use the funding for the provision of mobility-related services. Also, on a national level, the Netherlands has a smart-card system for traveling on public transport, as well as a platform consolidating planned and real-time schedules and an app to plan a trip. The national government itself is also active in developing MaaS in the country. It has selected seven experiments (one in the Amsterdam region) in which public and private parties work together.

When compared to the elements of MaaS mentioned above, the implementation is building on existing functions of identification and payment in the national travel smart card system. This is atypical for an integrated MaaS platform, which generally has its own identification and payment functions. In addition, it uses the existing national platform for trip planning, which provides flexibility to tie it into other planning apps and tools. Moreover, the implementation includes, in a single contract between the authority and the operator, both the integrating platform role (with functions like planning and paying) and the provision of services role (providing the mobility services, public transport, taxis, and bikes, to the traveler). This moves away from the narrow interpretation of MaaS mentioned before, which focuses on just the platform role. Also, the governance of public transport is shifting to make room for MaaS in a very specific way: as a part of existing concessions. The transport authority is taking more control over scheduled services and is pushing public transport operators to start delivering more integrated services to the traveler.

Birmingham in the United Kingdom

Transport for West Midlands (TfWM) is the transport authority for Birmingham and its surroundings and is also a cooperation of the municipalities in the region. It has a far less controlling role in public transport provision, as operators in the country (with the exception of London and Manchester) can independently initiate public transport services. The authority does play a role in consolidating the travel information of all the operators in the area, and it also manages the travel smart card (Swift). With the open market in the United Kingdom, bus operators are relatively autonomous in defining their services. This could lead to the provision of a fragmented set of services. In the Birmingham region, in 2015, the Bus Alliance started, in which the operators work together with the authority and others to improve the integration of the various bus services. After a study in 2015 initiated by the cooperating councils in the area, they saw the potential in improving the relatively fragmented landscape of services in the area. TfWM approached MaaS Global to see what they could offer. National Express (a major bus operator in the area), Gett taxis, NextBike, and Enterprise rent-a-car joined the initiative. They aimed at bringing Whim, the MaaS Global platform, to the region. It was set up to work with the existing smart card and was open to other operators to join in. Also, data sharing was part of the agreement. The pilot started with 500 users.

As the project commenced, it became clear that implementation was tricky. The way in which fares were structured, with large freedom for the operators, proved to be a barrier for the smooth implementation of MaaS, where the platform needed to integrate that variety. In addition, the market is open, which means that operators can start and stop operations whenever they please. That variety and those dynamics were somewhat harnessed by the existing cooperative governance bodies in the region, namely TfWM and the Bus Alliance. They are more stable institutions in the region and have already worked on integration of ticketing, through the smart-card system, and scheduling of services, through its role on travel information. Also, the authority is rethinking how it can further integrate through Swift and providing travel information, beyond the Whim initiative.

The national government in the United Kingdom is playing a limited role in relation to MaaS, mostly acting as a knowledge broker. It has been pushing for smart cards for ticketing on the national rail network. But compared to the other two countries discussed here, smart cards in the United Kingdom are not widespread, despite the London Oyster card being in use since 2003. The privatization and related franchising of public transport has limited the coordinating role of the government over the last decades, while in other countries governments have developed a major role in providing integrated ticketing and pricing. Currently, discussions are underway to bring the rail services back into the public sector. Even though there is a national system for through fares, there are no public transport wide national platforms for travel planning nor for payment systems.

When looking at the elements of MaaS in this case, with the privatized context and a relatively limited and recent introduction of smart-card payment (without integrated pricing), the added value of a MaaS platform could be the most substantial of the three cases discussed here. However, the existing governance seems not to be helpful. The open market service provision in the United Kingdom outside London and Manchester can be very fluid. It is not clear to authorities which operators will be running services in the area in the future, as they are free to enter and exit the market. Linking the platform to local operators requires robust and (inter)national standardization for the platform to be able to buy and plan trips for its MaaS customers. This could be done through either regulation or general willingness of the competing operators to open up to MaaS, but both seem unlikely in this case given the hands-off approach of the national government and the limited willingness of operators to step in. The regional Bus Alliance could play a role here, as it has set goals towards integration of services but isn't currently playing a part in this. MaaS in the West Midlands is not the expected integrating layer over all modes, but rather a cooperation between the authority, the platform provider, and a group of mobility service providers. Although open in its set up, involvement of service providers is necessary in this implementation to provide a real integrated alternative.

Helsinki in Finland

HSL is the public transport authority in the Helsinki area. It covers a select part of the greater Helsinki region with, as of January 2018, 7 of the total of 14 municipalities in the area cooperating. HSL has a policy-oriented role, with preparation of the Transport System Plan for the wider cooperation of jurisdictions on urban planning and transport policy in the area. HSL also plans the services and procures services from operators, both private and publicly owned companies, to carry them out.

Helsinki is seen as the birthplace of MaaS. MaaS Global, a key player in the space, has really pushed the original ideas of mobility as a service, showing the extent to which mobility services are fragmented and the potential of a platform solution for this problem. But even here, a key problem surfaced for MaaS service providers. Many operators provide month or year passes at a subsidized price compared to daily and single trip tickets. HSL would allow MaaS Global to buy single tickets for their customers, but not the reduced-price passes. This meant relatively high prices for users of the MaaS platform, making it viable only for specific high-income customers seeking high levels of service.

On a national level, the government is seeking ways to support MaaS platforms by introducing legislation that would simplify access to the market for MaaS service providers and their platforms. So the Finnish national government is focusing on market opening, differently from the other two examples in this chapter. The Dutch national government has focused on facilitating local and national experiments in MaaS service provision, after a history of breaking down barriers in public transport ticketing, pricing, and planning. The United Kingdom's national government, which is only taking the first steps towards further integrating ticketing and planning, is focusing on a role as a knowledge broker, rather than taking an active role intervening in the market.

Case Evaluation

We see different approaches in the three cases described above. On a regional level, the introduction of MaaS is dependent on the role of government and market players in the region. This is conditioned by (supra)national regulation and set in a broader governance framework. In Amsterdam, the government is procuring MaaS in its tendering of concessions, with MaaS being the shared delivery of the concession's public transport operator, which invited a global MaaS service provider as a partner. The platform is closed, but all regional mobility services in the concessions are included. In Birmingham, the government is only playing a facilitating role in bringing local mobility providers together with a global MaaS service provider, providing integration for the travelers of a limited subset of operators, who see it as a competitive advantage to join. The platform is open, but only a subset of the mobility services in the region is included. In Helsinki, the global MaaS service provider is only able to provide services at high costs, which provides a barrier for the integration, as it comes at a high cost to the MaaS operator and consequently the traveler. The authority is not buying integrated services for a regional concession, like in Amsterdam, that could include shared modes. Rather, the authority is buying separate services from operators, that it integrates itself through integrated pricing and ticketing, scheduling, and planning. Table 12.2 summarizes the differences between the cases.

On a national level, again three different approaches are observable in the cases. The Finnish government seems to be focusing on allowing private MaaS service providers to develop their role, independent of authorities, public transport operators, and mobility providers, by regulating open access to pricing and ticketing. The Dutch national government on the other hand is letting these parties work together to develop solutions in specific situations in which MaaS could offer additional value, and changes governance according

Table 12.2 Roles of actors and context of MaaS implementation compared

		Amsterdam, the Netherlands	Birmingham, United Kingdom	Helsinki, Finland
Roles	*National government*	Experiments starter	Knowledge broker	Market opener
	Regional government	Procurer of integrated mobility services	Facilitator of MaaS platform for the region	Facilitator to MaaS service provider
	MaaS service provider	Platform provider to public transport operator	Platform provider to regional public transport operators	Platform provider next to regional authority and transport operators
	Public transport operator	Provider of integrated mobility services procured by regional government with support of MaaS service provider	Provider of non-integrated mobility services, opting in to MaaS platform to add a layer of integration	Provider of services to MaaS service provider, legally forced to cooperate
Innovation	*The MaaS platform*	Single closed platform for the mobility provider in the region	Open platform with voluntary cooperation of mobility providers in the region	Supplemental platform with required cooperation of mobility providers in the region
	MaaS and existing public transport	MaaS absorbed in existing public transport provision	MaaS added on top of existing public transport provision	MaaS accepted next to existing public transport provision
	Existing multimodal integration environment	Mature: national integrated ticketing, pricing, and trip planning	Immature: regional development of integrated ticketing underway, with pricing and planning lagging behind	Semimature: regional integrated ticketing, pricing, and trip planning

to their findings. In addition, the national government has pushed for integration of ticketing and planning in public transport earlier, with the operators developing the platforms. Both approaches are hands-on, in different ways. In the United Kingdom, the national government is more hands-off; while some initiatives are being put in place to provide better integrated ticketing and pricing for rail services, in the field of MaaS the government is mainly operating as a knowledge broker.

SOCIETAL IMPLICATIONS

In the current state of the innovation, it is hard to predict the societal impact of MaaS. Utriainen and Pöllänen (2018) argue, based on several pilots, that MaaS will play an important role in changing travel behavior. But how is still rather unclear. Smith et al. (2018) expect that MaaS will have a positive effect for society. They expect access to mobility options will improve, as MaaS will make the use of shared and public modes easier. They expect the position of public transport to be strengthened, as public transport services can be oriented more to profitable areas and times, with shared modes available for less profitable areas and times, and with shared modes providing easier access to public transport. However, Jittrapirom et al. (2018) pose that it is still not clear to what degree MaaS is in fact able to reduce car use and ownership. MaaS and its technological, market, and institutional contexts still has so many moving parts that it is unclear what the effect will be, beyond that it could create a modal shift to public transport or to private (or shared) cars, with the well-understood effects of that. Here we distinguish a few mechanisms that could drive the development towards specific impacts, derived from the cases above and the literature. Those mechanisms provide us a glance of the possible futures of MaaS.

First, let's start with the obvious. In much of the MaaS literature, the landscape of mobility services is described as being very fragmented. However, this is only true when we go beyond the private car. Public infrastructure for the car is so good and so ubiquitous that owning the vehicle provides a highly integrated mobility system: a single mode for every trip from shopping at the supermarket to crossing the continent. The expansion of public infrastructure has long been the obvious way of improving mobility but is under scrutiny in many countries. Its high costs, financial and societal, and its limited benefits because of the induced demand slowing traffic down again swiftly after infrastructure expansion (Noland, 2001; Hymel, 2019) all make infrastructure expansion a less obvious solution. But to the user of mobility options the fragmentation is only a problem in those contexts where the car doesn't work. That can be in dense urban areas, or in circumstances where mobility poverty means a driver's license or a car is out of reach for many inhabitants.

Beyond the private car, public transport has indeed long been a fragmented landscape with modal (train vs bus), regional (jurisdiction vs jurisdiction), and organizational (operator vs operator) islands in terms of planning of and paying for trips. However, public transport has been performing better than the private car on safety, spatial efficiency, and emissions, making it an attractive policy alternative or addition to the private car (see also Veeneman, 2012). Consequently, policy makers have been supporting public transport, especially in those situations where the private car can be problematic from a mobility policy perspective, like in cities and for vulnerable groups. Part of making it attractive has been battling the fragmentation. In several cities and countries, that has already led to the development of easy physical transfers (like integrated service design in places like Canton Zurich (Veeneman, 2002)), integrated ticketing and pricing schemes (like the Hong Kong Octopus card and later the London Oyster card (Veeneman et al., 2018)), and standardization and integration of digital schedule information for all modes (like the Dutch OV9292 and later Google Transit (Veeneman et al., 2018)). Fragmentation is a real thing in public transport and it has been getting attention for quite some time.

And then there are the new shared modes that are being rolled out. This chapter has already discussed how these shared modes have become an alternative for both private and public transport, increasing the fragmentation. Karlsson et al. (2020) argue that governments need to give space to shared mobility and MaaS, as these new mobility options will undoubtably create societal value. Veeneman et al. (2020) argue that far more than just giving space, the better role for government is probably a more active one in bringing shared modes and public transport closer together, as users of public space and infrastructure, and to harvest their potentially synergetic relation and harness a possibly erosive relation. In line with Chapter 2 of this book, this requires a long-term involvement of key parties and co-development of more than just a platform or shared modes, but also the market, governance, and institutional context.

Consequently, even though efforts have been made to reduce fragmentation within public transport (through integrated payment and planning systems), the rise of shared modes has driven that fragmentation up again. This is where MaaS can really play a role in tying those modes, public and shared, together. In addition, private cars and public transport have their strengths and weaknesses in various contexts. Connecting the shared modes to the bigger system can allow for a far more appropriate use of modes in particular contexts. That can be done, for example, by making it simpler to transfer between modes and include individual shared modes in those areas where, and at those times when, scheduled public transport is too expensive to provide. And that can be done by supporting the transfer to public transport when and where car travel into the city is far too costly to a person or to society. That could have positive effects on the sustainability and efficiency of mobility (Cruz and Sarmento, 2020). That is a clear promise of MaaS.

However, whether that promise materializes depends heavily on the way the governance is developing. This brings us to the second mechanism. MaaS services are provided through a set of transactions between four stakeholders (see also Jittrapirom et al., 2020):

- Travelers,
- Mobility service operators,
- Government authorities, and
- MaaS service providers.

Generally, in any jurisdiction a tripartite relation exists between the first three stakeholders mentioned: travelers (as consumers and citizens), mobility operators, and government. Now a fourth party, an often private MaaS service provider, is entering and changing this into quadrilateral relation, with the potential for it to monopolize the relation with the traveler through its control over the single interface. The key incentive to the private MaaS service provider must come from one of the other three players (see also Cruz and Sarmento, 2020). When the incentive comes mostly from travelers as consumers, plans and planning from the app could favor travel speed or comfort as the key values, which could lead to more use of individual modes of transport, like shared cars, with possible negative societal impacts. When the incentive to the MaaS service provider comes mostly from the mobility operators, the operator with the deepest pockets could push their mode on the platform. This also has the potential to favor individual modes, that arguably can be more profitable, again with possible negative societal impacts. When the incentive to the MaaS service provider comes from governments, the expectation can be that MaaS platforms could favor modes supportive of public values.

In all three scenarios, the MaaS service provider must of course create value for all the three other stakeholders and cater to their interests. However, the monopolistic position of the MaaS service provider in a specific region once the platform has captured a substantial part of the market allows it to skew service provision towards the stakeholder providing the largest financial incentive (Frenken et al., 2020), which could be none of the other three but rather the main investor in the platform. A parallel here is how Google, through YouTube, is providing real value to its users, both content producers and consumers, but the incentive for Google in terms of how that value is provided is clearly to focus on catering for their advertisers (see for example Ruckenstein and Granroth, 2020), as they need to be profitable for their investors.

Whether MaaS will have positive societal implications depends heavily on the way in which it will be implemented, and in particular the way in which the governance changes will institutionalize and incentivize the relations between the key stakeholders. From the case studies of early MaaS implementations in Amsterdam, Birmingham, and Helsinki earlier in this chapter, we see a major role for public authorities. With their role in infrastructure, public space, and public transport, and because of the need for them to regulate private transport because of its externalities, they are already key players in the field. And because of that we see only limited implementations that favor the private side, by either private operators or MaaS service providers, in ways that could be at the detriment to the public side: the societal value of the innovation seems to be well secured. That could mean that the role of the MaaS service provider will end up rather regulated, maybe even as a procured separate service element. This could especially be the case in those instances where public transport authorities haven't been successful themselves in integrating key elements of mobility services, like planning trips and paying for services (see for example Veeneman et al., 2018). There is clearly a risk that the competitive push driving the innovation will weather away with this public control, leaving the mobility landscape still fragmented, as the public authorities regulating or procuring the MaaS services still generally represent a subset of the mobility services a traveler in the area could take. The Amsterdam example shows that authorities could decide to include more shared services, simplifying MaaS as the addition of more modes provided to the traveler in an integrated way by the public transport authority.

Finally, it is obvious that the traveler holds the key. This is an ecosystem change (Cruz and Sarmento, 2020) with all players realigning themselves in the short term with possibly changing mode choices, and in the longer term with possibly new mobility patterns and the related changes to spatial patterns. A new equilibrium might appear, but only if significant numbers of travelers want the change. It also is a change that is not suitable for a classic model of policy problem, assessment, and implementation, it being far more a system transformation and institutional change. Also, we see that a possible disruption that might have been expected seems to be unlikely, due to the classic mechanisms of transitions (see Geels, 2005).

In the big picture, a case could be made that MaaS service providers will become superfluous, as stakeholders other than MaaS service providers are providing the traveler with the tool to integrate all the services. Already now, their mobile phone provides travelers with an integrated tool for planning, reserving, identification, and payment for most of the modes in Table 12.1, all through separate apps. One could argue that the phone is providing the traveler with the most flexible tool for all elements of traveling, enabling the inclusion or exclusion of services through adding and removing apps. Obviously, this phone as an integrator is not for all, but it could limit the role of MaaS to very particular niches – for example where MaaS services can be helpful for specific vulnerable groups or in changing the modal split in a specific situation – rather than being a generic tool for the defragmentation of the mobility landscape. The mobile phone with its apps could already be the key mobility integrator.

SUCCESS/FAILURE FACTORS

Looking through the chapter so far, there are several factors that could drive the success of MaaS in the extent to which it will create societal value. First, the *existing level of integration of mobility services*. This has to be a key characteristic driving the potential of MaaS, but in two directions. When integration is already available for several of the functional elements of MaaS, the traveler's need for MaaS will be less, as probably will be the inclination of that traveler to sign up to a MaaS service. In other words, when it is easy to book a bike, scooter, or taxi at a station there is no need for a MaaS service provider to make that simpler. Paradoxically, when integration of several of the functional elements of MaaS is already available (for example when open public transport planning tools exist and ticketing and pricing systems are open for many modes), MaaS implementation will indeed be simpler (see also Veeneman et al., 2018). So, mature and open multimodal planning and paying for services reduces the need for MaaS for travelers but makes its implementation simpler. This brings us to the second factor.

Second, *existing planning, booking, pricing, and payment system variety*. MaaS service providers operating globally have to deal with a huge variety of systems for planning of and paying for mobility services. To those currently controlling the local systems (mostly transport authorities and operators), there is an interest in keeping control over those systems and in keeping their current direct link to the traveler and not relinquishing that to the MaaS service provider. Standardization of those systems for planning and paying can help enormously in the roll-out and scaling up of MaaS-providing platforms. Google has shown how that can work with the way in which their inclusion of public transport in Google Maps has kickstarted the standardization around scheduled service data communication in various General Transit Feed Specification (GTFS) formats (see also Veeneman et al., 2018). Similarly, payment systems like, again, Google or Apple Pay could help standardize payment for travelers. Likewise, smartphones allow for identification, which also can be used globally. However, standards for reserving a seat or a vehicle and standards for bulk capacity pricing are still missing. If these have to be implemented in various ways with a wide variety of system interfaces, this will make it harder to roll out MaaS globally. This is illustrated by the roll-outs with a stronger regional character that we have been seeing in this chapter. This also bring us to the third factor.

Third, *open pricing schemes*. A key challenge for global MaaS service providers is being able to buy and resell capacity in public transport and shared modes at reduced costs (see also Jittrapirom et al., 2020). Not only are standards missing there, but also the incentive for those controlling the pricing schemes to open the schemes up are limited. Existing pricing schemes and the value of direct sales for public transport operators and shared mode operators makes these operators hesitant to allow MaaS service providers to resell capacity. In the Helsinki case above, we saw how the Finnish government is pushing legislation to open pricing up, to allow MaaS service operators to thrive. In the Dutch case, it was solved altogether very differently, as the government was procuring the service as part of a larger package, asking the cooperating operators and service providers to come up with a shared pricing scheme. If open pricing schemes do not develop, the costs to the traveler of a package from a MaaS service provider will obviously be very high and only attractive to a small group of service-oriented, cost-indifferent travelers.

Fourth, *government and operator attitude towards MaaS as a risk*. A further factor is the way in which governments will position themselves towards MaaS (see also Jittrapirom et al., 2020). Obviously, it is to be expected that authorities value the defragmentation of the mobility landscape in their jurisdiction for their constituents, because it is a relatively simple and cheap way of improving the service the authority provides. On the other hand, they often have a stake in the game by having agency over specific modes (see Veeneman and Mulley, 2018), modes they prefer to strengthen or protect. And there can be a fear with governmental actors that MaaS service providers will not have the public interest at heart (van Waes et al., 2018), which they could see as a risk. We have seen responses in line with more hands-off approaches in Finland (in line with Pel's "conventional views" – see Chapter 2) and in line with more hands-on approaches in the Netherlands (in line with a "transition perspective"). If governments choose the latter, MaaS implementation is more likely, as the downsides for the existing actors in the market can be overcome. When involved these existing actors are likely less prone to fend off (niche) challenges to their position in the regime (see Geels, 2005). But, as with any transition, the character of MaaS will change accordingly, probably being less disruptive and less global. For some actors, like the entrepreneurial MaaS service providers, that will make MaaS less of a success; for others, like local governments, it will be seen as more successful.

Fifth, *potential customer base*. A final and obvious factor in the success of MaaS service providers to gain customers is whether their propositions are attractive to large groups of travelers (see also Jittrapirom et al., 2020). In the narrow form (a commercial provider of MaaS on top of existing services), the potential of this MaaS approach to be successful is highly dependent on the existing integration of services and the propensity of travelers to integrate services themselves through means of their smartphones. In the broader form (less fragmentation in the supply of mobility services), MaaS could become part of the mainstay of mobility provision in a region, like when a transport authority asks a MaaS service provider to bring all modes together into one – in this case governmentally controlled – app. This could help significantly in gaining customers.

Whichever way the MaaS concept will develop, narrow or broad, it has already had a major impact on the world of mobility. This impact is that the problem of the fragmented landscape of mobility services is higher up the agenda than ever and that the sector has a clear perspective in terms of reducing that fragmentation. Mobility service providers, policy makers, and academics are spending a great deal of time on the problem of fragmented mobility services, and many smaller repairs are being made because of that heightened interest, with or without the label of MaaS attached. In that respect MaaS, as the concept of integrating services rather than as the specific form of global mobility service providers, has been an innovation success.

RESEARCH CHALLENGES

In the perspective of this chapter, the research challenges for MaaS can be found in four areas. First, while research has already been carried out to understand the *general attitude* of travelers towards using MaaS services (for example Liljamo et al., 2020), this has little basis in empirical "real-world" cases and analyses. Any survey on the general characteristics of a mobility service like MaaS obviously provides a limited representation of the real-world mobility choices once the system is up and running. And indeed, in an environment where a few travelers are using MaaS or where various levels of integrative systems for planning and paying for mobility exist, the attitude is probably different than measured in that research. The dependencies between context variables and the willingness to use MaaS are still unclear. In the new transitional paradigm, the research problem here is that of experimenting and evaluating, rather than that of a priori analysis.

A second challenge is the effect of MaaS service provision on *travel behavior*, given specific service characteristics. This area is still not very well developed, with the first examples of research only recently being published (for example Alonso-González et al., 2020). It has to deal with the same challenges as the research mentioned above, with additional complexities. An attitude is directly measurable, even though the respondent might find it hard to understand the service, but the effect on travel behavior is dependent on much more than attitudes only, and is generally a result of a longer process of decision-making by the traveler. This is further compounded by the fact that most existing transport models still do a poor job when it comes to including shared modes, let alone integrating various modes. In the new transitional paradigm, this research could take off and develop quickly.

A third challenge is to understand the possible effect MaaS services will have on key *public values* related to mobility, like accessibility, sustainability, safety, health, inclusiveness, and more. There is the belief that MaaS can support a number of those public values (see also Veeneman et al., 2006) like sustainability, accessibility, and others, with expectation being positive on higher collective efficiency (both spatial and temporal) with faster travel and less use of space. Travelers could access shared modes more easily through MaaS, with the possible gain of reduced use of space compared to full private modes and more direct routes and hence faster travel than public transport can sometimes offer. Those gains could be added to the gains of easier use of public transport in terms of spatial and temporal performance, by quicker access to public transport through shared modes and simpler payment and planning of trips. Or it could help financial efficiency, for example by using demand-responsive services in times of low demand, with the traveler being supported through MaaS. However, MaaS could also increase car use through growing shared car availability, with expected effects on congestion, health, safety, and emissions. Obviously, with traveler attitudes and behavior still being unclear, estimating the effects of MaaS on key policy outcomes is highly challenging. Again, in the transitional paradigm, it will be important to keep track of these effects while experimenting and reworking the solutions.

Finally, different *business and governance models* are still in development, with no clear winner yet, as the case descriptions in this chapter showed. These different models will have different value for the different stakeholders, which will provide different incentives to those involved, which in turn will drive different focuses in the way that the services will be provided to the traveler (for an example see Hirschhorn et al., 2019). And this makes the circle round. If we do not understand the way in which the governance of MaaS will develop, it will be hard to predict the exact form in which MaaS will be implemented. This makes it hard to let users experience it and develop a mature attitude towards it, and following from that, to establish what the best estimate of usage is and what the effects on various public values will be. And at the same time, the most appropriate governance, more specifically the role of governmental control of MaaS, is highly dependent on what can be expected in terms of outcomes in use and how they relate to public values.

More real-world empirical analysis on all four aspects is needed, preferably in experimental and transitional settings, as well as the development of better multimodal modeling that includes shared modes in a mature way (see also Kamargianni et al., 2019). In line with Chapter 2 of this book, the complexity of and interdependence between these aspects asks for a transitional approach, with lots of experimentation and the involvement of relevant stakeholders, including the research community that has developed around the topic.

CONCLUSIONS

The expected global reach of Mobility-as-a-Service providers, building a world-wide platform for the integration of more or less all mobility options, has not materialized (yet). Globalizing a platform for hotel bookings, taxis, and food deliveries has proven to be more straightforward than globalizing a platform for mobility service provision. Part of that challenge has been that many transport operators are government-controlled monopolies, and many of them have localized ticket and payment systems. Whereas competition between hotels, taxis, and restaurants drove them to work with platforms to gain a competitive advantage over the others, a monopolistic transport operator can choose to ignore MaaS service providers.

For MaaS, there are so many moving parts right now that clarity about the outcome is hard to come by. This chapter describes several of those moving parts. However, there is a great deal of potential in reducing the fragmentation of existing and new mobility options. The best way forward is probably through experimental forms of layering MaaS in fitting ways on existing services.

We have shown in our cases that implementation of MaaS in various countries involves governmental players and that the earlier expected narrow "pure market" layer over all mobility services is not producing the major disruption to mobility that some would have anticipated. However, these examples are not representative of the entire MaaS landscape, now or in the future. It is still highly uncertain how MaaS will find its place and what effects it will have in the mobility markets. That outcome will be highly dependent on whether the implementation is more oriented towards traveler comfort, operator profit, governmental influence, or MaaS service provider control. With an imbalance in these, MaaS implementations that attract many travelers could have an extreme effect. For example, an implementation could drive people out of public transport as individual modes are more profitable to a monopolistic MaaS service provider, resulting in more congestion and negative outcomes on many public values. These can be desirable from the perspective of travelers or MaaS service providers, but are problematic from a societal perspective. At the other end of the spectrum, it could highly integrate various transport services, reducing the need for public transport services in areas where, and at times when, they are not optimal and at the same time simplifying and maximizing public transport use where and when its use provides major gains both for the individual traveler and for society, as it supports key public values. Both of these extremes in scenario still seem to be within the realm of possibility.

At the same time, there is a growing understanding that the fragmentation in mobility services should be addressed, fueled by the promising possibilities of MaaS. With that promise, and given the concerns, it makes sense to go through the transition together as travelers, MaaS service providers, governments, and operators of mobility services, by experimenting with implementing MaaS. Part of that transition would also entail changing regulation around MaaS, first in terms of creating good and open systems for planning, paying, booking, and the wholesale of transport capacity to resellers for a wide range of mobility services, and second in terms of limiting the possible negative effects on public values through optimizing services for the community, as opposed to allowing the monopolization of mobility provision and focusing on optimizing for the mobility provider or the individual traveler.

When smartphones allow for the integration of mobility services by simply adding apps, to what extent will travelers prefer a single fixed contract with a MaaS service provider to a set of flexible contracts with any provider of mobility of their choosing, all set up through a few clicks on an app on their phone? Maybe the innovations that make the MaaS platform possible are the exact innovations that make the MaaS platform and its service provider less necessary, if not redundant. Like with other innovations, there remain niches in which MaaS can still play a major role, like for companies wanting to provide an integrated travel solution to their employees or for regions wanting to provide an integrated solution to tourists visiting the area.

Introducing MaaS is a transition. The attention on MaaS has put the poor performance of existing systems – public, shared, and private transport – for both society and the traveler front and center. MaaS promises that facilitating the use of a wider variety of mobility options, including shared modes, can improve the performance of the total mobility system for all. It also shows that the complexity of such a transition requires us to revisit simplistic ideas of interventions and reconsider codeveloping this innovation so that it is a success for society as well as the traveler, for the public as well as the private.

NOTE

1. This section is based on Hirschhorn et al. (2019).

REFERENCES

Alonso-González, M.J., Hoogendoorn-Lanser, S., van Oort, N., Cats, O., and Hoogendoorn, S. (2020). Drivers and barriers in adopting Mobility as a Service

(MaaS) – a latent class cluster analysis of attitudes. *Transportation Research Part A: Policy and Practice, 132*, 378–401.

Atasoy, B., Ikeda, T., Song, X., and Ben-Akiva, M.E. (2015). The concept and impact analysis of a flexible mobility on demand system. *Transportation Research Part C: Emerging Technologies, 56*, 373–392.

Becker, H., Balac, M., Ciari, F., and Axhausen, K.W. (2020). Assessing the welfare impacts of Shared Mobility and Mobility as a Service (MaaS). *Transportation Research Part A: Policy and Practice, 131*, 228–243.

Cox, N.C.J. (2015). *Estimating Demand for New Modes of Transportation Using a Context-Aware Stated Preference Survey* (Doctoral dissertation, Massachusetts Institute of Technology).

Cruz, C.O., and Sarmento, J.M. (2020). "Mobility as a Service" platforms: A critical path towards increasing the sustainability of transportation systems. *Sustainability, 12*(16), 6368.

Durand, A., Harms, L., Hoogendoorn-Lanser, S., and Zijlstra, T. (2018). *Mobility-as-a-Service and Changes in Travel Preferences and Travel Behaviour: A Literature Review*. The Hague: Ministry of Infrastructure and Water Management.

Enoch, M.P. (2015). How a rapid modal convergence into a universal automated taxi service could be the future for local passenger transport. *Technology Analysis and Strategic Management, 27*(8), 910–924.

Finger, M., Bert, N., and Kupfer, D. (2015). Mobility-as-a-Services: From the Helsinki Experience to a European Model. *European Transport Regulation Observer*, (2015-01), 1–13.

Frenken, K., van Waes, A., Pelzer, P., Smink, M., and van Est, R. (2020). Safeguarding public interests in the platform economy. *Policy and Internet, 12*(3), 400–425.

Geels, F.W. (2005). Processes and patterns in transitions and system innovations: Refining the co-evolutionary multi-level perspective. *Technological Forecasting and Social Change, 72*(6), 681–696.

Hietanen, S. (2014). "Mobility as a Service" – the new transport model? *Eurotransport. ITS and Transport Management Supplement, 12*(2), 2–4.

Hirschhorn, F., Paulsson, A., Sørensen, C.H., and Veeneman, W. (2019). Public transport regimes and Mobility as a Service: Governance approaches in Amsterdam, Birmingham, and Helsinki. *Transportation Research Part A: Policy and Practice, 130*, 178–191.

Holmberg, P.E., Collado, M., Sarasini, S., and Williander, M. (2016). *Mobility as a Service – MaaS: Describing the Framework*. Gothenburg: RISE/Viktoria.se.

Hymel, K. (2019). If you build it, they will drive: Measuring induced demand for vehicle travel in urban areas. *Transport Policy, 76*, 57–66.

Jittrapirom, P., Caiati, V., Feneri, A.M., Ebrahimigharehbaghi, S., Alonso González, M.J., and Narayan, J. (2017). Mobility as a Service: A critical review of definitions, assessments of schemes, and key challenges. *Urban Planning, 2*(2), 13–25.

Jittrapirom, P., Marchau, V., van der Heijden, R., and Meurs, H. (2018). Dynamic adaptive policymaking for implementing Mobility-as-a Service (MaaS). *Research in Transportation Business and Management, 27*, 46–55.

Jittrapirom, P., Marchau, V., van der Heijden, R., and Meurs, H. (2020). Future implementation of Mobility as a Service (MaaS): Results of an international Delphi study. *Travel Behaviour and Society, 21*, 281–294.

Kamargianni, M., Yfantis, L., Muscat, J., Azevedo, C.L., and Ben-Akiva, M. (2019). Incorporating the Mobility as a Service concept into transport modelling and simulation frameworks. In *Special Report-National Research Council, Transportation Research Board*. Transportation Research Board.

Karlsson, I.C.M., Mukhtar-Landgren, D., Smith, G., Koglin, T., Kronsell, A., Lund, E., ... and Sochor, J. (2020). Development and implementation of Mobility-as-a-Service – a qualitative study of barriers and enabling factors. *Transportation Research Part A: Policy and Practice, 131*, 283–295.

Liljamo, T., Liimatainen, H., Pöllänen, M., and Utriainen, R. (2020). People's current mobility costs and willingness to pay for Mobility as a Service offerings. *Transportation Research Part A: Policy and Practice, 136*, 99–119.

Noland, R.B. (2001). Relationships between highway capacity and induced vehicle travel. *Transportation Research Part A: Policy and Practice, 35*(1), 47–72.

Ruckenstein, M., and Granroth, J. (2020). Algorithms, advertising and the intimacy of surveillance. *Journal of Cultural Economy, 13*(1), 12–24.

Smith, G. (2020). Making Mobility-as-a-Service: towards governance principles and pathways (PhD thesis, Chalmers University of Technology).

Smith, G., and Hensher, D.A. (2020). Towards a framework for Mobility-as-a-Service policies. *Transport Policy, 89*, 54–65.

Smith, G., Sochor, J., and Karlsson, I.M. (2018). Mobility as a Service: Development scenarios and implications for public transport. *Research in Transportation Economics, 69*, 592–599.

Sochor, J., Arby, H., Karlsson, I.M., and Sarasini, S. (2018). A topological approach to Mobility as a Service: A proposed tool for understanding requirements and effects, and for aiding the integration of societal goals. *Research in Transportation Business and Management, 27*, 3–14.

Utriainen, R., and Pöllänen, M. (2018). Review on Mobility as a Service in scientific publications. *Research in Transportation Business and Management, 27*, 15–23.

van Waes, A., Farla, J., Frenken, K., de Jong, J.P., and Raven, R. (2018). Business model innovation and socio-technical transitions. A new prospective framework with an application to bike sharing. *Journal of Cleaner Production, 195*, 1300–1312.

Veeneman, W.W. (2002). *Mind the Gap: Bridging Theories and Practice for the Organisation of Metropolitan Public Transport*. Delft: Delft University Press.

Veeneman, W.W. (2012). La mobilité alternative et les transports publics aux Pays-Bas. *Droit et gestion des collectivités territoriales, 32*, 279–289.

Veeneman, W., and Mulley, C. (2018). Multi-level governance in public transport: Governmental layering and its influence on public transport service solutions. *Research in Transportation Economics, 69*, 430–437.

Veeneman, W.W., Van de Velde, D.M., and Schipholt, L.L. (2006). The value of bus and train: Public values in public transport. In *Proceedings of the European Transport Conference* (pp. 1–12). Strasbourg, France, 18–20 September 2006 (Association for European Transport).

Veeneman, W., van der Voort, H., Hirschhorn, F., Steenhuisen, B., and Klievink, B. (2018). PETRA: Governance as a key success factor for big data solutions in mobility. *Research in Transportation Economics, 69*, 420–429.

Veeneman, W.W., Van Kuijk, J.I., and Hiemstra-van Mastrigt, S. (2020). Dreaming of the travelers' experience in 2040: Exploring governance strategies and their consequences for personal mobility systems. In B. Müller and G. Meyer (eds), *Towards User-Centric Transport in Europe 2* (pp. 225–239). Cham, Switzerland: Springer.

Wong, Y.Z., Hensher, D.A., and Mulley, C. (2020). Mobility as a Service (MaaS): Charting a future context. *Transportation Research Part A: Policy and Practice, 131*, 5–19.

13. E-shopping, travel behavior, and society: a multi-level perspective on sustainable transitions

Kunbo Shi, Long Cheng and Frank Witlox

1. INTRODUCTION

The past decades have witnessed a continuous increase in internet use. According to ITU (2020), only 17 percent (1.1 billion persons) of the total worldwide population were internet users in 2005. The share grew to 51 percent (4.0 billion persons) by the end of 2019. The use of the internet enables people to partake in shopping activities without any trips, that is, so-called e-shopping or online shopping. Particularly during the COVID-19 pandemic, travel-based in-store shopping has been considered risky behavior, thus being not recommended, and even not allowed in many countries. As an alternative to in-store shopping, online shopping is expected to see a marked increase (van Wee and Witlox, 2021). In 2020, online retail sales around the world amounted to over 4 trillion US dollars, accounting for 18 percent of all retail sales worldwide (Statista, 2021a, 2021b).

Shopping travel is one of the major components of transportation systems due to its high share in daily travel. In Europe, for example, shopping trips account for around 20 percent of daily trips (Rosqvist and Hiselius, 2016). In China the share is approximately 19 percent (Baidu, 2019). As an alternative to shopping travel, e-shopping can be understood as a significant innovative transition for transportation systems. This transition is a consequence of various macro socio-technical contexts (e.g., the widespread use of the internet, the COVID-19 outbreak) and will in turn bring changes in multiple systems (e.g., urban, transport, and social systems). Given this situation, this chapter will apply the multi-level socio-technical transition theory to understand the emergence and growth of e-shopping (Section 3).

Successful innovations are usually defined as ones that not only are implemented in the real world, but also have net benefits to society in the sense of contributing toward overcoming the limitations of current systems or addressing current problems (e.g., congestion, carbon emissions). The transition process from in-store shopping to online shopping is reshaping the way people participate in shopping activities, thus leading to considerable changes in socio-technical factors (e.g., a reduction in shopping trips and social connections). These changes are not always beneficial for sustainability. Therefore, it is hard to simply conclude whether e-shopping is a successful or failed innovation. In this chapter, from a behavioral perspective, we will discuss both successes and failures of e-shopping in multiple dimensions (Section 4).

From a viewpoint of transportation, built environment interventions could be effective practical measures to facilitate the successes and particularly mitigate the failures of online shopping. Therefore, we then discuss whether and how the built environment influences online shopping, which is expected to help yield policy recommendations (Section 5). Furthermore, current research limitations and challenges are presented in Section 6. The chapter ends with conclusions in the final section.

2.   DEFINITION OF E-SHOPPING

E-shopping can be simply understood as a process undertaken by consumers to purchase goods or services via the internet. In academia, however, its definition is a bit more complicated. In an early study by Mokhtarian (2004), the definition of e-shopping is considered to originate from that of *teleshopping*, which refers to "the use of ICT [i.e., information and communication technologies] to obtain information about or purchase consumer goods: pre-internet services such as home-shopping channels on cable television, specialized early computer-based systems …, and even telephone orders from a catalog mailed to the home can be placed in this category" (see Mokhtarian, 2004, p.259). Accordingly, Mokhtarian defined e-shopping as the use of the internet to acquire product information or purchase products. In this definition, e-shopping is understood as having two meanings, that is, internet-based browsing and internet-based purchasing (Rotem-Mindali and Weltevreden, 2013). This definition is widely acknowledged in the existing literature and is thus adopted in this chapter.

In this book, transport innovation is defined as "new elements of the transport system that are implemented in the real world" (see Chapter 1). By providing people with a novel way to visit stores without trips, e-shopping has been widely adopted across the world. Given this, e-shopping is usually understood as a type of virtual mobility (Ozbilen et al., 2021), that is, a new transport element, which can be certainly treated as a transport innovation.

Furthermore, there is a need to clarify whether e-shopping is a radical or incremental innovation. A radical innovation normally leads to radical changes in systems, while an incremental one does not. In addition, whether an innovation is radical or incremental depends on the considered stakeholders (e.g., consumers, companies, retailers, and urban systems) (Wu and Hisa, 2004). As mentioned above, both online searches and online purchases are considered e-shopping behaviors, but they are expected to have different impacts on transportation systems. Online searches without ordering products that are delivered to the home can be seen only as a new channel for consumers to acquire product information. After online searching, consumers still need to visit physical stores to purchase and collect products. Current evidence shows that online searching behavior tends to result in more frequent and longer one-way trips to physical stores, because the internet can help consumers extend their search spaces and acquire much more product information that consumers would otherwise not have been aware of (Shi et al., 2020, 2021a, 2021c; Xi et al., 2020b). However, these transport changes can hardly be considered radical. Therefore, internet-based searching behavior is seen as an incremental innovation here.

Compared to online searching behavior, online purchasing behavior usually involves more transport changes in at least three aspects. First, products purchased online are normally delivered to consumers. People do not need to travel to physical stores. Therefore, online purchases are usually considered a substitute for in-store shopping. Second, due to the wide adoption of online purchases, delivery demands are substantially increased (Weltevreden and Rotem-Mindali, 2009), thus leading to a great extension of delivery systems. Existing delivery companies have had to raise their investment in infrastructures to provide more delivery services (e.g., increasing logistics chains and warehouse spaces). New logistics systems are even built up specifically for online business, for example Amazon Prime and Jingdong Express. Compared to traditional delivery services, the services for online orders are usually faster and more time-reliable, for example the same day delivery of goods ordered online. Third, because of online purchases, self-service parcel lockers have been widely provided by delivery companies or a third party to cope with a high failure rate of home deliveries of online-ordered products (Shao et al., 2022), for example bpost parcel lockers provided by a delivery company in Belgium and Hive-box lockers provided by a third party in China. In sum, internet-based purchases provoke substantial changes in transportation systems and therefore online purchasing is treated as a radical innovation. Notably, as a radical innovation, online purchasing will receive more attention in this chapter, mainly because of its greater transport effects.

3. UNDERSTANDING E-SHOPPING FROM A MULTI-LEVEL PERSPECTIVE

In this section, we use a multi-level perspective (MLP) to understand e-shopping as a sustainable transition. According to Geels (2011), MLP is a theoretical framework including three dimensions (see Figure 13.1): 1) socio-technical landscape, which mainly refers to macro-level factors that put pressure on the existing regime, opening up windows of opportunity for novelty (e.g., e-shopping); 2) socio-technical regime, meaning a dynamically stable socio-technical environment surrounding novelty; and 3) niche-innovations, which refers to the emergence process of a novelty. An MLP framework is commonly employed to explain radical innovations (Geels, 2011, 2012). Therefore, here we will focus on online purchasing behavior rather than online searching behavior without purchases.

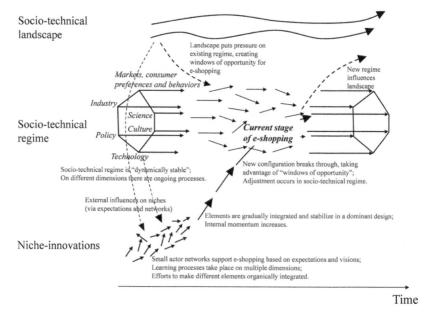

Source: Adapted from Geels (2011).

Figure 13.1   The development of e-shopping from a multi-level perspective

Socio-Technical Landscape

Five factors can be seen as a combined mechanism creating an opportunity window for the emergence of e-shopping. First, the internet has been used worldwide by the public since the 1990s, which provides a basic condition for the development of e-shopping. Second, thanks to highly effective production systems in past decades, overcapacity is becoming an issue for many manufacturers across different industries. They urgently need to seek new distribution channels like e-commerce to help them out. Third, rapid urbanization with a great extension of urban areas in recent decades makes people travel farther and suffer more from severe transportation congestion. Meanwhile, because of rapid economic development particularly since the age of globalization, people are subject to a faster pace of life, which makes them pursue quick ways to meet daily needs like shopping. Apparently, e-shopping – which can satisfy people's shopping demand without shopping trips – is an ideal innovation. Fourth, a high unemployment rate is a long-standing problem around the world. Innovation is widely considered the dominant factor driving economic growth (Hanusch and Pyka, 2007), thus mitigating unemployment issues. As a form of commercial innovation, e-commerce is therefore encouraged in many countries. Fifth, the COVID outbreak is a more recent booster. In-store shopping has been widely considered risky during the COVID-19 pandemic and is thus not recommended or even not allowed in many countries. As an alternative to in-store shopping, e-shopping has seen a large increase. For instance, the share of e-retail sales in total retail sales between the first and second quarters of 2020 has risen from 11.8 percent to 16.1 percent in the United States and from 20.3 percent to 31.3 percent in the United Kingdom. In China, the share increased from 19.4 percent to 24.6 percent between August 2019 and August 2020 (UNCTAD, 2020).

Socio-Technical Regime

Socio-technical regime or context is commonly interpreted in six aspects, including industry, markets and consumer preferences and behaviors, science, culture, technology, and policy (Geels, 2011) (see Figure 13.1). The left-hand side of Figure 13.1 represents the past socio-technical context that had been well established for traditional in-store shopping with rather limited consideration of e-shopping. For example, manufacturers provided their products only to traditional retailers who only ran in-store businesses. Meanwhile, relevant industrial supports (e.g., logistics and warehouses) were almost absent for e-retailing. Consumers had a quite low acceptance level for online shopping. In-store shopping was still a dominant cultural phenomenon. Technical supports (e.g., e-payment and parcel-track systems) had not been fully considered in the sciences. Governments had not issued any policies supporting and regulating the development of e-commerce markets. At the same time, urban (transportation) planning policies (e.g., land-use policies) were mostly designed and implemented given the assumption that people only met their shopping demands by traveling to physical stores. The changes in needs for in-store visits and logistics that would be caused by e-shopping had not been considered. The past socio-technical regime usually acts as a barrier toward sustainability at the beginning of an innovation transition (e.g., the shift from in-store shopping to e-shopping). To achieve a sustainable transition, these contextual factors need systematical changes to be compatible with the e-shopping age. This is a long-term process, which often needs 30–40 years.

Niche-Innovations

As an innovation, e-shopping has been widely adopted and is in the growth phase currently (see Figure 13.1). In this phase, to orient and coordinate e-commerce markets, the socio-technical regime is changing right now in the real world. For instance, many manufacturers and providers have started their online businesses. It has also been commonly seen that physical retailers open an online channel that allows people to order products via the internet. New and specific logistics chains have been developed (e.g., Amazon Prime and Jingdong Express). Various e-payment methods have emerged and become safer and faster (e.g., PayPal, AliPay, Apple Pay, and Amazon Pay). There is a considerably high level of acceptance of e-shopping by consumers. Researchers have paid scholarly attention to e-shopping and its transport and social implications. Many governments have issued laws or policies to manage e-commerce markets. These contextual changes encourage a continuous and rapid expansion in e-retail markets. As shown in Figure 13.2, worldwide e-retail sales amounted to 4,280 billion US dollars in 2020, while it was only 1,336 billion US dollars in 2014. The penetration shares – which are indicated by the percentages of online retail sales in total retail sales – have also contin-uously grown, from 7.4 percent in 2015 to 18.0 percent in 2020. Meanwhile, however, there are still many conflicts between e-commerce markets and the old socio-technical regime in terms of industry, markets and consumer prefer-ences and behaviors, science, culture, technology, and policy. For example, the security of e-payment is still a concern for consumers (Kim et al., 2010), and online shopping systems are not always friendly to all people (e.g., the elderly usually experience difficulty in the adoption of online shopping). These con-flicts are reasonable because the transformation of the socio-technical regime is a long-term process.

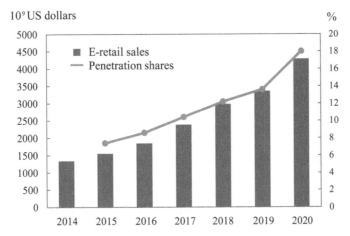

Sources: Statista (2021a, 2021b).

Figure 13.2   Global e-retail sales and penetration shares

4.   SUCCESS AND FAILURE OF E-SHOPPING

In the growth phase, the changes in socio-technical factors that are taking place to coordinate e-commerce markets have had wide-ranging consequences. Some of them are positive for the sustainable development of transportation and society while others are negative, and these can be therefore treated as the success and failure of e-shopping, respectively. Concentrating on changes in consumers' preference/behavior, this section will discuss these successes and failures from a behavioral perspective. The behavioral transition from in-store shopping to online shopping is considered the primary trigger that leads to success and failure in both transportation and social systems.

Success and Failure in Transportation Systems

Reducing travel demand and car use is widely accepted as one of the effective strategies to mitigate traffic congestion and transport-related carbon emissions. This will be used as a basis to assess whether e-shopping is a success or a failure for transportation systems.

In theory, researchers assume that e-shopping may have four possible impacts on travel demand (Salomon, 1985, 1986; Mokhtarian, 1990, 2002): 1) substitution – shopping trips are replaced by online shopping; 2) complementarity – additional shopping trips are generated by online shopping; 3) modification – the characteristics of shopping travel (e.g., destinations, routes, timing, and mode choices) are altered by online shopping; and 4) neutrality – online shopping has negligible effects on shopping travel.

Empirical evidence shows that the e-shopping impacts on travel behavior largely depend on how e-shopping is defined. When considering online purchasing behavior, it is often found that e-shopping tends to result in a reduction in shopping trips (i.e., substitution effect), therefore alleviating traffic congestion (Shi et al., 2019; Weltevreden, 2007; Weltevreden and van Rietbergen, 2009; Weltevreden and Rotem-Mindali, 2009; Xi et al., 2020a). In this sense, e-shopping seems a success for transportation systems. However, there is another possibility, that e-shoppers may make additional trips for shopping (i.e., complementarity effect) (Cao et al., 2010, 2012; Ding and Lu, 2017; Etminani-Ghasrodashti and Hamidi, 2020; Lee et al., 2017; Zhen et al., 2016) or for other purposes because they will have more spare time thanks to the use of e-shopping. In addition, there may simultaneously exist two plausible but conflicting situations where e-shopping influences modal split by replacing specific shopping trips. On the one hand, in order to save time, people may be more likely to substitute online purchases for long shopping trips compared to short ones. Therefore, e-shopping seems to particularly reduce the use of motorized modes (e.g., car, bus, metro, tram) that are usually for long trips. On the other hand, car users or people with high accessibility to public transit (e.g., bus, metro, tram) normally feel it is effortless to transport goods from stores to home compared to their counterparts in transport poverty. Therefore, they may have lower likelihoods of reducing shopping trips (Shi et al., 2021d). In this situation, online shopping will less likely influence the use of motorized modes but will more likely reduce the use of active modes (e.g., walking, cycling).

When only considering online searching behavior, it is almost certain that e-shopping will lead to extra shopping trips because via the internet consumers can acquire massive amounts of product information that they would have not been aware of otherwise (Shi et al., 2021c; Xi et al., 2020b). Meanwhile, because online searching is less spatially constrained, people can have a greater search space for consumption destinations, and consequently travel farther and use motorized modes (e.g., transit and car) more frequently to visit these destinations (Shi et al., 2020; Shi et al., 2021a). In this aspect, e-shopping is not necessarily nudging the transportation system toward sustainability.

Success and Failure in Social Systems

Social inclusion
Accessibility – which is traditionally understood as the potential of opportunities for interaction (Hansen, 1959; Handy and Niemeier, 1997) – is an important factor relating to the issue of social inclusion (Lucas, 2011; Preston and Rajé, 2007). The adoption of e-shopping creates a new concept – virtual

accessibility, which can be defined as the potential of opportunities for participation in shopping activities via the internet (Lavieri et al., 2018; Shao et al., 2022). The emergence of virtual accessibility is reshaping the landscape of social inclusion (Kenyon et al., 2002, 2003). Virtual accessibility adds value to improve the aggregate level of social inclusion (van Winden, 2001), because people can partake in shopping activities through not only the traditional channel (i.e., in-store shopping) but also the online channel (i.e., online shopping). For example, the disabled usually have more difficulties traveling for in-store shopping. Via the internet, they can access and order products as easily as others. From a geographical perspective, people in distant areas (e.g., exurban and rural areas) may have the same level of accessibility to consumer goods on the internet as their urban counterparts. On this point, e-shopping is a successful innovation.

However, online shopping cannot always be considered successful, but rather has the potential to generate new social inequality. Because of the difference in the availability of the internet or in the ability to use the internet among the population, the level of virtual accessibility varies largely between different groups. This leads to inequality in opportunities to partake in e-shopping. For instance, less-educated people and the elderly are usually less likely to shop online, because they have a lower ability to use the internet and limited acceptance of e-shopping. Meanwhile, new spatial inequality may also emerge. For example, because of the digital gap between urban and rural areas, rural people usually have lower accessibility to the use of the internet, thus limiting their participation in online shopping. Such spatial inequality also exists between countries. As shown in Table 13.1, the top 10 countries with the largest e-retail sales account for 88.6 percent of worldwide e-retail sales, while their total population is only 48.7 percent of the worldwide population. Among the top 10 countries, most are highly developed, such as the United States, the United Kingdom, Japan, South Korea, Germany, France, and Canada. As developing countries, India and Brazil also contribute considerable e-retail shares, which is mainly due to their large populations. Interestingly, China with a population share of 18.5 percent accounts for more than half of worldwide e-retail sales. In addition to its large population, rapid urbanization and economic growth in the last three decades have opened up a bigger opportunity window for e-shopping development.

Besides, in the long run, the possibly reduced visits to physical stores caused by e-shopping will lead to a decline in the number of these stores (Helm et al., 2020; Shi et al., 2019; Weltevreden, 2007). The disappearance of physical stores will raise the difficulty of in-store shopping for non-e-shoppers, who are often the elderly, low-income groups, and children. This is another potential failure of e-shopping that will aggravate the issue of social exclusion.

Table 13.1   *Top 10 countries with the largest shares of e-retail sales in*
              2021

Countries	E-retail shares/%[a]	Population shares/% (2020)[b]
China	52.1	18.5
US	19.0	4.2
UK	4.8	0.9
Japan	3.0	1.6
South Korea	2.5	0.7
Germany	2.1	1.1
France	1.6	0.8
India	1.4	17.7
Canada	1.3	0.5
Brazil	0.8	2.7
Total	88.6	48.7

Notes: [a] E-retail data are derived from von Abrams (2021); [b] population data are collected from the United Nations (2019).

Social connection

Even in the information era, face-to-face contact is still important because it can benefit people in various aspects such as recreational and socializing functions and non-financial incentives (Storper and Venables, 2004). For example, making trips to visit and shop at physical stores is helpful for consumers to moderate and even combat social isolation to some extent. In addition, recreational services (e.g., theme restaurants, live music performances) are usually available at shopping centers. In-store shopping can stimulate consumers to use these entertainment services (Mokhtarian, 2004). Apparently, online shopping lacks these social profits that in-store shopping can offer, thus possibly increasing social isolation (Xi et al., 2021). This is a failure of e-shopping. On the other hand, it can be argued that online shopping can help people save time, with which people can make trips to participate in other (social) activities in physical spaces. In this situation, e-shopping seems beneficial for social connections.

Wellbeing

Wellbeing is a broad construct that is commonly measured both psychologically and physically in the field of transportation (Chatterjee et al., 2020). According to current evidence, it seems that e-shopping can impact both dimensions. Psychologically, at least two advantages of online shopping can positively contribute to individuals' wellbeing. First, e-shopping usually helps consumers achieve an efficient consumption experience. This advantage can particularly

Table 13.2   *Success and failure of e-shopping in transportation and social systems*

Category	Transportation system	Social system
Success	• Reducing travel demand for in-store shopping • Reducing the use of motorized modes for shopping trips	• Improving the aggregate level of social inclusion • Encouraging social connections • Increasing the efficiency of consumptions • Diverse options of consumer goods
Failure	• Increasing travel demand for in-store shopping or non-shopping purposes • Reducing the use of active modes • Increasing the one-way distance of travel • Encouraging the shift from non-motorized modes to motorized modes	• Increasing the inequality in shopping opportunities between e-shoppers and non-e-shoppers • Increasing the spatial inequality in shopping opportunities due to digital gaps • Discouraging social connections • Indebtedness issues led by overconsumption online • Obesity and sedentary-related health problems

increase the wellbeing of those who have a fast pace of life. Second, people can access a great variety of product information via the internet, thus substantially increasing consumption options (Castellacci and Tveito, 2018; Guillen-Royo, 2019; Sabatini, 2011). However, overconsumption online possibly leads to an indebtedness issue, which will detract from individuals' psychological wellbeing (Guillen-Royo, 2019). In terms of physical wellbeing, frequent online shopping will lead to a longer period of indoor sedentary activities, which may thus make people obese (Aghasi et al., 2020) and have other health problems such as impaired eyesight and cervical spondylosis (Zheng et al., 2016). Obviously, the impacts of e-shopping on wellbeing consist of both success and failure.

Finally, Table 13.2 summarizes the successes and failures of e-shopping in transportation and social systems. It should be noted that some of them are contradictory. For example, as assumed above, e-shopping has the potential to either partially replace shopping trips or encourage additional shopping trips. These contradictory situations may be present separately or coexist in different socioeconomic, geographic, and cultural contexts.

5.   BUILT ENVIRONMENT AND E-SHOPPING

Different policies could be used to enhance the successes and particularly alleviate the failures of e-shopping. From a viewpoint of transportation, built environment interventions are usually considered effective in managing people's behavior and thus implemented in planning practice. In this chapter, we aim to focus on the relationship between the built environment and online shopping. In doing so, it will help to understand whether and how the built environment, as one element of the existing socio-technical regime, needs to be changed to steer online shopping toward sustainability.

The built environment is usually characterized by "five-Ds": density (e.g., population density, employment density), design (e.g., block size, proportion of four-way intersections), diversity (e.g., land-use mix), destination accessibility (e.g., accessibility to shopping centers), and distance to transit (e.g., shortest routes to the nearest transit stops) (Ewing and Cervero, 2001, 2010). In addition, neighborhood types (e.g., urban or rural neighborhood), which are closely correlated with the abovementioned five-Ds, are often considered a proxy variable of the built environment.

At the beginning of the use of ICT, some scholars made an announcement of "the death of distance" (e.g., Cairncross, 1997). This suggests that ICT-based activities are rarely influenced by the spatial constraint imposed by the surrounding environment. In other words, they assume that the built environment has negligible influences on online activities. However, a number of researchers argue that the statement of "the death of distance" is exaggerated, and geography still does matter in the information age (e.g., Morgan, 2004). Mainly focusing on e-shopping in recent years, transportation researchers and geographers have conceptually or/and empirically examined the relationship between the built environment/spatial attributes and e-shopping.

Anderson et al. (2003) formulated two hypotheses, namely the innovation diffusion hypothesis and the efficiency hypothesis, to explain how spatial factors influence online shopping behaviors. The former states that people in urban areas are more likely to conduct e-shopping because they are usually well-educated to use the internet actively for various purposes (e.g., shopping). The latter postulates that non-urban residents are more likely to shop online because they usually have lower levels of in-store shopping accessibility and thus benefit more from the access to a wide variety of products via the internet. Following the work by Anderson, researchers empirically examine the two conflicting hypotheses, supporting either or both (e.g., Cao et al., 2013; Ren and Kwan, 2009; Shi et al., 2019, 2021e).

In analogy with the principle of the five-Ds, some studies further examine the role of specific built environment elements in e-shopping. Similarly, inconsistent results are revealed. The link between accessibility to physical stores and online shopping is explored most frequently. As expected, some studies suggest a positive association of lower in-store shopping accessibility with online shopping (Loo and Wang, 2018; Ren and Kwan, 2009; Shi et al., 2021b). Meanwhile, some researchers find an insignificant relationship between in-store shopping accessibility and online shopping (Ding and Lu,

2017; Lee et al., 2017). On the other hand, the study by Cao et al. (2013) indicated that people with higher accessibility to in-store shopping opportunities have a higher likelihood of adopting e-shopping, particularly in urban areas. In addition to neighborhood type and accessibility to physical stores, other built environment elements are also examined in a few studies. It is found that, for instance, lower accessibility to metro stations, higher residential density, higher employment density, and higher white population density are positively correlated to online shopping (Loo and Wang, 2018; Ren and Kwan, 2009; Shi et al., 2021b, 2021d). Notably, some studies revealed that there are no significant correlations between a wide variety of built environment elements (e.g., transit accessibility, employment density, residential density, street design, and land-use diversity) and online shopping (e.g., Etminani-Ghasrodashti and Hamidi, 2020; Lee et al., 2017).

Although the majority of the literature supports that spatial constraints imposed by the built environment still matter in online shopping, is the announcement of "the death of distance" totally wrong? The answer seems to be "no". It can be theoretically assumed that – unless the internet is unavailable – a person can purchase online wherever he/she is. This means a limited importance of proximity as a spatial organizing principle of e-shopping. Some empirical studies seem to support this assumption, because they indicate that the accessibility to in-store shopping opportunities or transit services is significantly correlated with shopping trips, but is insignificantly related to online shopping (e.g., Ding and Lu, 2017; Etminani-Ghasrodashti and Hamidi, 2020). Therefore, it could be argued that the associations of the built environment with online shopping still exist but may be weaker than those with shopping travel.

Additionally, the associations between the built environment and shopping travel may be attenuated by online searching behavior. It is traditionally accepted that the built environment is closely correlated with one-way distances of travel. Briefly speaking, people in sparse areas tend to travel farther for shopping because of limited shopping opportunities compared to those in dense areas. However, it may not be the case when people search for information online about travel destinations (e.g., a restaurant) before departing for a consumption trip. In principle, a person can acquire information about all destinations of interest within a city via the internet. In other words, online searching is rarely spatially constrained by the surrounding built environment. Correspondingly, the trip made to visit a consumption destination searched online is also less constrained by the built environment. Therefore, the correlation between the built environment and the one-way distance of the shopping trip is weakened because of online searching behavior (Shi et al., 2020). This situation seems to suggest that the use of the internet could be challenging traditional knowledge about the relationships between the built environment and travel behavior.

Given the complex transport and social impacts of e-shopping and the mixed understanding of the relationship between the built environment and e-shopping, it is quite hard to point out a clear policy direction for land-use planning toward sustainability. This may be one of the major reasons why built environment interventions coordinating e-shopping have been rarely considered in planning practice.

6.   CURRENT RESEARCH LIMITATIONS AND CHALLENGES

It has been evident that online shopping has wide (both positive and negative) implications for transportation and social systems and meanwhile has associations with the built environment. With respect to these topics, however, there still exist some research limitations and challenges needing to be addressed in future research.

First, there is still a lack of clear understanding of e-shopping's implications for transportation systems, even after extensive empirical research (Le et al., 2022):

1.   Prior research mainly explores how e-shopping impacts shopping trip frequency (i.e., complementarity or substitution effects), while little considers the influences of e-shopping on shopping trip distance, duration, and mode choices (i.e., modification effects).
2.   Although online purchases and online searches are found to have different and even opposite impacts on travel, few researchers assess the net travel effects of both behaviors.
3.   The extent to which online shopping impacts travel also differs by type of product (Zhen et al., 2016). However, most scholarly attention is paid to some specific goods. It remains unknown what the overall impacts of e-shopping is on travel for all types of products.
4.   E-shopping can help save time and then possibly make consumers relocate their time used for other outdoor activities (e.g., recreational activities). Therefore, in theory, e-shopping has potential spillover effects on travel for other purposes. However, this issue is mostly ignored in previous studies.

Second, although it is theoretically expected that e-shopping has broad social implications, empirical studies are overall bare in this regard, and especially in relation to the following two aspects:

1. There is a lack of studies with sufficient empirical evidence examining whether and how e-shopping influences social inclusion and inequalities, particularly for vulnerable groups (e.g., suburban and rural residents, low-income people, the disabled, and the elderly).
2. Many empirical studies have explored the influence of general internet use on social connections or psychological and physical wellbeing. However, the investigation into the effects of e-shopping on these points is still in an early stage.

Third, like the interaction between the built environment and travel behavior (Cao et al., 2009), a residential self-selection issue may exist in the association between the built environment and online shopping as well. There may be either causality or correlation between the built environment and e-shopping due to the role of the attitude toward (e-)shopping or internet use in the following four scenarios (see Figure 13.3). In Figure 13.3a, the attitude plays an antecedent role in the relationship. This means that someone with a high preference for shopping may not only choose to live near the city center for more in-store shopping opportunities but also partake in online shopping frequently. Alternatively, someone who has a strong preference for e-shopping or the internet – regardless of accessibility to in-store shopping opportunities – tends to reside in a neighborhood with more convenience for e-shopping (e.g., fast access to the internet, sufficient parcel collection points) and make frequent online purchases. In both situations, there is a spurious causality between the built environment and online shopping. As displayed in Figures 13.3b and 13.3c, there are two indirectly causal directions between the built environment and e-shopping mediated by the attitude. Figure 13.3d shows a directly causal direction from the built environment to e-shopping. In this situation, the attitude seems less relevant. For example, one may live in a neighborhood with low accessibility to in-store shopping opportunities. Here, this person is forced to purchase online frequently regardless of whether he/she likes e-shopping or not. Alternatively, one may reside in an area with dense parcel collection points, which may stimulate the adoption of e-shopping. The self-selection issue may potentially exist but has rarely been considered in previous studies. Therefore, we need to be very careful when previous findings are used to inform policy. Researchers are strongly encouraged to take this issue into consideration in future research by implementing some strategies. For example, longitudinal data that are usually considered effective to address such an issue are encouraged in order to examine the causality direction between the built environment and e-shopping. When longitudinal data are unavailable, the attitude toward (e-)shopping or the internet should at least be taken into account in a cross-sectional analysis (Zhen et al., 2016).

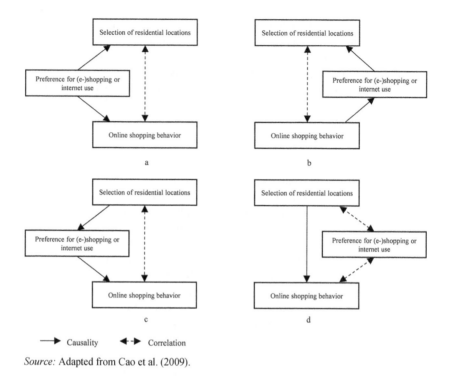

Source: Adapted from Cao et al. (2009).

Figure 13.3   *Potential associations between the built environment,*
                   e-shopping behavior, and attitudes

Fourth, as theoretically assumed in Section 5, the built environment may play
a weaker role in online shopping compared to that in shopping travel, and
the role of the built environment in consumption travel may be attenuated by
online searches. However, only limited studies have explicitly examined the
two assumptions. This knowledge gap needs to be filled by providing additional empirical evidence.

Fifth, the COVID-19 outbreak enlarges the opportunity window for the
development of e-shopping, which has accelerated the adoption of e-shopping
(van Wee and Witlox, 2021). Therefore, it is almost inevitable that e-shopping
will have stronger transport and social implications. Meanwhile, its relationship with the built environment may have also been reorganized. For example,
people residing in dense areas may be more worried about the pandemic risk
and are therefore more likely to substitute online shopping for shopping travel
to avoid physical contact with others. These changes may weaken the value of
knowledge contributed by previous studies before the pandemic. Against this

background, there is a need to reexplore the abovementioned topics during and post the COVID-19 pandemic.

Finally, we want to highlight the importance of geographical heterogeneity. Different countries and regions have a great variety of social, economic, and cultural backgrounds. These various contexts may be – to greater or less extent – shaping and reshaping people's e-shopping, travel behavior, and beyond. Because of this, theoretical or empirical research from multiple countries and regions is always encouraged, so that we can establish an in-depth and comprehensive understanding of e-shopping and its implications.

7.   CONCLUSIONS

As an alternative channel of shopping travel, e-shopping can be considered an innovation for transportation systems. From a multi-level perspective on sustainable transitions, this chapter conceptually analyzes the behavioral transition from in-store shopping to online shopping, and then identifies the possible success and failure of this transition for the sustainable development of transportation and social systems. Furthermore, we discuss how and whether the built environment is associated with online shopping, which helps yield policy recommendations for land-use interventions to achieve a sustainable transition. Finally, we present current research limitations and challenges. The main findings are as follows.

According to the multi-level socio-technical transition theory, the transition from in-store shopping to online shopping is induced by various macro socio-technical landscapes, mainly including wide use of the internet, over-capacity of industries, rapid urbanization, fast pace of life, unemployment issues, and the COVID-19 outbreak. Meanwhile, the behavioral transition has been in a growth phase. In this phase, the adoption of e-shopping provokes changes in or has conflicts with the old socio-technical regime/context, further generating wide-ranging consequences for transportation and society. These consequences are not always beneficial, but are harmful in some situations. For example, e-shopping may benefit transportation systems because it has the potential to reduce travel demand for shopping activities. However, due to the decreased shopping trips, e-shoppers may experience a reduction in social connections. Therefore, it can hardly be concluded that e-shopping is a successful or failed innovation. In addition, the built environment is found to be correlated with online shopping. However, the correlation needs to be carefully reexamined in future research, because there may be spurious causality when the attitude toward online shopping or internet use plays an antecedent role (i.e., a residential self-selection issue). Therefore, it remains unclear whether and how built environment interventions are useful to accommodate e-shopping so as to mitigate or even eliminate its failures for transportation and society.

REFERENCES

Aghasi, M., Matinfar, A., Golzarand, M., Salari-Moghaddam, A., and Ebrahimpour-Koujan, S. (2020). Internet use in relation to overweight and obesity: A systematic review and meta-analysis of cross-sectional studies. *Advances in Nutrition*, 11(2), 349–356.

Anderson, W.P., Chatterjee, L., and Lakshmanan, T.R. (2003). E-commerce, transportation, and economic geography. *Growth and Change*, 34(4), 415–432.

Baidu (2019). *Report on China Urban Transport 2018*. Available at: https://huiyan.baidu.com/reports/2018annualtrafficreport.html (accessed on 5 May 2021).

Cairncross, F. (1997). *The Death of Distance: How the Communications Revolution will Change Our Lives* (Vol. 302). Boston, MA: Harvard Business School Press.

Cao, X., Chen, Q., and Choo, S. (2013). Geographic distribution of e-shopping: Application of structural equation models in the Twin Cities of Minnesota. *Transportation Research Record*, 2383, 18–26.

Cao, X., Douma, F., and Cleaveland, F. (2010). Influence of e-shopping on shopping travel: Evidence from Minnesota's Twin Cities. *Transportation Research Record*, 2157, 147–154.

Cao, X., Mokhtarian, P.L., and Handy, S.L. (2009). Examining the impacts of residential self-selection on travel behaviour: A focus on empirical findings. *Transport Reviews*, 29(3), 359–395.

Cao, X.J., Xu, Z., and Douma, F. (2012). The interactions between e-shopping and traditional in-store shopping: An application of structural equations model. *Transportation*, 39(5), 957–974.

Castellacci, F., and Tveito, V. (2018). Internet use and well-being: A survey and a theoretical framework. *Research Policy*, 47(1), 308–325.

Chatterjee, K., Chng, S., Clark, B., Davis, A., De Vos, J., Ettema, D., Handy, S., Martin, A., and Reardon, L. (2020). Commuting and wellbeing: A critical overview of the literature with implications for policy and future research. *Transport Reviews*, 40(1), 5–34.

Ding, Y., and Lu, H. (2017). The interactions between online shopping and personal activity travel behavior: An analysis with a GPS-based activity travel diary. *Transportation*, 44(2), 311–324.

Etminani-Ghasrodashti, R., and Hamidi, S. (2020). Online shopping as a substitute or complement to in-store shopping trips in Iran? *Cities*, 103, 102768.

Ewing, R., and Cervero, R. (2001). Travel and the built environment: A synthesis. *Transportation Research Record*, 1780, 87–114.

Ewing, R., and Cervero, R. (2010). Travel and the built environment: A meta-analysis. *Journal of the American Planning Association*, 76(3), 265–294.

Geels, F.W. (2011). The multi-level perspective on sustainability transitions: Responses to seven criticisms. *Environmental Innovation and Societal Transitions*, 1(1), 24–40.

Geels, F.W. (2012). A socio-technical analysis of low-carbon transitions: Introducing the multi-level perspective into transport studies. *Journal of Transport Geography*, 24, 471–482.

Guillen-Royo, M. (2019). Sustainable consumption and wellbeing: Does on-line shopping matter? *Journal of Cleaner Production*, 229, 1112–1124.

Handy, S.L., and Niemeier, D.A. (1997). Measuring accessibility: An exploration of issues and alternatives. *Environment and Planning A*, 29(7), 1175–1194.

Hansen, W.G. (1959). How accessibility shapes land use. *Journal of the American Institute of Planners*, 25(2), 73–76.

Hanusch, H., and Pyka, A. (2007). Principles of neo-Schumpeterian economics. *Cambridge Journal of Economics*, 31(2), 275–289.

Helm, S., Kim, S.H., and Van Riper, S. (2020). Navigating the "retail apocalypse": A framework of consumer evaluations of the new retail landscape. *Journal of Retailing and Consumer Services*, 54, 101683.

ITU (2020). The number and percentage of worldwide internet users. Available at: https://www.itu.int/en/ITU-D/Statistics/Pages/stat/default.aspx (accessed on 8 May 2021).

Kenyon, S., Lyons, G., and Rafferty, J. (2002). Transport and social exclusion: Investigating the possibility of promoting inclusion through virtual mobility. *Journal of Transport Geography*, 10(3), 207–219.

Kenyon, S., Rafferty, J., and Lyons, G. (2003). Social exclusion and transport in the UK: A role for virtual accessibility in the alleviation of mobility-related social exclusion? *Journal of Social Policy*, 32(3), 317–338.

Kim, C., Tao, W., Shin, N., and Kim, K.S. (2010). An empirical study of customers' perceptions of security and trust in e-payment systems. *Electronic Commerce Research and Applications*, 9(1), 84–95.

Lavieri, P.S., Dai, Q., and Bhat, C.R. (2018). Using virtual accessibility and physical accessibility as joint predictors of activity-travel behavior. *Transportation Research Part A: Policy and Practice*, 118, 527–544.

Le, H.T., Carrel, A.L., and Shah, H. (2022). Impacts of online shopping on travel demand: A systematic review. *Transport Reviews*, 42(3), 273–295.

Lee, R.J., Sener, I.N., Mokhtarian, P.L., and Handy, S.L. (2017). Relationships between the online and in-store shopping frequency of Davis, California residents. *Transportation Research Part A: Policy and Practice*, 100, 40–52.

Loo, B.P., and Wang, B. (2018). Factors associated with home-based e-working and e-shopping in Nanjing, China. *Transportation*, 45(2), 365–384.

Lucas, K. (2011). Making the connections between transport disadvantage and the social exclusion of low income populations in the Tshwane Region of South Africa. *Journal of Transport Geography*, 19(6), 1320–1334.

Mokhtarian, P.L. (1990). A typology of relationships between telecommunications and transportation. *Transportation Research Part A: General*, 24(3), 231–242.

Mokhtarian, P.L. (2002). Telecommunications and travel: The case for complementarity. *Journal of Industrial Ecology*, 6(2), 43–57.

Mokhtarian, P.L. (2004). A conceptual analysis of the transportation impacts of B2C e-commerce. *Transportation*, 31(3), 257–284.

Morgan, K. (2004). The exaggerated death of geography: Learning, proximity and territorial innovation systems. *Journal of Economic Geography*, 4(1), 3–21.

Ozbilen, B., Wang, K., and Akar, G. (2021). Revisiting the impacts of virtual mobility on travel behavior: An exploration of daily travel time expenditures. *Transportation Research Part A: Policy and Practice*, 145, 49–62.

Preston, J., and Rajé, F. (2007). Accessibility, mobility and transport-related social exclusion. *Journal of Transport Geography*, 15(3), 151–160.

Ren, F., and Kwan, M.P. (2009). The impact of geographic context on e-shopping behavior. *Environment and Planning B: Planning and Design*, 36(2), 262–278.

Rosqvist, L.S., and Hiselius, L.W. (2016). Online shopping habits and the potential for reductions in carbon dioxide emissions from passenger transport. *Journal of Cleaner Production*, 131, 163–169.

Rotem-Mindali, O., and Weltevreden, J.W. (2013). Transport effects of e-commerce: What can be learned after years of research? *Transportation*, 40(5), 867–885.

Sabatini, F. (2011). Can a click buy a little happiness? The impact of business-to-consumer e-commerce on subjective well-being (No. 12/2011). EERI Research Paper Series.

Salomon, I. (1985). Telecommunications and travel: Substitution or modified mobility? *Journal of Transport Economics and Policy*, 19(3), 219–235.

Salomon, I. (1986). Telecommunications and travel relationships: A review. *Transportation Research Part A: General*, 20(3), 223–238.

Shao, R., Derudder, B., and Witlox, F. (2022). The geography of e-shopping in China: On the role of physical and virtual accessibility. *Journal of Retailing and Consumer Services*, 64, 102753.

Shi, K., Cheng, L., De Vos, J., Yang, Y., Cao, W., and Witlox, F. (2021a). How does purchasing intangible services online influence the travel to consume these services? A focus on a Chinese context. *Transportation*, 48(5), 2605–2625.

Shi, K., De Vos, J., Yang, Y., and Witlox, F. (2019). Does e-shopping replace shopping trips? Empirical evidence from Chengdu, China. *Transportation Research Part A: Policy and Practice*, 122, 21–33.

Shi, K., De Vos, J., Yang, Y., Li, E., and Witlox, F. (2020). Does e-shopping for intangible services attenuate the effect of spatial attributes on travel distance and duration? *Transportation Research Part A: Policy and Practice*, 141, 86–97.

Shi, K., De Vos, J., Cheng, L., Yang, Y., and Witlox, F. (2021b). The influence of the built environment on online purchases of intangible services: Examining the mediating role of online purchase attitudes. *Transport Policy*, 114, 116–126.

Shi, K., De Vos, J., Yang, Y., Xu, J., Cheng, L., and Witlox, F. (2021c). Does buying intangible services online increase the frequency of trips to consume these services? *Cities*, 119, 103364.

Shi, K., Shao, R., De Vos, J., Cheng, L., and Witlox, F. (2021d). Is e-shopping likely to reduce shopping trips for car owners? A propensity score matching analysis. *Journal of Transport Geography*, 95, 103132.

Shi, K., Shao, R., De Vos, J., and Witlox, F. (2021e). Do e-shopping attitudes mediate the effect of the built environment on online shopping frequency of e-shoppers? *International Journal of Sustainable Transportation*, 1–11.

Statista (2021a). Retail e-commerce sales worldwide from 2014 to 2024. Available at: https://www.statista.com/statistics/379046/worldwide-retail-e-commerce-sales/ (accessed on 6 November 2021).

Statista (2021b). E-commerce share of total global retail sales from 2015 to 2024. Available at: https://www.statista.com/statistics/534123/e-commerce-share-of-retail-sales-worldwide/ (accessed on 6 November 2021).

Storper, M., and Venables, A.J. (2004). Buzz: Face-to-face contact and the urban economy. *Journal of Economic Geography*, 4(4), 351–370.

UNCTAD (2020). *COVID-19 and E-commerce: A Global Review*. Available at: https://unctad.org/webflyer/covid-19-and-e-commerce-global-review (accessed on 5 May 2021).

United Nations (2019). World Population Prospects 2019, Online Edition. Department of Economic and Social Affairs, Population Division. Available at: https://population.un.org/wpp/Download/Standard/Population/ (accessed on 6 November 2021).

Van Wee, B., and Witlox, F. (2021). COVID-19 and its long-term effects on activity participation and travel behaviour: A multiperspective view. *Journal of Transport Geography*, 95, 103144.

Van Winden, W. (2001). The end of social exclusion? On information technology policy as a key to social inclusion in large European cities. *Regional Studies*, 35(9), 861–877.

Von Abrams, K. (2021). These are the top global ecommerce markets. Available at: https://www.emarketer.com/content/top-global-ecommerce-markets (accessed on 6 November 2021).

Weltevreden, J.W. (2007). Substitution or complementarity? How the Internet changes city centre shopping. *Journal of Retailing and Consumer Services*, 14(3), 192–207.

Weltevreden, J.W., and Rotem-Mindali, O. (2009). Mobility effects of B2C and C2C e-commerce in the Netherlands: A quantitative assessment. *Journal of Transport Geography*, 17(2), 83–92.

Weltevreden, J.W., and van Rietbergen, T. (2009). The implications of e-shopping for in-store shopping at various shopping locations in the Netherlands. *Environment and Planning B: Planning and Design*, 36(2), 279–299.

Wu, J.H., and Hisa, T.L. (2004). Analysis of e-commerce innovation and impact: A hypercube model. *Electronic Commerce Research and Applications*, 3(4), 389–404.

Xi, G., Cao, X., and Zhen, F. (2020a). The impacts of same day delivery online shopping on local store shopping in Nanjing, China. *Transportation Research Part A: Policy and Practice*, 136, 35–47.

Xi, G., Cao, X., and Zhen, F. (2021). How does same-day-delivery online shopping reshape social interactions among neighbors in Nanjing? *Cities*, 114, 103219.

Xi, G., Zhen, F., Cao, X., and Xu, F. (2020b). The interaction between e-shopping and store shopping: Empirical evidence from Nanjing, China. *Transportation Letters*, 12(3), 157–165.

Zhen, F., Cao, X., Mokhtarian, P.L., and Xi, G. (2016). Associations between online purchasing and store purchasing for four types of products in Nanjing, China. *Transportation Research Record*, 2566, 93–101.

Zheng, Y., Wei, D., Li, J., Zhu, T., and Ning, H. (2016). Internet use and its impact on individual physical health. *IEEE Access*, 4, 5135–5142.

14. Identifying disruptive innovations in transport: the case of the Hyperloop[1]

Yashar Araghi and Isabel R. Wilmink

1. INTRODUCTION

Innovations in transport modes and services enable people to travel more kilometres for work or leisure (Walton, 2006) compared with even a decade ago (except for the COVID-19 pandemic years). However, current motorized transport modes are either using fossil fuels directly or using electricity mainly generated by such fuels. This contributes to global CO_2 emissions, one of the leading causes of adverse climate change. Transport research, as well as practice, has focused on innovating cleaner and safer transport modes that can positively contribute to people's lives and health and cover their mobility needs while avoiding further negative impacts on climate by the transport sector. Undoubtedly, this calls for outside-the-box thinking and being innovative.

Innovations can be categorized into incremental or disruptive.[2] Incremental innovations are generally improvements to existing products and services. They have a limited impact on society, whereas disruptive innovations have a high impact on society or even disrupt the current market and offer many new technological features (Hopp et al., 2018b; Kylliäinen, 2019). An example of an incremental innovation in transport could be the addition of winglets at the tip of an aircraft's wings, which reduces fuel consumption by 2–5 per cent (Takenaka et al., 2008) and thus leads to a small improvement in aircraft emissions. But some innovations have disruptive impacts, such as container shipping (more than 60 years ago), which allowed for large-scale production and distribution of goods globally (Levinson, 2006; Notteboom and Rodrigue, 2008). Another example is Netflix, which started as a simple mail-order company but grew up to put Blockbuster out of business due to its convincing business model, and utilization of internet and online streaming technology for delivering media content (Hopp et al., 2018a).

Disruptive innovations involve unexpected trends in innovation pathways and often require new areas of research and development (R&D), the creation of new ways of production, and new markets. They can lead to sectoral transformations and the displacement of incumbent companies (Christensen et al., 2015).

Since disruptions often bring many changes to society, they can be challenging if citizens and governments are unprepared (Chen, 2018). This could leave policymakers, people and even businesses exposed to hard choices and even lead to closures of long-running businesses. Thus, it is crucial to grasp what makes an innovation disruptive in the transport domain.

To the best of our knowledge there are limited studies in the transport domain that focus on how innovations in transport can become disruptive and what are the underlying conditions and requirements for disruptions to occur. Therefore the focus of this study is to examine the disruption phenomenon in the context of transport and introduce a method for identifying the disruptive potentials of innovations in the transport sector. For this purpose, we have developed a framework that can help transport planners and researchers assess if a given innovation can be potentially disruptive.

We will further apply this framework to discuss the case of the Hyperloop to see if it could be a disruptive innovation in the transport sector in the coming decades. The Hyperloop has been chosen as a case study for our framework since there is a heated debate among academics and practitioners on how disruptive this new mode will be for the current transport sector.

The Hyperloop is poised to offer several societal advantages and appealing features such as ultra-high speeds (1200 km/hr) using only electricity (i.e. fewer emissions), much-reduced noise or vibration due to technical novelties (Mitropoulos et al., 2021), and being weatherproof (Dudnikov, 2017). Together, these attributes can be considered significant changes from the existing modes in the transport sector. Some even call it the 'fifth mode of transport' (Jacob et al., 2017; Armağan, 2020).

There are claims that the Hyperloop will reshape future travel and be a competitive mode to high-speed trains (HSTs) and air travel (Jia et al., 2019; Armağan, 2020; Mitropoulos et al., 2021). But the critics argue that the Hyperloop's construction costs would be very high and its effectiveness in increasing the overall welfare of society is doubtful (Hansen, 2020).

We emphasize that our goal in this chapter is not to forecast if and when an innovation (such as the Hyperloop) will be disruptive. Instead, we aim to explore when innovations could be potentially disruptive.

2.   AN OVERVIEW OF THE LITERATURE

Disruptive Innovations

Many authors have tried to explain what disruptive innovations are and describe the nature of these innovations (for instance, see Marsden and

Docherty, 2013; Hopp et al., 2018b; Si and Chen, 2020; Petzold et al., 2019; Martínez-Vergara and Valls-Pasola, 2021). However, in this chapter we use the definition provided by Christensen et al. (2018): disruptive innovations are those products or services that initially start at the bottom of the market with simple applications. They are normally less expensive and more accessible than the incumbent products or services. Eventually the simple, cheap and accessible product or service manages to aggressively move upmarket and force the incumbents out of the market or even put them out of business.

Tait and Wield (2021) point to the importance of policymakers paying extra attention towards disruptive innovations, since such innovations offer 'potential gains for a national economy that hosts the next generation of disruptive innovations' (Tait and Wield, 2021, p. 316).[3] The authors refer to the policies put in place by the UK government to support these innovations by implementing support systems, measuring the success or failure of policy initiatives that support disruptive innovations, and offering a regulatory system that enables the adoption of the innovative technology (Tait and Wield, 2021).

Finally, Millar et al. (2018) identify an important gap in the literature, namely research that addresses 'the impact of disruptive technology [innovation] on macro-systems such as societies or ecosystems, or on specific strategies and instruments to leverage, mitigate, or ameliorate systemic disruption'(p. 256). Therefore, the impact of disruptive innovations in the transport sector is of significant importance and needs to be carefully examined and researched to reveal the full potentials (or even threats) to society.

The Hyperloop: An Innovation with Great Potential and Challenges

The academic discussions on this innovative mode in the transport sector have recently intensified, with 95 per cent of published studies dated after 2016 (Mitropoulos et al., 2021). The original idea of the Hyperloop, where a pod travels at high speed through a vacuum tube, dates back to the early 20th century when Robert Goddard designed magnetic floating trains that in vacuum tubes could reach 400 km/hr (Salter, 1972; van Goeverden et al., 2018).

The discussions about the Hyperloop were revived after the publication of a white paper on the Hyperloop by Elon Musk and his group of engineers at SpaceX company in 2013 (SpaceX, 2013). In his 'Hyperloop Alpha' white paper, Musk provided some technical and economic features for a possible 'fifth mode of transport' (SpaceX, 2013). To demonstrate the potential of the Hyperloop, Musk chooses the example of the journey between Los Angeles and San Francisco, a roughly 610 km trip (each way) that by his estimate 7.4 million passengers take every year. Using the Hyperloop, the travel time would be slashed from 75 minutes by air travel (airport to airport) to 35 minutes by the Hyperloop (station to station), with a $20 (€15.3 in the year 2013) ticket cost. The $6 billion[4] (€4.6B in 2013) capital investments would be repaid in 20 years, assuming 40 pods operating every 2 minutes and each pod carrying 28 passengers (SpaceX, 2013; Rajendran and Harper, 2020).

The Hyperloop is designed to offer services for passengers and freight, possibly mixed, in tubes that are expected to be around 3.3 metres in diameter. However, most Hyperloop proposals indicate that pods carry only passengers or only freight, using the same tubes and infrastructure, since this would be beneficial from an operational perspective (Doppelbauer, 2013; van Goeverden et al., 2018).

Van Goeverden et al. (2018) conclude that the Hyperloop could have positive societal impacts by increasing accessibility to points of interest and job markets, and further it could be environmentally beneficial (e.g. less energy consumption, no emissions of greenhouse gases (GHGs), and much-reduced noise), if the promised technical and safety features are fulfilled. However, unlike Musk, they indicate that the drawback for the Hyperloop is that the construction and operational costs do not suggest a strong financial performance. The Hyperloop requires a revenue of more than €0.30 per passenger-km to cover the annual costs for operations and amortization of line infrastructure capital investments[5] compared to €0.17 per passenger-km for high-speed rail and €0.18 per passenger-km for air travel for a distance of 600 km (van Goeverden et al., 2018).

On the issue of construction costs, there are more recent studies that provide different estimates. For instance, in a relatively recent feasibility study conducted for Transport Canada, Stubbin et al. (2020) estimate construction costs of $56.4M (€49.0M in the year 2020) per km for a distance of 500 km in Canada. They further provide a range of estimates from other feasibility studies, varying between $37.8M (€32.9M) and $52.6M (€45.7M) per km for Stockholm to Helsinki (500 km), Abu Dhabi to Dubai (150 km), and Toronto to Windsor (350 km). These cost estimates all tend to contradict the appraisal of Musk (SpaceX, 2013), which was around $19M (€16.5M) per km for the Los Angeles to San Francisco line.

We will use the Hyperloop as a case from the transport sector when applying our disruptive innovation framework. Applying the framework to the Hyperloop provides an opportunity to debate its disruptive potential based on those above-mentioned (potential) societal advantages and possible financial difficulties.

3.   CONCEPTUAL FRAMEWORK FOR DISRUPTIVE INNOVATIONS

Bower and Christensen (1995) investigated why the leaders of companies, which lead the market, often failed to react or respond to a challenge arising from a technological change or innovation that was initiated by new entrant(s) with far less resources (Bower and Christensen, 1995; Christensen et al., 2015; Hopp et al., 2018b).

Initially, Christensen and his colleagues investigated how 'disruptive technologies' are developed, but later Christensen and Raynor (2003) switched to the term 'disruptive innovations' since there are cases where the technological part of the innovation is limited. However, there are other reasons why innovation can still be disruptive. For instance, the business model and the enabling value network can be highly influential in bringing disruptive innovation to the market (Hopp et al., 2018a). In later publications, Christensen et al. (2018) provide more detailed features and characteristics of disruptive innovations. They explain that disruptive innovations '... are NOT breakthrough technologies that make good products better; rather, they are innovations that make products and services more accessible and affordable, thereby making them available to a larger population' (Christensen et al., 2018).

We will now introduce our framework, which applies Christensen's theory to the transport sector. While other frameworks, such as the 'innovation feasibility' framework of Feitelson and Salomon (2004) (see Annema, Chapter 6 in this volume), the 'sociotechnical transition pathways' of Geels and Schot (2010), and several others as introduced in Part I of this book, have already explained how innovations are adopted in the transport sector, few studies address the disruptive aspects of innovations. The Disruptive Innovations Theory of Christensen seems to help shed light on how some innovations become disruptive.

As described by Christensen et al. (2015), disruption begins when an entrant firm with fewer resources challenges incumbent(s) who have considerable resources and revenues. Incumbents, however, try to improve their products and services for their most demanding and profitable clients. Incumbents tend to ignore or pay less attention to the needs of others (i.e. non-consumers, or the low end of the market). Entrants try to target those overlooked segments, and gain a foothold among them by offering suitable functionalities, usually at a lower price. Incumbents do not respond appropriately to these moves by entrants. Then entrants grow and claim some of the incumbents' mainstream customers while keeping those features that made them successful. The disruption is completed when (or if) the mainstream customers start adopting the entrants' products on a large scale.

This chapter puts Christensen's theory into practice with the help of a conceptual framework (see Figure 14.1). The conceptual framework aims to help researchers explore if a given innovation can potentially disrupt the market/sector (or not). The framework guides the researcher to consider different types of enablers that may help the innovation to be disruptive. These innovation enablers are categorized into supply related enablers and demand related enablers, as elaborated upon later in this section. Once these innovation enablers have been investigated, the researcher or the market analyst can gain a better outlook on the disruptive potential of the innovation under study.

The proposed conceptual framework starts by identifying the innovation. This framework can address diverse types of innovations in transport such as transport modes (e.g. the Hyperloop, flying car), transport services (e.g. Mobility as a Service (MaaS); see Veeneman, Chapter 12 in this volume), or technologies that can impact the way people may travel (e.g. Virtual Reality (VR), which might reduce the need to travel).

In the first part of the framework, the basic characteristics of the innovation are determined. The following questions could be helpful: what does the innovation do? Who will use this innovation (i.e. which user segments)? What are the investment costs? What are the operational costs? What are the key drivers and barriers of the innovation? What are the environmental impacts of the innovation?

The next two parts of the conceptual framework form its core, containing the two main types of innovation enablers: the *supply related innovation enablers* and the *demand related innovation enablers*. We explain these two in detail below.

The Supply Related Innovation Enablers

The second part of the framework focuses on the supply related *enablers*, and includes three essential enablers of an innovation which can make it a potentially disruptive innovation. They are called[6] 1) 'enabling technology', 2) 'business model' and 3) 'coherent value network'. Table 14.1 explains what is meant by each of these three innovation enablers.

According to Christensen, the three innovation enablers should be in place in order to make disruptive innovations available to consumers or users (Christensen et al., 2015, 2018). These enablers can be seen as the key ingredients that allow the innovating firm to supply (i.e. make available) the innovation to customers in collaboration with a network of partners and distributors. If the supply related enablers of an innovation are not in place, then it would be hard to make the innovation widely available to the public in a short time scale (i.e. few years).

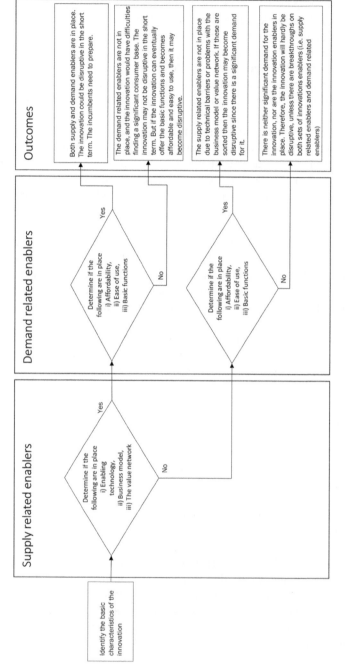

Figure 14.1   Conceptual framework for exploring the disruptive potential of an innovation

*Table 14.1   The three supply related innovation enablers and their
                description*

Supply related innovation enablers	Description
Enabling technology	An enabling technology that makes a product or a service more affordable and accessible to a broader population
Business model	A business model that targets new customers, non-consumers (who previously did not buy products or services in a given market) or low-end consumers (the less profitable customers)
Coherent value network	A network in which suppliers, partners, distributors and customers are each better off when the disruptive technology prospers

The first supply related enabler is the actual technology behind the innovation. This refers to the mechanism that enables the innovation to function. If the 'enabling technology' for a given innovation is not yet ready or mature (i.e. is still at lower TRL levels[7]) that means the technology (behind the innovation) is in its infancy or early conceptual phase (e.g. below TRL level 3 or 4). The technology exists only in labs or simulation environments (at these TRL levels). The prototypes or working models of the innovation[8] need to be built and tested. The safety features also need to be checked and verified.

The second supply related enabler is the business model for the innovation. Suppose the innovation's business model does not indicate a positive financial performance for the innovation; this means that the innovation is not yet feasible (financially) and requires (substantial) investment to take off. The cost of building the infrastructure for some innovations, or building the IT platforms for an innovative service, may be simply too high to amortize the capital costs for investors in a time frame set by them.

The third supply related enabler is the coherent value network. Often, the innovators are not independently able to bring the end product or service (i.e. the core of the innovation) to the users/customers. The firm needs suppliers of parts or raw materials, and it also needs distributors and partner companies who can help with the logistics. Without these networks of companies, the innovating firm will have difficulties in getting its product or service to its clients, and the innovation will probably not disrupt the market.

The Demand Related Innovation Enablers

The innovation must have features that will increase the utility of consumers by using or adopting it. Therefore, the needs and requirements of the consumer on the demand side are very important and are investigated as the third part of the conceptual framework. The demand related enablers include three ele-

Table 14.2   *Demand related enablers for a potentially disruptive*
                  passenger mode or service

Demand related innovation enablers	Description for passenger transport mode or service
Affordability	Travel costs are comparable to, or lower than, those of the existing modes or services
Ease of use	Availability, comfort and accessibility of the mode or service is attainable by consumers
Basic functions	The mode or service can be used safely by the travellers The travel time when using the innovative mode or service will be comparable to, or lower than, that when using existing modes or services

Table 14.3   *Demand related enablers for a potentially disruptive freight*
                  mode or service

Demand related innovation enablers	Description for freight transport mode or service
Affordability	Shipping costs are comparable to, or lower than, those of existing modes or services
Ease of use	Loading/unloading locations are accessible for shippers There is a suitable network connection so that the freight can be distributed smoothly
Basic functions	The mode or service can be used reliably, with a shipping speed that is comparable to, or higher than, that of existing modes and services

ments,[9] namely: 1) 'affordability', 2) 'ease of use' and 3) 'basic functions' of the product or service offered by the innovation. These elements in the demand related innovation enablers should be in place so that innovations can find a substantial consumer base and potentially gain a foothold among consumers and even attract non-consumers or the lower end of the market. This part is also essential for making an innovation potentially disruptive.

In Tables 14.2 and 14.3, the elements of demand related enablers are listed and further translated into specific key criteria for a potentially disruptive passenger and freight transport mode (or service), respectively.

The last (fourth) part of the framework provides a series of outcomes, informing the user of the framework about the status of the innovation and what changes are needed to enable a disruptive entry to the market. The two sets of innovation enablers can generate four different outcomes. Based on each of these four outcomes, the framework provides guidance regarding the disruptiveness of the innovation.

In the next section, we will apply the above-mentioned conceptual frame-work for the case of the Hyperloop and discuss the disruptiveness potential of this mode of transport.

4. APPLYING THE CONCEPTUAL FRAMEWORK FOR THE CASE OF THE HYPERLOOP

To make a fair judgement on the disruptiveness potential of the Hyperloop, one needs to implement a full-scale and up-to-date technical and financial feasibility study of a Hyperloop system in a potential corridor or a network of corridors to obtain a better picture of the Hyperloop's impact on the trans-port sector. We have not yet conducted such a comprehensive assessment. However, we rely on recent technical and economic feasibility studies, which will be the sources for our case study.

The first step of our conceptual framework is to identify the basic charac-teristics of the Hyperloop. The literature review article by Mitropoulos et al. (2021) is a solid point of departure in determining these basic features of the Hyperloop. The paper offers answers to the questions we proposed at the start of section 3. The authors list various studies conducted in different regions of the world (Europe, Asia and America) and address multiple technical aspects of the Hyperloop, for instance the pod, the power systems, the infrastructure, levitation, propulsion, and transport engineering and planning aspects.

For the second step of the framework, we reflect on the supply related ena-blers of the Hyperloop (as discussed in section 3), which are the following: 1) the enabling technology, 2) the business model for the Hyperloop and 3) the coherent value network that needs to be created for the Hyperloop.

The Enabling Technology Needed for the Hyperloop

The Hyperloop is, on many levels, a highly advanced piece of technology that needs to be studied at multiple levels (van Goeverden et al., 2018):

- pod technology
- levitation and propulsion technology
- the infrastructures (i.e. the tubes, the pylons, stations, high-speed switches)
- the vacuum environment
- the stations and loading and unloading of passengers and cargo to the pods
- the safety and regulatory processes and guidelines
- communication, traffic monitoring and control systems.

Based on the discussion provided by van Goeverden et al. (2018), Stubbin et al. (2020), Rajendran and Harper (2020) and Mitropoulos et al. (2021), among other authors, the technical features of the Hyperloop are in the 'design' or 'define' phase or 'in the very early stages of development'. Much more data needs to be gathered from different pilots or test environments (where actual-size pods are put on test tracks, in tubes) to make a definite conclusion on the technological readiness of the Hyperloop.

There are uncertainties about the pods' comfort level, including in relation to seat and thermal comfort, amount of walking space within the pods, and vibration and noise levels at such high speeds. Amenities such as WiFi connection, luggage compartments and lavatories are also essential for the Hyperloop's competitiveness with air travel or HSTs. There are also issues about maximum thrust forces and acceleration or deceleration that can be applied on pods with passengers onboard, which must be tested (Mitropoulos et al., 2021).

Assuming all these technical barriers and uncertainties are solved, the Hyperloop could provide a 'good alternative' to air travel for 'medium to long-distance travel' (van Goeverden et al., 2018).

The Business Model for the Hyperloop

Since there are not yet any commercially available Hyperloop lines, there is no market in place to allow us to assess the current business model for the Hyperloop. However, according to Christensen's theory, a business model of disruptive innovation needs to target non-consumers or new customers from the low end of the market. The low end of the market is where the disruption of the market often occurs (Christensen et al., 2018).

The developers of the Hyperloop expect it to capture some part of the market for air travel and HST, or even disrupt these markets. Hence, the Hyperloop business model needs to be comparable with these two incumbent modes in terms of capital investments, operational costs and revenues. That is why some studies, including those by van Goeverden et al. (2018), Rana (2020) and Rajendran and Harper (2020), compare financial indicators of these two modes with that of the Hyperloop.

The business model of the Hyperloop, if it wants to be disruptive and capture a significant market share from other modes, needs to cover the operational and overhead costs and repay the capital costs over a period of time. At the same time, the travel costs for the passengers and the freight fares per ton-km needs to be competitive with the existing modes so that reasonable modal shifts can be observed over time.

The study by van Goeverden et al. (2018) estimated that the travel time for a 600 km distance would be 40 minutes for the Hyperloop, compared to 98 minutes for air travel and 139 minutes by HST. The energy consumption per passenger would be 3.3 times less for the Hyperloop and HST than for air travel, and similarly, the GHG emissions[10] would also be three times less. This means that there would be passengers who would be willing to pay a premium to travel on the Hyperloop for reasons of either speed or low emissions (or both), which means the Hyperloop could have higher ticket prices than its competitors. However, asking for a premium ticket price for the Hyperloop would make it impossible to capture the low end of the market.

Stubbin et al. (2020) have reviewed the feasibility studies[11] of several Hyperloop projects and collected the capital costs of different potential corridors. The capital costs for infrastructure are in the range[12] of \$37.8M to \$56.4M per km (in 2020 \$) in these feasibility studies (Stubbin et al., 2020), €33.5M[13] per km in the article by van Goeverden et al. (2018), and €38.9M per km in the article by Maja et al. (2020). Here, the type of soil where the pylons for the tubes need to be built and the geography of the corridor play a significant role in making estimates of capital costs. Based on the corridor's terrain, the tubes may need bridges or tunnels, which would increase the costs per km and would be highly challenging from a safety point of view.

In addition to the capital costs, the operational and overhead costs also need to be covered by the Hyperloop revenues. Providing estimations for these costs is extremely difficult since it depends on the demand size, the frequency of pods, the time saving per corridor and the freight capacity that could be handled in a corridor. Yet, a few studies have provided helpful revenue estimates translated into ticket costs. Based on certain assumptions regarding the indicators that we just mentioned for the operating costs, van Goeverden et al. (2018) has estimated that €0.30 per passenger-km would at least mean the project would break even. Stubbin et al. (2020) have used the example of MagLev trains as a gauge for the Hyperloop and estimated a unit ticket price of \$0.26 to \$0.28 per person-km. Maja et al. (2020) calculated that it is possible to envision a Hyperloop service between Rome and Milan with a ticket price of €0.23 per passenger-km if the project is allowed to amortize its capital costs over 44 years.

These price estimates per passenger-km are often comparable to or higher than legacy air carriers and certainly higher than the low-cost air carriers. These ticket cost estimates make the Hyperloop affordable for the high end of the market rather than the lower end, which means it is unlikely that the innovation will be disruptive.

The Value Network that Needs to Be Created for the Hyperloop

Several private Hyperloop developers such as Hyperloop Transportation Technology (HTT), Hyperloop Italia, Virgin Hyperloop One, Zeleros Hyperloop (UPV), Delft Hyperloop, DGW Hyperloop, and The Boring Company are actively studying, experimenting with and testing different concepts and technological hurdles of Hyperloop.

The European Hyperloop Centre is a not-for-profit and open innovation initiative in the Netherlands. It is a public–private partnership between a provincial body and two Dutch Ministries,[14] and a group of private companies trying to support the R&D on Hyperloop transportation.

The formation of several private companies and public–private partnerships as well as various publicly funded feasibility studies on the Hyperloop signal increasing interest for the Hyperloop. However, hurdles remain regarding both technical matters, such as the promised speeds and the acceleration and deceleration challenges, and safety and health aspects, all of which need to be overcome. There needs to be much more industrial collaboration from different stakeholders to overcome the technical and safety barriers. These collaborations are essential to acquire the approval of regulators that will allow the Hyperloop to be put into public use.

Next to technical challenges, there are some legal challenges regarding the land, both private and publicly owned, that needs to be acquired (i.e. right of way) for building the Hyperloop infrastructure (Stubbin et al., 2020). There are plausible concerns about the visual pollution caused by the tubes placed over pylons in urban and suburban areas, which might lead to resistance from citizens. Therefore, in addition to technical collaboration, there needs to be stakeholder collaboration on legal and societal issues for the Hyperloop to be built in urban and suburban areas. Without the collaborations among stakeholders, the network of suppliers and regulatory authorities, it is unlikely that even a technically ready Hyperloop would be allowed to operate.

As a next step in assessing the conceptual disruptive innovation framework, we focus on the demand related enablers of the Hyperloop. This part looks from the user's perspective. In the following subsections, we discuss if the Hyperloop is affordable, easy to use, and has the basic functions a mode needs for passengers and cargo movements.

Affordability

According to Christensen, disruptive innovations need to address the low end of the market (Christensen et al., 2015). But there are examples of innovations that were initially expensive and eventually disrupted the mobility market, for instance the car following mass production by the Ford company. In the car example, the price of a car was too high for the lower end of the market, but Ford managed to reduce the cost significantly so that it was no longer a luxury product for the very rich. Ford managed to make cars accessible to broader sectors of the market, and with added competition from other carmakers and economics of scale, the car eventually became available to the middle classes.

As discussed in the business model section, previous research and feasibility studies have put the Hyperloop ticket price at between $0.26 and $0.36 (€0.23

to €0.31 in the year 2022) per person-km to break even. Air travel is expected to be a strong competitor for the Hyperloop. We have taken the average cost of a one-way plane ticket from many connections (ranging from 350 km to 1200 km) in different parts of the world (i.e. developed and developing countries). We arrived at prices between $0.06 and $0.12 (€0.05 to €0.10 in the year 2022) per passenger-km for the low-cost carriers and between $0.11 and $0.35 (€0.09 to €0.24 in the year 2022) per passenger-km for the legacy carriers (for a one-way economy class seat). However, many legacy carriers sell their return tickets (for many connections) with a price only 10 to 30 per cent higher than their one-way tickets. Hence, the travel cost per passenger-km for a return ticket would be between $0.05 and $0.18 (€0.04 to €0.16 in the year 2022) per passenger-km for the legacy carriers. The price for HST tickets, depending on the distance and the location, is in the same order of magnitude as air travel. Van Goeverden et al. (2018) put the HST price at $0.20 and air travel at $0.21 per passenger-km, which is in line with our observation of the current situation (in 2021).

The conclusion is that given the current travel cost estimations for the Hyperloop, it would not compete with air travel or HST, purely based on costs.

Ease of Use

When assessing the 'ease of use' of a transport mode we consider three attributes, namely 1) availability, 2) comfort and 3) accessibility, that are highly relevant to travellers when selecting a travel mode.

Comfort
Walker (2018) has defined a few criteria for passengers' comfort in the Hyperloop, which include acceleration/deceleration, seat comfort, thermal comfort, crowdedness, psychological distress, noise, motion sickness and access to facilities.

To ensure passenger comfort, maximum acceleration and deceleration needs to be strictly limited to 0.1G and 0.3G, respectively. However, motion sickness, which can be anticipated with the presence of lateral gravity when travelling around bends and curves at high speeds, is yet to be studied (Walker, 2018).

Given the experience with MagLev trains, Hyperloop passengers are less likely to sense much noise or feel much vibration since the driving mechanism for pods is relatively similar, except pods travel in a more controlled vacuum environment (Walker, 2018). However, the actual noise and vibration levels need to be demonstrated at speeds of 1200 km/hr, which has not yet been achieved either in practice or in the test phases.

Another important comfort factor is that passengers will not be able to enjoy the view of the surroundings while travelling in the tube. This may need to be compensated by onboard amenities such as video or visual features in the Hyperloop pods.

Although Elon Musk did not include lavatory facilities in his design, it seems essential for trips longer than 10 minutes. Therefore, to compete with air travel and HST, the Hyperloop pods need to be equipped with lavatories. Additionally, the passenger compartment needs to be easily walkable with enough luggage room, especially for trips of medium to long distance.

Given what has been said above, it is clear that the level of comfort for the Hyperloop is not yet certain. Further data from test cases and pilots are required so that there is enough evidence to make a sound analysis of the comfort levels.

Availability

Various feasibility studies (e.g. Stubbin et al., 2020; NOACA et al., 2019; Rajendran and Harper, 2020) assume that the number of pods travelling on a given corridor (e.g. 600 km) per hour would be anything from 15 to 40, or even 50, with a passenger capacity of 28 to 44 and an operation time of 15 to 18 hours per day. SpaceX has estimated the frequency of pods to be one pod per 2 minutes, and van Goeverden et al. (2018) have used a frequency of one pod per 5 minutes. Maja et al. (2020) have used two scenarios of 4 minutes in peak times and 8 minutes in non-peak times.

Given all these estimates, we can conclude that the Hyperloop would have a robust competitive edge when it comes to availability, since HST and air travel, even for busy point-to-point destinations, cannot compete with such high frequencies. However, the Hyperloop and HST are bound to the corridors built for them and do not have the flexibility available to air travel. No corridor or infrastructure other than the airport itself is required for air travel, and air travel therefore wins in terms of availability for less densely populated areas.

Accessibility

If we define accessibility in terms of passengers boarding or disembarking the pods, then the Hyperloop seems to be comparable to HST and would therefore be a relatively passenger-friendly mode (Walker, 2018; Jia et al., 2019). Reaching the pods at the station would be relatively quick, requiring less screening and waiting times than air travel. Therefore, the time lost in Hyperloop stations is thought to be somewhat less than air travel (Jia et al., 2019).

If the accessibility of passengers is defined in terms of accessing the station itself, then the Hyperloop would be similar to air travel, given that the Hyperloop stations would most likely be positioned on the outskirts of cities, with good access/egress connections. Assuming that the Hyperloop stations could be located in or near urban centres, they would be more accessible than airports and comparable to trains. Spatial design topics play an important role in determining the optimal location for the Hyperloop stations and whether the Hyperloop stations should be built underground or above ground with tubes connecting them.

Given some safety concerns due to the vacuum environment and the high speeds of the pods, the Hyperloop stations might need a special design to allow the vacuumed and normal pressure areas to be split (Mitropoulos et al., 2021) and might have to be located outside of built-up areas (Jia et al., 2019).

In conclusion, boarding the Hyperloop would, at best, be similar to HST in terms of waiting time at the station and accessing the station. But it would be similar to air travel should passengers have to go to the outskirts of cities to get to the stations, which would require good access/egress links. Therefore, the Hyperloop would not be much better or worse than the incumbent modes in terms of accessibility.

We emphasize that there are many socio-economic factors involved in choosing the location of the Hyperloop stations. These factors include accessing workplaces in the metropolitan areas, increased land value, increased economic activities and enhanced accessibility to other metropolises. Therefore the optimal choice of location for a Hyperloop station, and its consequential impacts, requires detailed research.

The Hyperloop could be an instrumental mode in making two metropolitan areas accessible to each other. The high frequency of pods is an important advantage of the Hyperloop over HST and local air travel. The high capacity offered by pods travelling at high frequencies (e.g. every 2 to 5 minutes) between two metropolitan areas would give the Hyperloop an unmatchable edge in connecting two places, increasing the accessibility of people wanting to reach those areas that are linked to the Hyperloop network. This might also generate new trips, for instance people working in Paris but deciding to live in Rotterdam.

Basic Functions

A fundamental goal of transport modes is to move people (and cargo) safely at an acceptable cost, in relative comfort, and within a reasonable travel time (Mokhtarian, 2019). We call these 'basic functions' of a transport mode. Comfort and cost have already been discussed in previous sections, and here we focus on the other two basic functions: safety and travel time. To be commercially successful, the Hyperloop needs to be safe and preferably have lower or similar travel times compared to the existing travel modes.

Safety

Around a third of previous research on the Hyperloop has paid some attention to safety issues (Mitropoulos et al., 2021). This volume of research reflects the level of concern about the safety of the Hyperloop.

Since the Hyperloop aims to travel at very high speeds in tubes, it needs to overcome many technical challenges before it can be considered safe by regulators for public use. The near-vacuum operating environment, the frequent pressurization and depressurization of pods, air cracks and fire hazards in pods and tubes, and emergency exits in different parts of the corridor (in tunnels, over bridges, over water), are among the topics which should be addressed for safety (Gkoumas and Christou, 2020). Therefore, Hansen (2020) considers 'safety constraints' to be the 'most serious barrier' to 'successful commercial operations' (p. 817).

Hyperloop developers are promising that the highest safety standards and stringent regulations are being pursued in the design of the systems. However, these claims remain to be substantiated during the pilot phases. Statistically, the Hyperloop needs to be at least as safe as air travel, given that air travel would be one of the main competitors of the Hyperloop. The safety record of air travel is well documented by the International Air Transport Association (IATA), which has recorded 1.3 to 1.7 accidents per 1 million flights (IATA, 2022). These statistics would be the benchmark during the initial and pilot phases of the Hyperloop.

Assessing and proving the safety of a novel concept like Hyperloop can be deemed a difficult challenge. The mapping of risks and hazards in the state-of-the-art literature (e.g. railway, aviation, automotive) is often done on the basis of statistics. Although the Hyperloop concept consists of many subsystems which have been proven to be safe in various existing applications (e.g. magnetic levitation, airlocks, vacuum systems), the statistics on the risks and hazards of these components rely on the application context. Moreover, when dealing with such a novel concept, an additional set of complications is induced by the lack of existing application-specific norms or standards (Arup et al., 2017).

In order to assess the safety of Hyperloop in its early development stages, in the absence of application-specific norms/standards and/or statistics, novel and potentially tailored approaches may have to be designed. These approaches should accommodate interactive discussions between the applicant and the assessor on notions such as, for example, acceptable risk (Arup et al., 2017).

In this chapter we will not go deeply into the safety aspects and concerns of the Hyperloop; we refer readers to Gkoumas and Christou (2020) for a deeper discussion regarding the safety aspects of the Hyperloop.

Travel time

Travel time is another essential aspect for any transport mode, not just disruptive ones. Due to the levitation of pods and reduced air resistance in the tubes, the pods are poised to travel at very high speeds, thereby drastically reducing travel times over short to medium distances compared to incumbent modes. If these promised low travel times are realized in the real world, the Hyperloop can claim to have a significant advantage over its competitor modes, that is, air travel and HST.

The perception of time or the value of time (VOT) is not consistent among different user groups, such as commuters, business travellers and leisure travellers (Horowitz, 1978). Establishing the VOT and willingness to pay for such low travel times among different segments of the Hyperloop users requires further research (Mitropoulos et al., 2021).

BAK Economics AG (2020) has discussed the time aspects for the Hyperloop and concludes that even considering the access/egress time and the time needed for the security screening, the boarding process and baggage handling, the Hyperloop will still be considerably faster than competing modes, and will probably have a significant impact on the Basel–Paris corridor of 415 km, which was taken as an example. Furthermore, the Hyperloop is poised to have a winning edge over air freight in the high-speed cargo market, which accounts for 2 per cent of total ton-miles but covers 40 per cent of total freight value (BAK Economics AG, 2020).

In the two years preceding the writing of this chapter in 2022, many business trips have been scrapped and replaced by online meetings due to the COVID-19 pandemic. Going forward, many business travellers will likely continue to reduce their amount of travel compared to before the pandemic. Some researchers claim that this reduced work-related travel may stay for the long term (at least for high-income individuals), even after the pandemic is over (Brough et al., 2021). This would cause severe doubts as to the travel demand for the Hyperloop, and would weaken one of the best advantages that the Hyperloop has to offer in comparison to its competitor modes.

5. DISCUSSING UNCERTAINTIES REGARDING THE HYPERLOOP

The conceptual disruptive innovation framework provides a tool to explore different criteria of innovations and swiftly check if an innovation (mainly in the transport sector) has the potential to be disruptive or not and under which conditions.

We used the case of the Hyperloop to validate the framework and to see if the Hyperloop could potentially disrupt the transport sector. The findings of recent research on the Hyperloop was used as input for our assessment.

Table 14.4   *Index for determining the disruptiveness of each element*

Highly unlikely to be disruptive	Unlikely to be disruptive	Neutral	Likely to be disruptive	Highly likely to be disruptive
$--$	$-$	0	$+$	$++$

Tables 14.5 and 14.6 provide a summary of the two central components of the framework (i.e. the supply related enablers and the demand related enablers), which will guide us to the outcome of the framework. An index for determining the disruptiveness of each element discussed in these tables is given in Table 14.4.

Framework Outcome

Based on the above discussions, one can conclude that some supply related enablers for the Hyperloop are in place, some are uncertain, and some are not in place. Similarly, some demand related enablers for the Hyperloop are present, and some are yet to be proven. Therefore, given the current level of uncertainty, it would be challenging to consider the Hyperloop as a disruptive mode of transport in the short term. This finding is based on what has been revealed by the currently available academic literature on Hyperloop assessment and by independent feasibility studies. However, the future may look different for the Hyperloop given other circumstances. Next, we offer some possible future developments that could help the Hyperloop disrupt the transport sector or conversely make the Hyperloop uncompetitive in the future.

Possible Future Developments

The environmental pull: the need to reduce the carbon footprint of the transport sector
Due to the significant role of the transport sector in total emissions and the increased public sensitivity to climate change, the Hyperloop seems to be an attractive alternative to air travel for policymakers, and could replace many short to medium air trips between metropolitan areas. Aircraft currently use fossil-based fuels, which is highly unsustainable (Upham et al., 2012), and there are growing public concerns, reflected, for instance, by the flight shaming campaign (Flaherty and Holmes, 2020; Mkono, 2020).

There are additional reasons for policymakers to support the Hyperloop developments. Here we name just a few: the uncertainties in oil prices; the Paris Agreement and the follow-up obligations for CO_2 reduction; and the

Table 14.5   *Summary of findings for supply related enablers of the Hyperloop*

Supply related innovation enablers	Discussion	Potential for disruption	Assumption
Enabling technology	The technical features of the Hyperloop are highly uncertain. Some features are in the 'design' or 'define' phase, and some are in the 'early stages of development'. Therefore it is too early to say if the final product would be more affordable to a broader population than the incumbent modes.	0 (too early to say)	Technologies must be safe and cost-efficient to operate a network of pods.
Business model	Based on different scenarios on infrastructure costs and market size, researchers have generally evaluated that the ticket price of the Hyperloop will be around two times higher than HST and legacy air carriers (three times more costly than low-cost carriers) to cover the capital costs and operational costs. This will make it difficult for the Hyperloop to disrupt the market and target the low end of the market.	– –	Environmental drivers are not the main drivers for extensively adopting a mode (i.e. disruption). The ticket price and willingness-to-pay play a more significant role in determining the business model of the Hyperloop.
Coherent value network	There is a significant interest and excitement among several private parties to further push ahead with the technical development of the Hyperloop. There is also interest from the public sector to assess the potential of the Hyperloop, at least at the level of feasibility studies. Some public–private collaborations are underway, which creates the foundation of a value network. This factor could positively contribute to the Hyperloop's disruptiveness, provided that other aspects (technical and financial) are fulfilled.	+	Private–public collaborations will continue in future.

Table 14.6   Summary of findings for demand related enablers of the Hyperloop

Demand related innovation enablers	Discussion	Potential for disruption	Assumption
Affordability	Given the current estimations, the Hyperloop ticket price would be between $0.26 and $0.36 per person-km. In contrast, the low-cost carriers are at $0.06 to $0.12 per passenger-km, and the legacy carriers are between $0.11 and $0.35 per passenger-km for one-way economy tickets and even less for a return ticket. The Hyperloop could not compete with air travel or HST, purely based on ticket price.	– –	The ticket price for the Hyperloop is determined with the assumption that it only recovers the operational costs and initial capital investments with a discount rate of 2%.
Ease of use	*Availability:* for an established Hyperloop corridor, the frequency of the Hyperloop will be very high and much better than flight and HST frequencies. However, the air carriers are more flexible in choosing destinations and responding to the market demand. The Hyperloop, like HST, is bound to the existing tracks, making it difficult for the Hyperloop to be available at many locations, thus limiting its availability.	+ (for frequency)	Based on the frequency of one pod per 2 to 5 minutes.
		– (for network availability)	New routes would take a longer time to construct and considerable investment and depend on the demand.
	Comfort: Hyperloop developers claim that it would be a smooth ride, with less noise and vibration. The Hyperloop's comfort level is not yet certain and requires further proofs through test cases and pilots to gather enough evidence to make a sound analysis.	+ (with uncertainty)	Based on the validation of claims made by the Hyperloop developers.

Demand related innovation enablers	Discussion	Potential for disruption	Assumption
Ease of use (continued)	*Accessibility*: the Hyperloop would be (at best) similar to HST in terms of waiting time at the station and travelling to the station; otherwise, reaching a Hyperloop station would be similar to going to airports on the outskirts of cities, which require good access/egress links. Therefore, the Hyperloop would not be much better or worse in terms of accessibility than incumbent modes.	+	Based on screening at the Hyperloop stations not being intensive.
Basic functions	*Safety*: this is highly debated in the literature, and many technical and safety challenges are yet to be overcome. This factor provides the highest level of uncertainty for Hyperloop's deployment and possible disruption in the coming years.	0 (for safety)	High level of uncertainty for Hyperloop's deployment.
	Travel Time: if the promised speeds are fulfilled, the Hyperloop will have a significant advantage over its competitors. This creates potential disruption capabilities for the Hyperloop, assuming that there is sufficient demand for fast travel after the pandemic.	+ + (for travel time)	The Hyperloop can reach speeds of 1200 km/hr (as promised)

increasingly available electricity from renewable sources, which is more easily used in land-based transport than in air transport.

The need to reduce the carbon footprint of the transport sector may allow policymakers to justify spending public resources on the infrastructure costs of the Hyperloop. This would have a reinforcing effect on efforts to fast-track technological developments of the Hyperloop and to attract additional investment from the private sector, in order to overcome the technical and safety challenges.

There is the possibility that policymakers will opt for imposing environmental tax levies on the aviation sector, to force the aviation sector to innovate and reduce emissions and also to support the cleaner competing modes, such as the Hyperloop or HST. This would again be an advantage for the Hyperloop. However, considering the costs of tickets, it seems that without substantial increases in the price of flying, it will be challenging for the Hyperloop to compete and become disruptive.

Breakthrough in aviation emissions problem
There is also the possibility that the aviation sector might make a breakthrough in flying without the need for fossil fuels, by replacing them with electro-fuel and liquid hydrogen (Åkerman et al., 2021) or any other innovative technology. This scenario would create a very high entry barrier for the Hyperloop, given the high capital investment costs compared to air transport, which already has a lot of infrastructure (local and international airports) in place.

Breakthrough in reducing the Hyperloop costs
Currently, the most influential factor in determining the high ticket cost of the Hyperloop seems to be the capital investments needed for building the infrastructure (van Goeverden et al., 2018; Hansen, 2020). If there are breakthroughs in significantly reducing the infrastructure costs and building cost-efficient pods and operating systems, the disruptive potential of the Hyperloop may increase.

Furthermore, the reduced infrastructure costs would enable more countries to invest and collaborate in developing an international Hyperloop network (e.g. Amsterdam–Frankfurt–Brussels–Paris network). This scenario would work as a catalyst in fast-tracking the Hyperloop's entry into the market.

6.   CONCLUSIONS

This chapter introduces a conceptual framework to explore disruptive innovation in the transport sector. The framework is based on the Disruptive Innovation Theory of Christensen and could further help transport researchers and practitioners to put this theory into practice.

The framework is then applied to the case of the Hyperloop as a potentially disruptive mode. The supply related enablers of the Hyperloop – namely, the Hyperloop technology, business model and value network – are explored, along with the demand related enablers of the Hyperloop. Further, the demand related aspects – namely, affordability, availability, comfort, accessibility, safety and travel time – are discussed. This is done based on recent literature and feasibility studies on the Hyperloop.

The outcome of the conceptual framework is that it is unlikely that the Hyperloop will be disruptive in the short term (i.e. the coming decade), given the arguments provided in the discussion section and the current uncertainties on the technological readiness and safety issues. Moreover, the current estimated ticket costs for the Hyperloop would be acceptable only to the high-income groups of society. Therefore, policymakers could find it hard to justify the infrastructure costs to only benefit this small segment of society instead of investing in projects with more comprehensive and overall benefits for society as a whole.

Finally, we argue that the need to reduce the carbon footprint of the transport sector combined with possible breakthroughs in reducing the Hyperloop's infrastructure expenses (and subsequently reducing travel costs) may be helpful for the realization and potential disruption of the transport sector.

NOTES

1. The authors would like to thank TNO colleagues Eleni Charoniti and Sri Ramakrishnan Ganesan for their support and useful discussions on the overall Hyperloop research and also to Chris van der Ploeg from TNO's Integrated Vehicle Safety group for his insight into safety aspects of Hyperloop. Further, the authors are grateful for the funding provided by the Ministry of Economic Affairs and Climate Policy and the Ministry of Infrastructure and Water Management of the Netherlands.
2. In the innovation literature, there is another category of innovations, called radical innovations. Radical innovations also point to novel ideas or products and technological breakthroughs. However, there are some conceptual and fundamental differences between disruptive innovations and radical innovations. They are created by different mechanisms and the response of incumbent firms or organizations towards these two types of innovation should be different. We refer the interested reader to Hopp et al. (2018a) and Hopp et al. (2018b). In this chapter we focus on disruptive innovations and base our work on the literature related to such innovations.
3. Some of the potential gains are listed as creating new business models, value chains and transforming or propelling the sector (in which the disruption has occurred) to a leading position (Tait and Wield, 2021).
4. All the prices given in this chapter are in US Dollars ($) or Euros (€) unless otherwise mentioned.

5. Also assuming pods with a capacity of 28 passengers and a frequency of 12 pods/ hr each direction with at least a 15 hr/day operating time.
6. These terms are taken from Christensen's theory of Disruptive Innovations (Christensen et al., 2018).
7. TRL levels refer to the technology readiness levels explained by Mankins (1995).
8. In case of transport, this could be either a mode, a service or a new technical tool.
9. These terms are taken from Christensen's theory of Disruptive Innovations (Christensen et al., 2018).
10. If the electricity for the Hyperloop or HST comes from renewable sources then the GHG emission saving would be much larger compared to air travel.
11. These studies have been conducted by different commercial sector consultancy companies including, among others, SpaceX, KPMG, TEMS and Virgin Hyperloop.
12. There is a low estimate of $19M per km by SpaceX (2013), which seems to be an outlier, given the rest of the feasibility studies – see section 2.
13. Given that 50 per cent of the 600 km corridor has solid soil, 40 per cent weak soil and 10 per cent consists of tunnels.
14. The Ministry of Economic Affairs and Climate Policy, the Ministry of Infrastructure and Water Management, and the Province of Groningen.

REFERENCES

Åkerman, J., Kamb, A., Larsson, J., and Nässén, J. (2021). Low-carbon scenarios for long-distance travel 2060. *Transportation Research Part D: Transport and Environment*, 99, 103010.

Armağan, K. (2020). The fifth mode of transportation: Hyperloop. *Journal of Innovative Transportation*, 1(1), 1105.

Arup, BCI, TNO, and VINU (2017). *Hyperloop in The Netherlands*. Report no. 2017 R10715, TNO, The Hague.

BAK Economics AG (2020). *Hyperloop: A Breakthrough for Vacuum Transportation?* https://wissenschaftsrat.ch/images/BAK_2020_Hyperloop.pdf (accessed 30 November 2021).

Bower, J.L., and Christensen, C.M. (1995). Disruptive technologies: Catching the wave. *Harvard Business Review*, 73(1), 43–53.

Brough, R., Freedman, M., and Phillips, D.C. (2021). Understanding socio-economic disparities in travel behavior during the COVID-19 pandemic. *Journal of Regional Science*, 61(4), 753–774.

Chen, T.M. (2018, January). Impact of disruptive innovations on road transport strategy. Paper presented at the HVTT15 15th Symposium on Heavy Vehicle Transport Technology, Rotterdam, 2–5 October.

Christensen, C.M., McDonald, R., Altman, E.J., and Palmer, J.E. (2018). Disruptive innovation: An intellectual history and directions for future research. *Journal of Management Studies*, 55(7), 1043–1078.

Christensen, C.M., and Raynor, M.E. (2003). Why hard-nosed executives should care about management theory. *Harvard Business Review*, 81(9), 66–75.

Christensen, C.M., Raynor, M., and McDonald, R. (2015). What is disruptive innovation? *Harvard Business Review*, 93, 44–53.

Doppelbauer, J. (2013). *Hyperloop – An Innovation for Global Transportation*. EU Agency for Railways, Valenciennes, France.

Dudnikov, E.E. (2017, October). Advantages of a new Hyperloop transport technology. In *2017 Tenth International Conference Management of Large-Scale System Development (MLSD)* (pp. 1–4). IEEE.

Feitelson, E., and Salomon, I. (2004). The political economy of transport innovations. In M. Beuthe, V. Himanen, A. Reggiani, and L. Zamparini (eds), *Transport Developments and Innovations in an Evolving World* (pp. 11–26). Berlin: Springer.

Flaherty, G.T., and Holmes, A. (2020). Will flight shaming influence the future of air travel? *Journal of Travel Medicine*, 27(2), taz088.

Geels, F.W., and Schot, J. (2010). The dynamics of transitions: A socio-technical perspective: A multi-level perspective on transitions. In J. Grin, J. Rotmans, and J. Schot, *Transitions to Sustainable Development: New Directions in the Study of Long-Term Transformative Change* (pp. 18–28). London: Routledge.

Gkoumas, K., and Christou, M. (2020). A triple-helix approach for the assessment of hyperloop potential in Europe. *Sustainability*, 12(19), 7868.

Hansen, I.A. (2020). Hyperloop transport technology assessment and system analysis. *Transportation Planning and Technology*, 43(8), 803–820.

Hopp, C., Antons, D., Kaminski, J., and Salge, T.O. (2018a). What 40 years of research reveals about the difference between disruptive and radical innovation. *Harvard Business Review*, 9 April.

Hopp, C., Antons, D., Kaminski, J., and Salge, T.O. (2018b). The topic landscape of disruption research – a call for consolidation, reconciliation, and generalization. *Journal of Product Innovation Management*, 35(3), 458–487.

Horowitz, A.J. (1978). The subjective value of the time spent in travel. *Transportation Research*, 12(6), 385–393.

IATA (2022). IATA releases 2021 airline safety performance. IATA Press Release No: 11, 2 March. https://www.iata.org/en/pressroom/2022-releases/2022-03-02-01/ (accessed 15 March 2022).

Jacob, R., Phillip, R., Sunny, S., and George, J. (2017). Review of Hyperloop: The fifth mode of transport. *International Journal of New Technology and Research*, 3(3), 101–103.

Jia, P.Z., Razi, K., Wu, N., Wang, C., Chen, M., Xue, H., and Lui, N. (2019). *Consumer Desirability of the Proposed Hyperloop*. UC Santa Barbara, Department of Economics.

Kylliäinen, J. (2019). Types of innovation – the ultimate guide with definitions and examples, 4 October. https://www.viima.com/blog/types-of-innovation (accessed 15 March 2022).

Levinson, M. (2006). *The Box: How the Shipping Container Made the World Smaller and the World Economy Bigger*. Princeton, NJ: Princeton University Press.

Maja, R., Favari, E., and Mariani, C. (2020). Forecasting the success of hyperloop technology on Italian routes: A broad feasibility study. Conference paper, European Transport Conference 2020.

Mankins, J.C. (1995). Technology readiness levels. White Paper, 6 April.

Marsden, G., and Docherty, I. (2013). Insights on disruptions as opportunities for transport policy change. *Transportation Research Part A: Policy and Practice*, 51, 46–55.

Martínez-Vergara, S.J., and Valls-Pasola, J. (2021). Clarifying the disruptive innovation puzzle: A critical review. *European Journal of Innovation Management*, 24(3), 893–918.

Millar, C., Lockett, M., and Ladd, T. (2018). Disruption: Technology, innovation and society. *Technological Forecasting and Social Change*, 129, 254–260.

Mitropoulos, L., Kortsari, A., Koliatos, A., and Ayfantopoulou, G. (2021). The Hyperloop system and stakeholders: A review and future directions. *Sustainability*, 13(15), 8430.

Mkono, M. (2020). Eco-anxiety and the flight shaming movement: Implications for tourism. *Journal of Tourism Futures*, 6(3), 223–226.

Mokhtarian, P.L. (2019). Subjective well-being and travel: Retrospect and prospect. *Transportation*, 46(2), 493–513.

NOACA, HTT, and TEMS (2019). *Great Lakes Hyperloop Feasibility Study*. https://www.hyperlooptt.com/2019/great-lakes-hyperloop-feasibility-study/ (accessed 25 June 2022).

Notteboom, T., and Rodrigue, J.P. (2008). Containerisation, box logistics and global supply chains: The integration of ports and liner shipping networks. *Maritime Economics and Logistics*, 10(1), 152–174.

Petzold, N., Landinez, L., and Baaken, T. (2019). Disruptive innovation from a process view: A systematic literature review. *Creativity and Innovation Management*, 28(2), 157–174.

Rajendran, S., and Harper, A. (2020). A simulation-based approach to provide insights on Hyperloop network operations. *Transportation Research Interdisciplinary Perspectives*, 4, 100092.

Rana, Y. (2020). On the feasibility of the Hyperloop concept. Doctoral dissertation, Massachusetts Institute of Technology.

Salter R.M. (1972). *The Very High Speed Transit System*. RAND Corporation, Santa Monica.

Si, S., and Chen, H. (2020). A literature review of disruptive innovation: What it is, how it works and where it goes. *Journal of Engineering and Technology Management*, 56, 101568.

SpaceX (2013). *Tesla Hyperloop Alpha Document*. https://www.tesla.com/sites/default/files/blog_images/Hyperloop-alpha.pdf (accessed 12 November 2021).

Stubbin, E., Charette, G., and Ashraf, K. (2020). *Preliminary Feasibility of Hyperloop Technology (Final)*. AECOM for Transport Canada, Ontario, Canada, Tech. Rep. RFP-T8080-180829, 2020-08.

Tait, J., and Wield, D. (2021). Policy support for disruptive innovation in the life sciences. *Technology Analysis and Strategic Management*, 33(3), 307–319.

Takenaka, K., Hatanaka, K., Yamazaki, W., and Nakahashi, K. (2008). Multidisciplinary design exploration for a winglet. *Journal of Aircraft*, 45(5), 1601–1611.

Upham, P., Maughan, J., Raper, D., and Thomas, C. (2012). *Towards Sustainable Aviation*. London: Earthscan.

Van Goeverden, K., Milakis, D., Janic, M., and Konings, R. (2018). Analysis and modelling of performances of the HL (Hyperloop) transport system. *European Transport Research Review*, 10, 41.

Walker, R. (2018). *Hyperloop: Cutting through the Hype*. https://trl.co.uk/uploads/trl/documents/ACA003-Hyperloop.pdf (accessed 3 December 2021).

Walton, J. (2006). Transport, travel, tourism and mobility: A cultural turn? *Journal of Transport History*, 27(2), 129–134.

15. Mission-oriented innovation policy: the case of the Swedish "Vision Zero" approach to traffic safety[1]

Jannes Craens, Koen Frenken and Toon Meelen

MISSION-ORIENTED INNOVATION POLICY

Innovation policy, in part, consists of mission-oriented policies that direct the efforts of multiple actors towards reaching a concrete goal. The notion of mission refers to NASA's mission to put a man on the moon in the 1960s. The achievement of this goal could obviously not be left to the market nor to scientists alone. Instead, a dedicated organization needed to be created bringing together a diverse set of expertise with strong political backing. A similarly successful example of mission-oriented innovation policy has been the development of high-speed trains by the French government in the late 1970s.

Mission-oriented innovation policy is currently making a revival (Mazzucato, 2018; Schot and Steinmueller, 2018), including in the context of transport innovation (Bugge et al., 2021). This revival is motivated by the need to address today's grand societal challenges, notably global warming, biodiversity loss, and ageing populations. Contrary to the past missions which were characterized by a technological challenge, today's missions are motivated by societal challenges. Most prominently, global warming is threatening the livelihoods of millions of people. Similarly, biodiversity loss and chemical pollution is affecting food production and human health in a myriad of ways. To tackle these challenges, production and consumption will need to change drastically. Apart from challenges of an environmental kind, there are other societal challenges often noted by politicians, including cybersecurity, obesity, ageing, and mental health.

While these societal challenges are all quite different, they have in common that the problems that need to be tackled are "wicked" (Wanzenböck et al., 2020). Wicked problems are characterized by:

- *contestation*, resulting from divergent claims, values, and framings;
- *complexity*, resulting from the multitude of relevant actors and geographical scales (local, national, global) causing a "problem of many hands" (Thompson, 1980);
- *uncertainty*, resulting from limited knowledge to develop effective policies.

Traditional innovation policies are considered to be unfit to deliver solutions to wicked problems. In most high-income countries, traditional innovation policies focus on supporting innovative firms, in particular by subsidizing R&D personnel, stimulating collaboration with universities, and granting the firms patent protection (Schot and Steinmueller, 2018; Wanzenböck et al., 2020). However, as there is no well-developed market for societal problems (being externalities or insufficient public goods), firms are unlikely to develop effective solutions to them. Instead, government itself may have to take the lead.

For governments to be effective in mission-oriented innovation policy, they must avoid a number of pitfalls. In this context, Weber and Rohracher (2012) argued that a new type of innovation policy is needed that is "transformative". Rather than leaving it to firms to develop innovation within the context of existing markets and regulations, government should take the lead in providing "directionality" to innovative activities – including those by firms – so as to transform society. One type of transformative innovation policy is the use of missions: setting a bold and well-defined goal and providing funds and a policy mix to reach such a goal. A policy mix here refers to a combination of policy instruments. The articulating of what is called a "mission" can guide the creativity and investments of many different actors in a particular direction. The key example today is innovation policy in the energy domain, following the target set for countries in the Paris Agreement from December 2015 (to limit global warming to below 2 degrees Celsius, preferably to 1.5 degrees Celsius).

To tackle societal challenges, it is not just about coming up with innovative solutions, but also about having these solutions diffuse within society. Hence, mission-oriented innovation policy involves a much broader policy mix than traditional innovation policy as it focuses not only on innovation but also on diffusion, involving stimulating new markets (e.g., by public procurement or product subsidies), the development of new regulations and standards (e.g., regarding the phasing out of harmful technologies or minimum performance standards), and inducing behavioral change (e.g., through information campaigns and training schemes). As so many policies are needed at the same time, a key aspect of mission-oriented innovation policy is to provide coordination among ministries and between government levels (municipality, province, state, Europe) (Hekkert et al., 2020; Larrue, 2021).

While there is an increasing consensus among policy makers and academics that mission-type policies are needed to tackle the grand societal challenges of our time (Schot and Steinmueller, 2018; Wanzenböck et al., 2020; Larrue, 2021), there is little experience in actually carrying out such ambitious policies. While there has been some experience with technological missions – such as the aforementioned man-on-the-moon and high-speed train projects – the lessons learned from technological challenges do not necessarily carry over to societal challenges. Technological challenges are not so wicked as societal challenges, as there is a clear technological goal for which a dedicated organization can be created under government control. Instead, in the case of missions for societal challenges, the articulation of the mission itself is an important process in its own right. The articulation of the mission, and follow-up policies, should then mobilize a variety of actors in a distributed manner rather than within a single government organization. If this process of "demand articulation" involves relevant actors, has broad political support, and is clearly defined and measurable, the formulation of the mission reduces the wickedness. The policy can then set in motion a coordinated and reflexive process among various actors (Hekkert et al., 2020; Wanzenböck et al., 2020).

While mission-oriented innovation policy towards societal challenges has emerged very recently, we may still learn from policies in the past that – with hindsight – can be understood as policies with a societal mission. In the Netherlands, for example, the persistent challenge to avoid massive flooding has led to a new type of water policy that involves controlled flooding of designated areas (van Staveren et al., 2014). Other examples include re-structuring polluted areas (Coenen et al., 2018) and anti-smoking policies (Wanzenböck et al., 2020).

Here, we report on Sweden's ambitious traffic safety policy known as "Vision Zero". We consider this policy as a mission-oriented innovation policy towards a societal challenge, as it started from the articulation of a bold, societal goal (zero traffic deaths) and involved all sorts of innovations from a variety of actors (public, private, and professional organizations). This chapter first explains what the Vision Zero policy entails and then investigates the factors that made it a success as well as the policy failures that were not overcome. We then draw lessons for the development of new mission-oriented innovation policies to address societal challenges in transport. The research was based on 14 interviews with key people involved in the policy over the years, including employees at Trafikverket, the governmental body that works on Swedish road traffic infrastructure and safety, employees in other governmental bodies, and employees at private firms connected to Vision Zero. As part of the interview series, we also had conversations with two fellow academics from Sweden with expert knowledge on Vision Zero.[2]

VISION ZERO

Vision Zero is a traffic safety policy introduced by the Swedish government in 1997 (Ministry of Transport and Communications, 1997). The objective of the Vision Zero policy holds that "eventually no one will be killed or seriously injured within the road transport system" (Tingvall and Haworth, 1999, p. 1). While a specific time-span is not provided to reach the goal of zero, the vision was introduced with a long time-span in mind (Belin et al., 2012). Its long-term orientation, as well as ambitious aim to bring fatalities to zero, set the policy apart from different traffic safety policies (Johansson, 2009). Before Vision Zero, traffic safety policies were built around reducing fatalities or centered around new innovations that could prevent them. As a defining characteristic, the Vision Zero policy does not start from the supply of possible solutions (thus asking "what *can* be done?"), but rather starts from the demand articulation of zero deaths (asking then "what *should* be done?"). With its use of a mission for dealing with societal problems, Vision Zero is considered not just a new traffic policy, but also a policy innovation as such (Belin et al., 2012; Belin and Tillgren, 2013).

The use of explicit quantitative goals was not new in Swedish traffic safety policy. From 1982, goals were established by the Swedish government regarding fatalities in road traffic accidents (Belin et al., 2010). The main policy change of Vision Zero in this regard was its ethical basis. The ethical principle underlying Vision Zero holds that "It can never be ethically acceptable that people are killed or seriously injured when moving within the road transport system" (Tingvall and Haworth, 1999). The policy thus reoriented priorities from preventing accidents in general to preventing accidents resulting in serious injuries or deaths. This also means that investments in traffic safety policy are not evaluated using a cost–benefit analysis, where the return on investment is computed by aggregating all effects using monetary valuations, including the valuation of people's lives as well as of travel time (which may increase as a result of safety investment).

Two explicit ethical rules in Vision Zero have been highlighted (Tingvall and Haworth, 1999, p. 2). First, "life and health can never be exchanged for other benefits within the society". This principle breaks with cost–benefit analysis that treats traffic fatalities as any other externality so as to include the social costs of traffic fatalities in the total of costs and benefits of investments in road infrastructure. As a second rule, it is stated that "whenever someone is killed or seriously injured, necessary steps must be taken to avoid a similar event". This principle ensures that traffic safety policy is a continuous process of learning.

Alongside the ethical approach of the policy, the responsibility of the road users and designers of the road systems was changed in three main aspects that differ greatly from other road traffic systems (Tingvall and Haworth, 1999, p. 1):

- "The designers of the system are always ultimately responsible for the design, operation and use of the road transport system and thereby responsible for the level of safety within the entire system."
- "Road users are responsible for following the rules for using the road transport system set by the system designers."
- "If road users fail to obey these rules due to lack of knowledge, acceptance or ability, or if injuries occur, the 'system designers' are required to take necessary further steps to counteract people being killed or seriously injured."

The key change here is that the responsibility for safety is not fully centered on the road user, as in traditional road safety systems, but that "system designers" also carry responsibility. System designers are defined as organizations that have responsibilities related to the design and maintenance of the elements of the road system (such as vehicles and roads), as well as to the support systems enabling safe road traffic, such as regulation and education (Fahlquist, 2006; Rosencrantz et al., 2007). In particular, they should design traffic systems in such a way that road users' mistakes do not result in serious or fatal injuries. Hence, this change does not mean that less responsibility is placed on road users. Rather, given that road users – being human – make mistakes, system designers carry the responsibility to make sure that such mistakes do not lead to major injuries.

This engineering challenge was further elaborated by setting the physical abilities of humans to withstand crash impact central in the design of the traffic system. These factors are taken up in the ability of a vehicle to withstand an accident and the forgivingness of road infrastructure. The combined scientific knowledge on these aspects was used as a starting point for the design of traffic systems, in particular road infrastructure and car design.

One example of an innovation that resulted from these principles is the "2 + 1 road", with a barrier which is now a standard design in Sweden on new roads where speeds are over 70km/h (Figure 15.1). The barrier is erected in the middle of a 2 + 1 road, which has two lanes in one direction and one in the other. This wired barrier prevents head-on collisions between opposing drivers, which often have severe consequences. It is considered extremely effective, lowering fatalities by up to 80 percent (Johansson, 2009). Besides safety improvements, this type of road is also a less expensive alternative to four lane roads, while enabling almost the same traffic flow.

Source: Skvattram (2007).

Figure 15.1   Example of 2 + 1 road in Sweden

Another practical outcome is the use of roundabouts. Roundabouts are usually not effective in reducing the number of accidents, but ensure that when one occurs, the impact is much lower. Because of lower speed levels they can ultimately contribute to a reduced number of fatalities (Belin et al., 2012). Besides these measures, attention has been paid to improving safety conditions for unprotected road users, leading to widespread implementation of separated bike lanes among other measures. Another important focus of improvements is on vehicle design so as to address irresponsible driver behavior. One of the innovations is the "alcolock", a device which ensures the driver can only start a vehicle when sober. This solution is especially used with people who are professional drivers, such as taxi drivers and bus drivers, or have a history with intoxicated driving (Johansson, 2009). Furthermore, seatbelt reminders have been developed, which provide warnings to drivers when seatbelts are not worn. To stimulate vehicle innovation, the government takes a leading role by procuring cars that are equipped with the most recent safety features. Herewith, it contributes to faster and more widespread diffusion (Belin et al., 2012).

Another distinctive example is the use of so-called "safety cameras". These speed cameras are installed on roads which have a high record of injuries or fatalities but lack possibilities to take other measures such as median barriers. To encourage social acceptance of cameras, they were redesigned in the early 2000s, making them round and partly blue, the color used for traffic information. A concurrent campaign was launched on "Sweden's new lifesavers". The number of speed tickets issued is confined to a set yearly amount, which practically means that every camera is only operational for 10 percent of the time (Lindberg and Håkansson, 2017). While ticketing is limited, the cameras do generate continuous data, enabling road operators to learn lessons about effectiveness. These various aspects of camera implementation enabled lowering speeds on large stretches of roads, with high levels of public acceptance.

SUCCESS FACTORS

The Vision Zero policy can be seen as a radically new policy, starting from an ethical rather than from a cost–benefit perspective and introducing a new "system engineering" paradigm which leads to a host of innovations of various kinds. What is more, the policy is radical in its explicit and ambitious mission: zero deaths. This well-defined mission not only guides innovation efforts in a clear direction, but also allows – at least at the general level – for a straightforward evaluation of the effectiveness of policy by measuring the number of fatal accidents every year.

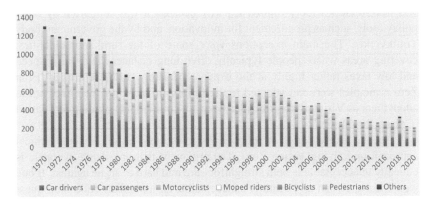

Source: Transport Analysis and Swedish Transport Agency (2021).

Figure 15.2   Traffic fatalities in Sweden

As is clear from Figure 15.2, the number of fatal accidents has indeed gone down remarkably. While other countries have managed to bring down the number of traffic deaths as well over this period, Sweden is ranked among the top countries in the world in terms of traffic safety (Mendoza et al., 2017). Recent data for 2020 show that the country experienced the lowest number of traffic deaths in Europe: Sweden lost 18 lives per million inhabitants against 42 lives per million inhabitants for the European Union on average (European

Commission, 2021). From a policy analysis point of view, then, one can ask why the Swedish Vision Zero policy has been such a success (Craens, 2019). Obviously, not all reduction in traffic deaths can be attributed to the Vision Zero policy, as exogenous factors (reduced alcohol consumption) and other policies (innovation from abroad) may also have contributed to the fall in fatalities.

Among the interviewees, the large majority indicated that the system design perspective was the most crucial success factor, referring to the design principle that traffic systems should be constructed in such a way that accidents never (or very unlikely) become fatal. Following this principle, many innovations were developed, tested, and implemented including the aforementioned 2 + 1 roads, roundabouts, and new speed limit system. What is more, Volvo and Saab intensified their safety innovation programs a few years after Vision Zero started.

The principle of system design was never implemented into legislation, meaning the policy was not binding for regional or local governments (Belin and Tillgren, 2013). This means that system designers and municipalities, in principle, could have resisted the Vision Zero principles. The system design was nevertheless widely followed, partly because it was supported by other policy tools, such as procurement for innovation, and by the government body Trafikverket. The main exceptions were some of the rural municipalities, covering areas where people typically drive long distances and value speed and low taxes rather highly at the expense of safety. In these areas, Vision Zero principles were not adopted to the full extent and they were free not to adopt them as Vision Zero principles were not codified into binding legislation (Craens, 2019).

A second aspect often mentioned as a success factor was the ethical approach underlying Vision Zero. The argument that any death in traffic is ethically unacceptable is both straightforward and hard to argue against. If one would disagree that any death is unacceptable, an immediate follow-up question is how many deaths one is willing to accept. Providing an exact positive number, then, is much harder than to argue that this number should be zero. This is even harder for a politician, who would have to argue that it is acceptable for a government to "kill" a particular number of people, while at the same time a government is expected to care about its citizens' wellbeing. Another reason why the ethical basis is compelling relates to people's personal experience. Most people have experience with traffic deaths or severe injuries in their personal sphere of family and friends. In all, the ethical basis of the policy is considered a key factor for politicians, experts, and the public in accepting the new policy. However, it must be noted that the ethical principles underlying Vision Zero are not uncontested (van Wee, 2011). In particular, as the policy continues, the marginal cost of preventing another traffic death is likely to increase. Given the rising investment to reduce the number of traffic deaths, the policy may become "counterproductive" in terms of overall mortality as an alternative investment may save more lives (Elvik, 1999).

A third success factor relates to learning. Given the long-term goal and wide support for Vision Zero, learning can take place over long periods of time and be based on precise data. Policies went hand in hand with scientific research, meaning that most measures were only taken if scientific evidence was available, while scientific research projects were, in turn, motivated by questions related to traffic safety. The nation-wide adoption of Vision Zero and the innovation coming out of the policy also help in collecting systematic evidence from many sites and for different circumstances.

A fourth success factor is the cooperation between different types of actors. Through intensive cooperation, a wide range of stakeholders are activated to work within the Vision Zero policy. This allows different actors, each with their own expertise and legitimate scope for action, to contribute to the shared goal in a complementary and coordinated manner. Examples of these collaborations include those between the automotive industry and Trafikverket. Knowledge and information about the development of infrastructure and about fatalities is shared, in order to be taken up in the designs of new vehicles and infrastructures. As a result, the government policy and the innovation strategies of Volvo and Saab became increasingly aligned and complementary. Another successful cooperation is the speed camera project, with a consultancy mediating between private companies and Trafikverket. In addition, cooperation was sought with the police to set up a new approach to ticketing offenders. A final example is the establishment of platforms – such as the Group for National Cooperation for Roads – holding regular meetings between automotive firms, the police, ministries, municipalities, and regions.

Finally, some interviewees emphasized the ambition underlying the target of zero deaths. The ambitious nature of the target helped in achieving goals and going beyond actions that would normally be taken. The level of ambition not only motivates stakeholders involved to reach ever lower numbers of deaths, but also creates a continuous pressure from outside as politicians expect improved results year after year.

LESSONS FOR MISSION-ORIENTED INNOVATION POLICY

Vision Zero marked a paradigm shift in traffic safety policy by taking an ethical principle as the starting point and putting system design central. It can be considered a successful policy program as the long-term trend of traffic deaths in Sweden has been going down since, and Sweden currently has the lowest number of traffic deaths per capita across Europe. Obviously, a zero number of deaths has not been achieved – and may never be achieved in the future – but the objective has remained and efforts are ongoing to get closer to this goal. The success of Vision Zero is further exemplified by similar programs being set up in different countries and cities worldwide. Although the exact policy process and implementation differs around the world, similar successes have been reported in bringing down traffic deaths.[3]

In the context of today's innovation policies oriented towards societal challenges, Vision Zero may serve as one example of how such an innovation policy can be conceived and implemented. In particular, Vision Zero can be understood as an instance of mission-oriented innovation policy *avant la lettre*. The mission character of the policy lies in the articulation of an ambitious and well-defined societal goal with broad support in politics and society at large (Larrue, 2021). Such bold objectives are now also being articulated in other domains, especially in the context of climate change (e.g., carbon neutral maritime transport in 2050) and healthcare (e.g., zero suicides). And, in its implementation, Vision Zero also resonates with the idea that mission-oriented innovation policy is about understanding innovation as "socio-technical", involving technological, behavioral, and regulatory changes (Wanzenböck et al., 2020), and about a broad policy that coordinates several policy instruments, actors, and scales in a coherent manner (Larrue, 2021).

While mission-oriented innovation policy has quickly gained in popularity across governments, mostly in the Global North, there is not much consensus yet on how such policies should be designed (Larrue, 2021). Nevertheless, some possible "failures" have been formulated by policy scholars, failures that one needs to try to avoid in mission-oriented innovation policy (Weber and Rohracher, 2012; Wanzenböck and Frenken, 2020). We list these in Table 15.1 and provide an explanation of each. We can then use these possible failures that can be encountered in mission-oriented innovation policy as a way to evaluate Vision Zero.

If we look back at 25 years of Vision Zero policy through the lens of the four possible failures, we can conclude from our interviews that this policy has – indeed – been able to avoid most of the failures listed in the table. Handling these challenges has greatly contributed to the success of the policy. Regarding directionality failures, the policy did not only benefit from a clear articulation of a well-defined goal, but also profited from the wide support among different stakeholders. The underlying ethical principle that any deaths in traffic are seen as societally unacceptable greatly contributed here. To maintain focus and consistency in directionality, the delegation of policy coordination to a specialized agency with sustained funding helped. This agency identified strongly with the policy and actively promoted the vision among stakeholders. Demand articulation failures were also properly addressed in the Vision Zero policy. Before the introduction of the policy, both automotive manufacturers

Table 15.1   *Four typical failures in mission-oriented innovation policy*

Directionality failure	Lack of shared vision regarding the goal and direction of the transformation process; Inability of collective coordination of distributed agents involved in shaping systemic change; Insufficient regulation or standards to guide and consolidate the direction of change; Lack of targeted funding for research, development, and demonstration projects and infrastructures to establish acceptable development paths.
Demand articulation failure	Insufficient spaces for anticipating and learning about user needs to enable the uptake of innovations by users; Absence of orienting and stimulating signals from public demand; Lack of demand-articulating competencies.
Policy coordination failure	Lack of multi-level policy coordination across different systemic levels (e.g., regional–national–European) or between technological and sectoral systems; Lack of horizontal coordination between research, technology, and innovation policies on the one hand and sectoral policies (e.g., transport, energy, agriculture) on the other; Lack of vertical coordination between ministries and implementing agencies which leads to a deviation between strategic intentions and operational implementation of policies; No coherence between public policies and private sector institutions; No temporal coordination, resulting in mismatches related to the timing of interventions by different actors.
Reflexivity failure	Insufficient ability of the system to monitor and to involve actors in processes of self-governance; Lack of distributed reflexive arrangements to connect different discursive spheres and provide spaces for experimentation and learning; No adaptive policy portfolios to keep options open and deal with uncertainty.

Sources: Weber and Rohracher (2012), p. 1045; Wanzenböck and Frenken (2020), p. 54.

(like Volvo) and Trafikverket thought that there would be no market for road safety. With the introduction of Vision Zero, they learned this was not the case, as Trafikverket actively created and enhanced the market for traffic safety. They did this by using procurement tools and using the government as a lead example for taking safety measures. As a consequence, manufacturers introduced more safety features. Furthermore, Trafikverket actively informed end-users on how to behave more safely and use safer equipment in traffic. Another example of a measure to stimulate the market for safer vehicles was the Euro NCAP crash test program. The Swedish government, jointly with the national governments of the Netherlands and the United Kingdom, took a leading role in setting up this European program to test the safety of cars and enhance standards (Hobbs and McDonough, 1998). The resulting crash test scores helped in creating further awareness of safety among potential car buyers.

Dealing with policy coordination proved more challenging in the Vision Zero policy. On the one hand, a broad mix of measures to enhance traffic safety was introduced in a generally well-coordinated way. Policy coherence was also addressed as stakeholders were actively flagging policies in other domains that could – unintendedly – worsen traffic safety. More generally, mission-oriented innovation policy is built on both developing new policies to achieve the mission and abandoning existing policies that are counterproductive in reaching the mission. This is in line with the broader idea of transitions studies, that transformations also involve "phasing out" technologies and regulations (see Pel, Chapter 2 in this volume). Yet the Vision Zero policy also struggled with coordination. While a paradigm shift has taken place in the whole traffic safety sector, the legislation concerning traffic safety has not been adapted to the Vision Zero policy, leaving the concept of system designers without a legal framework. The lack of legislation created problems with regard to implementation. At regional levels especially, this causes problems, since regional authorities have their own responsibility in infrastructure planning. As some of the regions favor speed over safety, some roads still have a high rate of severe and fatal accidents. Trafikverket has few legal measures to increase safety on these roads. Furthermore, coordination between European and national levels of legislation proved difficult. Sweden and Norway are the only countries that have implemented Vision Zero in such a rigorous manner. The development of new national measures is sometimes hampered as implementation of safety regulation is coordinated among countries at the European level. There were also challenges in horizontal coordination between different policy domains. The focus on a single goal, which is common in mission-oriented innovation policy, can lead to trade-offs with other policy areas. While actors involved in Vision Zero acknowledge the importance of coordinating with people involved in other policy domains (sustainability, privacy, etc.), concrete actions in this regard are limited. This also limits potential spillovers between the development of technical innovations in these domains (Langeland et al., Chapter 7 in this volume).

Finally, potential reflexivity failures are well addressed in the Vision Zero policy. One of the main principles of Vision Zero is having a learning attitude. This principle is exemplified in the evidence-based nature of the policy. Proposed improvements are first tested and evaluated before implementation. Insights from industry about quality control and continuous improvement have been translated for the Vision Zero policy. There is an ongoing search for "best practice measures" that can be taken in order to save lives in the traffic system. Yearly follow-up data is retrieved for multiple traffic safety indicators and processed in statistical models so as to get insight into the current safety situation. Additionally, platforms which include stakeholders such as the vehicle industry and non-governmental organizations (NGOs) have been established to discuss developments in traffic safety on a yearly basis.

The more general lessons that can be drawn from Vision Zero regarding mission-oriented innovation policy are threefold (Craens, 2019). For mission-oriented innovation policy to succeed, high-level political support is key. Given that, generally, missions cannot be completed in a few years, such a commitment needs to be rather independent from changing political coalitions in government cabinets, as these tend to change every few years. One way to ensure political commitment is to start from ethical principles that are both straightforward and widely shared. In transportation research, there is now a lively discussion about ethical principles such as equity and sufficiency, which could be drawn upon in developing new missions to address societal challenges related to transport (van Wee and Geurs, 2011; Verlinghieri and Schwanen, 2020). An example of such a mission would be: in 2050, everyone in the city should have access to all daily necessities within 15 minutes. As with Vision Zero, various types of innovations, such as new cycling infrastructure or Mobility as a Service (MaaS) technologies, could be combined to achieve the mission. A second way to ensure commitment to the mission-oriented innovation policy is to anchor parts of the policy in strict targets (such as the Paris agreement of CO_2 reduction in Europe) which can be translated into binding laws. Indeed, the Vision Zero case suggests that the policy could have been even more effective if principles were codified in legislation rather than being dependent on voluntary adoption by lower governmental levels. In the context of climate change, missions can be developed that combine a legal CO_2 reduction target with ethical principles such as equity. For example: all citizens should have access to affordable, zero-emission mobility in 2040.

A second key lesson that can be drawn from Vision Zero is the need to have the ability to measure progress along the way, and to attribute success and failure to particular measures. This allows stakeholders to evaluate progress on a regular basis as well as be adaptive in what policies and innovations to use. Collective learning is supported by a shared evidence base produced by an independent body. It should be noted that an evidence-based approach is highly established in transportation research and policy. Traffic measures are routinely evaluated by collecting large amounts of data on traffic participants. In policy fields such as transportation and healthcare, which have a longer tradition of evidence-based policy making, the learning component of mission-oriented innovation policy may be easier to implement. On the flipside, a challenge of current evidence-based approaches is measuring progress in terms of deeper-level institutional change (e.g., in terms of culture) (Pel, Chapter 2 in this volume). In addition, these types of changes can be important for achieving missions. A potential solution might be a more open strategy of evaluation which is sensitive to different types of impacts (including social impacts) and unintended effects. For example, early electric vehicle users joined online communities, which then influenced the development of new driving and charging habits (Meelen et al., 2019). For mission-oriented innovation policies, learnings about such unexpected effects could be applied in subsequent rounds of innovation experiments.

Finally, in terms of impacts, while mission-oriented innovation policy is motivated by societal challenges, it may also contribute to more classic economic objectives of innovation policy such as job creation or economic competitiveness. By actively sharing the vision with a broad range of actors, including industry, new products can be developed and new markets can be created. In this respect, governments can use public procurement for innovation, promote standard-setting, introduce clear regulations, and support university–industry collaboration to help make local firms more innovative. An ambitious policy at national or regional level can thus create a "testbed" for all kinds of innovations that may later on be exported in global markets. A recent example of this is the Norwegian strategy for reduced shipping emissions, which went hand in hand with the development of local industrial capabilities to produce electric ferries (Bugge et al., 2021). Yet economic objectives should not be considered a key objective of mission-oriented innovation policy, as the primary goal of mission-oriented innovation policy is to tackle a persistent societal problem which traditional policies have failed to solve.

NOTES

1. Koen Frenken benefitted from financial support from the INTRANSIT center funded by the Norwegian Research Council.
2. For further details, see Craens (2019).
3. The success of the Vision Zero policy has not remained unnoticed outside Sweden, and several governments have adopted similar policies including Norway, provinces in Australia, several large cities in the United States, and London (Craens, 2019). Moreover, in Sweden the Vision Zero style of policy has been developed in sectors other than traffic safety. In particular, challenging issues like fire prevention, suicide prevention, and patient safety in hospitals have been subjected to the Vision Zero approach.

REFERENCES

Belin, M.Å., and Tillgren, P. (2013). Vision Zero. How a policy innovation is dashed by interest conflicts, but may prevail in the end. *Scandinavian Journal of Public Administration, 16*(3), 83–102.

Belin, M.Å., Tillgren, P., and Vedung, E. (2010). Theory and practice in Sweden: A case study of setting quantified road safety targets. *Journal of Health Medical Information, 1*(1), 1000101.

Belin, M.Å., Tillgren, P., and Vedung, E. (2012). Vision Zero – a road safety policy innovation. *International Journal of Injury Control and Safety Promotion, 19*(2), 171–179.

Bugge, M.M., Andersen, A.D., and Steen, M. (2021). The role of regional innovation systems in mission-oriented innovation policy: Exploring the problem-solution space in electrification of maritime transport. *European Planning Studies,* https://doi.org/10.1080/09654313.2021.1988907.

Coenen, L., Campbell, S., and Wiseman, J. (2018). Regional innovation systems and transformative dynamics: Transitions in coal regions in Australia and Germany, in: Isaksen, A., Martin, R., and Trippl, M. (eds.), *New Avenues for Regional Innovation Systems: Theoretical Advances, Empirical Cases and Policy Lessons* (Berlin: Springer), pp. 199–217.

Craens, J. (2019). Mission-oriented innovation policy and societal challenges: Lessons from the Swedish Vision Zero program. MSc Thesis "Innovation Sciences", Utrecht University.

Elvik, R. (1999). Can injury prevention efforts go too far? Reflections on some possible implications of Vision Zero for road accident fatalities. *Accident Analysis and Prevention, 31*(3), 265–286.

European Commission (2021). Road safety: 4,000 fewer people lost their lives on EU roads in 2020 as death rate falls to all-time low, https://ec.europa.eu/commission/presscorner/detail/en/ip_21_1767.

Fahlquist, J.N. (2006). Responsibility ascriptions and Vision Zero. *Accident Analysis and Prevention, 38*(6), 1113–1118.

Hekkert, M.P., Janssen, M., Wesseling, J.H., and Negro, S. (2020). Mission-oriented innovation systems. *Environmental Innovation and Societal Transmissions, 34,* 76–79.

Hobbs, C.A., and McDonough, P.J. (1998). Development of the European New Car Assessment Program (EURO NCAP). Transport Research Laboratory, United Kingdom, Paper Number 98-S11-O-06, https://cdn.euroncap.com/media/1387/development-of-euro-ncap-1998-0-7498ffa2-36b6-4328-8b6e-f44fe5d35563.pdf.

Johansson, R. (2009). Vision Zero – implementing a policy for traffic safety. *Safety Science, 47*(6), 826–831.

Larrue, P. (2021). The design and implementation of mission-oriented innovation policies: A new systemic policy approach to address societal challenges. OECD Science, Technology and Industry Policy Papers, No. 100, OECD Publishing, Paris, https://doi.org/10.1787/3f6c76a4-en.

Lindberg, H., and Håkansson, M. (2017). *Vision Zero 20 Years.* http://www.afconsult.com/contentassets/8f0c19f4f7d24aa5bdbfd338128391ec/2017057-17_0194-rapport-nollvision-eng_lr.pdf.

Mazzucato, M. (2018). Mission-oriented innovation policies: Challenges and opportunities. *Industrial and Corporate Change, 27*(5), 803–815.

Meelen, T., Truffer, B., and Schwanen, T. (2019). Virtual user communities contributing to upscaling innovations in transitions: The case of electric vehicles. *Environmental Innovation and Societal Transitions, 31,* 96–109.

Mendoza, A.E., Wybourn, C.A., Mendoza, M.A., Cruz, M.J., Juillard, C.J., and Dicker, R.A. (2017). The worldwide approach to Vision Zero: Implementing road safety strategies to eliminate traffic-related fatalities. *Current Trauma Reports, 3*(2), 104–110.

Ministry of Transport and Communications (1997). En route to a society with safe road traffic. Selected extract from Memorandum prepared by the Swedish Ministry of Transport and Communications. Memorandum, DS.

Rosencrantz, H., Edvardsson, K., and Hansson, S.O. (2007). Vision Zero – is it irrational? *Transportation Research Part A: Policy and Practice*, *41*(6), 559–567.

Schot, J., and Steinmueller, W.E. (2018). Three frames for innovation policy: R&D, systems of innovation and transformative change. *Research Policy*, *47*(9), 1554–1567.

Skvattram (2007). 2+1 road with cable barrier on Road 34 near Linköping, Sweden [Photograph]. Wikimedia Commons. https://en.wikipedia.org/wiki/2%2B1_road#/media/File:MMLNorr1.JPG.

Thompson, D.F. (1980). Moral responsibility of public officials: The problem of many hands. *American Political Science Review*, *74*(4), 905–916.

Tingvall, C., and Haworth, N. (1999). Vision Zero – an ethical approach to safety and mobility. In *Sixth ITE International Conference Road Safety and Traffic Enforcement: Beyond 2000*, Melbourne, 6–7 September 1999. https://eprints.qut.edu.au/134991/.

Transport Analysis and Swedish Transport Agency (2021). Road traffic injuries. https://www.trafa.se/en/road-traffic/road-traffic-injuries/.

Van Staveren, M.F., Warner, J.F., van Tatenhove, J.P.M., and Wester, P. (2014). Let's bring in the floods: De-poldering in the Netherlands as a strategy for long-term delta survival? *Water International*, *39*(5), 686–700.

Van Wee, B. (2011). *Transport and Ethics: Ethics and the Evaluation of Transport Policies and Projects* (Cheltenham, UK and Northampton, MA, USA: Edward Elgar Publishing).

Van Wee, B., and Geurs, K. (2011). Discussing equity and social exclusion in accessibility evaluations. *European Journal of Transport and Infrastructure Research*, *11*(4), 350–367.

Verlinghieri, E., and Schwanen, T. (2020). Transport and mobility justice: Evolving discussions. *Journal of Transport Geography*, *87*, 102798.

Wanzenböck, I., and Frenken, K. (2020). The subsidiarity principle in innovation policy for societal challenges. *Global Transitions*, *2*, 51–59.

Wanzenböck, I., Wesseling, J.H., Frenken, K., Hekkert, M.P., and Weber, K.M. (2020). A framework for mission-oriented innovation policy: Alternative pathways through the problem–solution space. *Science and Public Policy*, *47*(4), 474–489.

Weber, K.M., and Rohracher, H. (2012). Legitimizing research, technology and innovation policies for transformative change: Combining insights from innovation systems and multi-level perspective in a comprehensive "failures" framework. *Research Policy*, *41*(6), 1037–1047.

Index